Portrait of Walton  **W9-DJM-842**

PORTRAIT OF
WALTON

Michael Kennedy

Oxford   New York
OXFORD UNIVERSITY PRESS
1990

Oxford University Press, Walton Street, Oxford OX2 6DP

Oxford New York Toronto
Delhi Bombay Calcutta Madras Karachi
Petaling Jaya Singapore Hong Kong Tokyo
Nairobi Dar es Salaam Cape Town
Melbourne Auckland

and associated companies in
Berlin Ibadan

Oxford is a trade mark of Oxford University Press

First published 1989
First issued (with corrections) as an Oxford University Press paperback 1990

British Library Cataloguing in Publication Data
Kennedy, Michael, 1926–
Portrait of Walton.—(Oxford lives).
1. English music. Walton, William, 1902–1983
I. Title
780.92
ISBN 0–19–282774–X

Printed in Great Britain by
Clays Ltd.
Bungay, Suffolk

TO
JOHN WARRACK

Preface

IN 1968 Sir William Walton, after reading my book on Elgar, asked that I should be his biographer. A contract was signed and Oxford University Press kindly accommodated my insistence that I would not write any biography while the subject was alive, although I contributed a brief essay to Stewart Craggs's catalogue in 1977. In the years between 1968 and 1983 a biography by Gillian Widdicombe was promised, to which Walton gave his blessing and much assistance while still sanctioning mine. After his death Lady Walton and Neil Tierney wrote books about him. In the belief and hope that there is still room for another, I now fulfil my twenty-year-old promise, not out of contractual obligation or a sense of duty, but out of admiration and affection for a fellow-Lancastrian, whose music has given me a lifetime of pleasure and whom I very much liked as a man.

I have called this book *Portrait of Walton* because I have approached the subject on roughly similar lines to those on which I approached Elgar. It is not an analytical probe into the construction and technique of the music such as wins favourable professorial reviews but bores and baffles many readers. In any case, Frank Howes's classic survey of the music, which is emphatically not boring, seems to me to do the job as well as it can be done.

My acknowledgements are principally due to Lady Walton, for permission to quote from Sir William's letters and interviews; to Dr Stewart Craggs, who has generously placed all his unrivalled researches at my disposal; to my dear friend Roy Douglas, for providing much help and many memories; to Alan Frank, former head of the music department of Oxford University Press, who not only read the book and made many helpful suggestions, but allowed his superb drawing of Walton by Gino Coppa to be used as an illustration; to the Britten–Pears Foundation, for permission to quote from the late Lord

Britten's letters and diary (which are the Foundation's copyright and may not be further reproduced without written permission); to the late Dora Foss; and to Gillian Widdicombe; Peter Heyworth; John Warrack; Andrew Potter; Angus Morrison; John Amis; Owain Arwel Hughes; Rosamund Strode; Diana Rix of Harold Holt Ltd.; Viscount Wimborne; André Previn and his secretary Fiona Page; EMI Records and the Decca Record Company. Mr Previn was particularly generous in providing me with a photostat of the autograph fragment of Walton's Third Symphony which is in his possession. I have been much fortified by the enthusiasm and encouragement of Bruce Phillips, of Oxford University Press, and Joyce Bourne has typed the book not once or twice but several times, *sans reproche*.

I gratefully acknowledge the respective publishers' permission to quote brief extracts from the following books: *Music—a Joy for Life*, by Edward Heath (Sidgwick and Jackson); *Malcolm Sargent, a Biography*, by Charles Reid (Hodder & Stoughton); and *A View from the Pit*, by Richard Temple Savage (David and Charles).

Walton was careless about grammar and punctuation in letters and never seems to have read them after finishing them to supply missing words or clarify the sense! In 95 per cent of extracts I have given them as he wrote them. In one or two cases, a little judicious editing has been necessary to save the reader incomprehension.

M.K.

Contents

PART I. WILLIE

1. Oldham and Oxford, 1902–20   3
2. London and Salzburg, 1920–3   18
3. *Façade*, 1921–79   27
4. With the Sitwells, 1923–9   37
5. Viola Concerto, 1928–9   47
6. *Belshazzar's Feast*, 1929–31   53
7. In Switzerland,1931–3   62
8. Unfinished Symphony, 1933–4   71
9. Finished Symphony, 1934–5   80

PART II. WILLIAM

10. *Crown Imperial*, 1936–7   89
11. Violin Concerto, 1938–9   96
12. War Films, 1939–42   108
13. *Columbus* and *Henry V*, 1942–5   119
14. Post–war Quartet, 1944–7   127
15. Birth of an Opera, 1947–8   136
16. Making an Opera, 1948–52   143

PART III. SIR WILLIAM

17. Coronation Interlude, 1952–3   159
18. Finishing an Opera, 1953–4   168

19. *Troilus and Cressida*, 1954–76   179
20. Cello Concerto, 1955–7   194
21. *Partita* and Second Symphony, 1957–60   205
22. *The Bear*, 1964–7   223
23. Britten and *Bagatelles*, 1969–72   241
24. Seventieth Birthday, 1972   250
25. Roaring Fanfares, 1972–7   257
26. The Haunted End, 1977–83   269
27. Epilogue   279

APPENDICES

I. Classified List of Works   290
II. *Troilus and Cressida* Synopsis   314
III. Discography: Walton conducts Walton   320
IV. Select Bibliography   325

Classified Index of Works   329

General Index   335

Illustrations

1. As a schoolboy (*Mrs C. Noel Walton*)
2. With Dora Foss at Symonds Yat (*Diana Foss Sparkes and Christopher Foss*)
3. Musical evening in Wimborne House, painted by Sir John Lavery (*Viscount Wimborne*)
4. Walton in 1949 (*BBC*)
5. Congratulating Sir Malcolm Sargent at a Royal Albert Hall Promenade Concert, 1 September 1965 (*The Hulton Picture Company*)
6. Drawing by Gino Coppa, 1975 (*Reproduced by permission of the artist. In the collection of Alan Frank*)
7. Susana and William Walton, 1971 (*The Hulton Picture Company*)
8. At the Ritz on his seventieth birthday, 29 March 1972 (*The Hulton Picture Company*)
9 and 10. The only surviving sketch of the Third Symphony (*André Previn*)
11. In his study in Ischia, 1981 (*Christopher Warde-Jones*)

PART I

Willie

1
Oldham and Oxford, 1902–20

WILLIAM WALTON'S is a classic example of the career of a British creative artist: at first regarded as subversive; then acclaimed as the answer to all prayers and absorbed into the body politic of music as part of the establishment; and finally treated as an institution, though one whose foundations are by then found to be suspect. He himself picked a careful and wary path through the minefields strewn around him at various points in his progress from *enfant* not so very *terrible* to unofficial laureate. He never lacked patrons, whether they were kindly clerics, titled ladies, or film producers. But none, except perhaps one, claimed his full allegiance. He had no close association with a locality, as Elgar had with Worcestershire and Britten with East Anglia, nor did his music take root at a specific festival like the Three Choirs or Aldeburgh; he owed no debts to any musical creed or dogma, as Vaughan Williams did to folk-song and others do to serialism; he was in the midst of the most fashionable intellectual literary circle of the 1920s yet continued, like Kipling's Cat, to walk by himself as far as his tastes and prejudices were concerned. He had the misfortune to compose an inimitable, unique masterpiece—*Façade*—at the start of his career, and although he wrote superb examples in the traditional forms of symphony, concerto, and cantata, he carried that early and deserved success (and notoriety) like an incubus for the rest of his life. It is no easy matter if one's first major work is regarded as a *non plus ultra*.

The character of Walton's music is as strong as that of the man it reflects; and the strength probably came from his being one of those Lancastrians in whom stubborn independence and musical sensitivity are often blended. If his career sometimes suggests a progress 'through gilded trellises'—and Walton himself was the last person to deny the element of luck—this is

perhaps because legends often arise from attractive superficial-
ities. It is unwise to underestimate the sheer grit which Walton
displayed in his boyhood in his determination to escape from
his native town of Oldham, and the acutely developed power
of self-criticism which resulted in a body of work of almost
uniformly high quality. There was, too, an element of insecurity, often present in artists whose powerful individuality
would seem to suggest the opposite. He was shy and reserved,
but a romantic. He would weep while listening to Puccini. He
kept the same short 'choirboy' haircut all his life. He never lost
his inoffensively grubby sense of humour, and lurking amid
the languid Oxford accent and economical mode of speech
were always the flattened vowel-sounds of his Lancashire
origin, just as Beecham's oratorical mannerisms never wholly
concealed the son of St Helens.

The musical and biographical similarities between Walton
and Elgar are striking. Both were born in provincial towns to
not-very-well-off parents with musical backgrounds, although
Worcester in the 1860s was a much pleasanter place in which to
spend one's childhood than the Lancashire mill town of
Oldham where Walton was born. Oldham bore the scars of the
Industrial Revolution—mill chimneys, terrace houses, drab
streets, extremes of wealth and poverty, and the dour climate of
a town on the edge of the Pennine Hills. Whereas Elgar's father
was a piano-tuner and organist, Walton's was a singing-
teacher, organist, and choirmaster. Charles Alexander Walton
was born in Hale, Cheshire, in 1867, the son of an Inland
Revenue official. He met his future wife, Louisa Maria Turner,
at a recital in Chorlton-cum-Hardy, Manchester. She was a
contralto. She was born in Stretford in 1866. They were
married at a Congregational church in Chorlton-cum-Hardy on
10 August 1898 and went to live in Oldham, where Charles
had a job in an ironworks. But he described himself on his
marriage certificate as a teacher of singing. Although twenty–
six, he had been one of the first intake in 1893 when the Royal
Manchester College of Music opened its doors with Sir Charles
Hallé as its first principal. He was a bass–baritone pupil of
Andrew Black (the first singer of the title-part in Elgar's
Caractacus in 1898) and during his three years as a student he

sang in oratorio, Mephistopheles in Gounod's *Faust* and Papageno in Mozart's *Die Zauberflöte*.

Charles and Louisa Walton's first son, Noel, was born in 1899. William was born at 93, Werneth Hall Road, on 29 March 1902, his sister Nora in 1908 and another brother, Alexander, in 1910. Charles taught music at Hulme Grammar School, Oldham, and for twenty-one years was organist and choir-master of St John's Church, Werneth. He also taught in Glossop, Hyde, Manchester, and Leeds. He was a strict disciplinarian, of violent temper, and his choir's attendance at church services and rehearsals was compulsory. Noel and William sang in it. William did not enjoy the experience because whenever he sang a wrong note he was rapped on the knuckles by his father's ring 'and it hurt'. Among works in which he and Noel took part were Haydn's *The Creation*, Mendelssohn's *Hymn of Praise*, Handel's *Messiah*, Gounod's *Messe solennelle*, and many anthems. As was the case with Elgar, a musical gift was soon apparent: William learned to play the piano and organ and had violin lessons until his father stopped them because he was lax about practising. Walton could never play the piano well and he was even worse on the violin. 'I could never organize my fingers properly and it sounded so awful.'[1] Here is part of Walton's telegrammatic third-person account of his early days from a letter to his publisher Hubert Foss in 1932: 'It is said that he could sing before he could talk (doubtless untrue).[2] Anyhow, he remembers making a scene (tears, etc.) because not allowed to sing a solo in local church choir when about the age of six. Won probationership to Ch. Ch. Cath. Choir (after being very sick on first long train journey).'

His father had seen a newspaper advertisement for a voice trial for scholarships for probationer choristers at Christ Church, Oxford. Application was made, and Mrs Walton and her son arrived at the college when the examinations had finished (they had missed the first train because the money for

[1] This and other direct quotations from Walton are from the 1981 television profile *At the Haunted End of the Day*.
[2] Bits of *Messiah*, so Mrs Walton told Osbert Sitwell: *Laughter in the Next Room* (London, 1949), p. 174.

their fares had been drunk by Charles Walton in the pub the previous evening, and had had to be borrowed from the greengrocer). Mrs Walton pleaded for her son to be heard and the organist, Dr Henry Ley, then aged twenty-five,[3] heard William sing Marcello's 'O Lord, our Governor' and asked him to sing the middle note of a five-note chord as an ear-test. He was accepted and left an Oldham board school (which had been 'a nightmare' for him) to become a boarder at Oxford, where, he recalled, his first term was 'odious' until he learned how to disguise his Lancashire accent.

Walton's brother Noel told the late Hugh Ottaway in 1970 that he believed his parents never intended William to have a musical career.[4] If the boy had returned home at the end of his time at choir school, he would have been sent to work in a cotton mill or a bank. Charles Walton was bitter about his own ill-fortune as a musician and was determined that his sons would not enter the profession.

The only reason my parents sent Willie to Oxford [Noel wrote] was the opportunity to give him a superior education (not all that 'superior' according to W!) at a comparatively low price, which at that time they could just afford, the fees being £30 per annum, plus extras . . . It is my definite opinion that my parents were unaware of Willie's musical talent. I think my father heard one or two of Willie's compositions by 'cat's whisker' radio.[5] Yet I think he was incredulous to the end.

(William Elgar had a similar sceptical opinion of his son Edward.)

For the next six years, 1912–18, Walton remained at the choir school. His letters to his mother show that he was regularly treble soloist in the anthems, although Dr Leslie Russell (a fellow-chorister who became head boy in 1914) wrote, in a letter to Stewart Craggs in 1976, that he remembered him as 'an uninteresting, quiet, unathletic boy with a *poor* voice'. He was quite athletic, in fact, being good at football and cox of his

---

[3] He had been appointed in 1909 while still an undergraduate.
[4] Walton wrote to Alan Frank in August 1975 that he suspected Ottaway's monograph was 'full of misinformation obtained from Noel who is more gaga, I gather, even than I am. Neither of us, I suspect, are to be trusted for accuracy—unless there's confirmation and I doubt if any exists.'
[5] Charles Walton died in November 1924.

college boat. The contrast of holidays in wartime Oldham after 1914 must have been considerable and certainly stiffened William's will to avoid the otherwise inevitable alternative of becoming an office-boy or a cotton-clerk there. Luck—in the person of Dr Thomas Strong, Dean of Christ Church—prevented this calamity when the outbreak of war dealt a financial blow to Charles Walton by reducing the numbers of his singing pupils. An obvious family economy was to bring William home, but Dr Strong himself paid the balance of the school fees (most of which were met by the scholarship). When William was confirmed in Christ Church Cathedral on 15 March 1916, Strong gave him a Bible inscribed to mark the occasion. William took this home to Oldham, where he left it. Noel Walton remembered Mrs Walton trying to persuade William to take it into his possession, but he refused to have it.

Strong's account of his interest in Walton was given in a letter he wrote to Hubert Foss on 8 January 1938, when 'my intimate acquaintance with W. is long past'. He continued:

In those days the senior boys used to come to my house every Sunday morning after Cathedral, i.e. about 11.30 a.m. It began by being a sort of little Bible class, but they gradually developed the habit of staying till 1 p. m. and messing about with my books etc. *I think*, but am not quite sure, that he used to strum on my piano . . . One Sunday, when he was in the choir, he brought with him a large bundle of music-paper covered with his compositions: he was then about 15. He asked if he might leave them for me to look at and dumped them on the table in the hall. It so happened that the examinations for music degrees were going on just then and Parry was staying with me. He picked up W.'s MSS and was interested. I remember his saying 'There's a lot in this chap, you must keep your eye on him!'

Walton's characteristically understated account of why he began to compose was: 'I thought "I must make myself interesting somehow or when my voice breaks I'll be sent back to Oldham. What can I do? Write music." So I did.' His description of his childhood compositions, in the 1932 letter to Foss from which I have already quoted, is:

First signs of composition 'Variations for violin and pf. on a chorale by J.S.B.' didn't progress (like his latest composition) more than a dozen bars. Not very interesting and wisely decided to stop. However broke loose again about 13 and wrote two 4-part songs, 'Tell me where is

fancy bred' and 'Where the bee sucks'. After that, fairly went in for it
and produced about 30 very bad works of various species, songs,
motets, Magnificats etc.

The manuscripts of two of these compositions, 'Tell me
where is fancy bred' and a choral prelude on 'Wheatley', dated
respectively July and August 1916, are in the British Library.
They explain the interest of the astute Hubert Parry, always
alert to give a new talent a helping hand, as does another piece
of Walton juvenilia, the *Litany*, a setting of Phineas Fletcher's
'Drop, drop, slow tears', dated Easter 1916, fluently and
expertly written for choir and a genuine Waltonian experience,
with surprising harmonic progressions. Most significant of all,
it is an astoundingly well-developed example of his characteristic bitter–sweet romanticism. His instruction was under
general supervision from Hugh Allen, who was in Oxford as
organist of New College until becoming Professor of Music in
1918. Walton recalled in later years that it was from Allen that
he 'obtained some insight into the mysteries of the orchestra, as
he could bring scores vividly to life by playing them on the
organ'.[6] Allen introduced Stravinsky's *Petrushka* to Walton by
playing it to him on the piano. William also spent hours in the
Ellis Library in the Radcliffe Camera which possessed a large
number of scores by Debussy, Ravel, Stravinsky, and Prokofiev. 'I fear I spent perhaps too much time there, to the
detriment of my scholarly studies in Latin, Greek, and algebra.'
His brother Noel remembered William on holiday in Oldham
'making horrible noises for hours, playing from scores of *Le
Sacre du Printemps* and Bartók—especially *Allegro barbaro*'. Henry
Ley, who taught him organ and piano, and his assistant Basil
Allchin introduced Walton to a wide range of styles in church
music. Ley became convinced of Walton's unusual promise
through an incident when William took him a choral composition for comment. While playing it through, Ley bowdlerized
one or two of the chords, only to be told at the end: 'But, sir,
that is not what I wrote.' Walton wrote his first March for Ley's
wedding, where it was played on the organ by Noel Ponsonby,
who was about to become organist of Christ Church. (His son
Robert was BBC Controller of Music from 1972 to 1985.)

[6] R. M. Schafer, *British Composers in Interview* (London, 1963), p. 74.

Walton's letters to his mother from Oxford give a picture of his progress at a point in his life about which little detail has hitherto emerged. Perhaps the six-part Motet was the work Ley bowdlerized:

[Undated, 1916] I finished my lectures with Dr Ernest Walker on harmony. Dr Iliffe says I have done wonderfully well at my counterpoint and thinks I shall easily be able to get through my 1st Mus. Bac. in May . . .

[Undated, 1916] H. G. Ley came back on Thursday. I showed him my six-part Motet. He said it had wonderful ideas in it. I showed him the others. Those were quite excellent, especially the Fantasia . . . A new Choral Prelude [probably 'Wheatley'] and two others didn't sound well on the organ but were fairly respectable on the piano. He is teaching me harmony free and is going over the motets . . .

[17 September 1916] On Tuesday I sent my compositions to Sir Hubert Parry . . . This morning Mr Ley *played my Choral Fantasia* after service. People said it was very fine, but I don't give my opinion.

[8 October] The Dean has been saying something to me about The Royal College of Music. He says it is unpatriotic to England to let slip such a musical brain . . . I expect to be able to do florid counterpoint in four parts before half term.

[26 November] I finished my composition on 'For all the Saints'.

[17 June 1917] This morning . . . I went to Dr Allen's to be introduced to Sir Hubert Parry but his car had bust down so I am going at 3.30.

[24 June] I went to see Sir Hubert Parry on last Sunday afternoon and had quite a long talk with him. He is an awfully jolly old person.

The extent to which Walton taught himself orchestration, by trial and error and instinct, is even more remarkable than in the case of Elgar. Walton claimed only to play the pianoforte very badly, and his practical musical work was chiefly as a chorister. Writing home in May 1918 he said: 'Next term I hope to be an organist at Brasenose College if I can learn enough by October. It will be rather good if it is possible to do it.' (He was organist there for three services until the regular organist returned from the army.) Elgar, on the other hand, was an accomplished violinist, organist, and woodwind player and had a long semi-professional apprenticeship as orchestral player and as amateur conductor. Even so, he was twenty-five before he composed the *Suite* for small orchestra and thirty-three when he wrote the overture *Froissart*. Walton composed *Portsmouth Point* when he was twenty-three and the Viola Concerto when he was twenty-six. He began the First Symphony when he was twenty-nine.

There had been a second threat of a permanent return to 93 Werneth Hall Road, Oldham, during the summer holiday of 1916, when William's voice broke and an illness of his mother made the Walton finances again precarious. A letter to Charles Walton from the Revd Edward Peake, headmaster of the choir school, on 6 September 1916, said:

I have seen the Dean [Dr Strong], and he says he does not think it right that we should let Billy go, and he will consult with the canons about settling the account. So I am sure you need not worry about it. I dare say this will be a comfort to Mrs Walton. I am so sorry that she has been ill . . . It is just such illness that makes one long for money most, isn't it?

Nor was this the end of Dr Strong's beneficence, for which English music should be truly thankful. Walton remained in the choir school for a further two years. He passed the first half of his Bachelor of Music examination on 11 June 1918 at New College, the examiners being Sir Walter Parratt, Dr Hugh Allen, and Dr Ernest Walker. Strong then wrote to Charles Walton:

He has, as you know, shown considerable gifts in the way of music and he is, he tells me, very anxious to follow music as his profession. I saw him a day or two ago and talked over matters with him, and I have also discussed his prospects with Dr Allen, the new Professor of Music. It seems that it is necessary now to take some decision as to his future, as he is now out of the choir and cannot go on indefinitely in the school. I am anxious that he should have the best chance he can get of doing well in music. I am venturing therefore to make the following suggestion: that he should join the College in October as an undergraduate and continue his musical studies under Dr Allen and the other members of the Faculty. If he did this, he would be able to obtain the degree of Bachelor of Arts side by side with that in music . . . I have the control (as Dean) of a Fund available for the assistance of undergraduates of the College who need it. Owing to the great lack of undergraduates by reason of the war, this fund is in a rather prosperous position and I could draw upon it in order to repay his university expenses. The boy is young to be an undergraduate, but he seems to me to be a very steady and trustworthy fellow, and I do not think he should come to any harm. Last year I took a boy of 16.3 years and he has done very well: he is aiming at a degree in medicine . . .[7]

[7] This last sentence seems to dispose of the oft-repeated statement that Walton was the youngest member of Oxford University since the reign of Henry VIII, since in October 1918 he was sixteen years and seven months.

William's letter to his mother on 16 July was more dramatic:

My fate hangs in the balance. The Dean is writing to dad to see whether I shall go into Ch. Ch. and get my Mus Bac and B.A. or go into an office. Mind Dad replies in the affirmative immediately. He will probably ask if I shall be able to get on after with piano, and you'd better say it would not be improbable that I should be able to get a post as organist at a public school.

To Walton himself Dean Strong wrote a few weeks later, on 12 August:

I gather there was some doubt as to whether you had got exemption from any part of Responsions. Your music examination does not excuse you, but I think some of the things you have passed through the Locals may help. Can you tell me exactly what *subjects* you passed in last summer? [for matriculation]. I suppose the list will be out soon for this year, and then I shall be able to see what you have done towards exemption. I think you have not taken Greek, so that there will certainly be this subject wanted . . . In a way it is rather a plunge from a small school into the University, and some boys might make a great mess of things, but I think this will not be the case with you.

Greek—and algebra—were indeed wanted but they were never forthcoming, with the result that William did not obtain a degree, either in music or arts, because he failed Responsions at three attempts, in June, September, and December 1919. Through Strong's influence, the governing body of Christ Church had granted Walton an in-College Exhibition of £85 a year for two years on 13 November 1918. At Michaelmas 1919, his fifth term of residence, against Walton's name was inscribed 'Pass or Go', and with his third failure in Responsions he went. He was not in residence for the Hilary and Trinity terms of 1920 but passed the second part of his B.Mus. on 8 and 9 June 1920, when the examiners were Allen, Percy Buck, and Vaughan Williams.[8] (In the History of Music paper, candidates

[8]   I am indebted to Dr J. F. A. Mason, former Librarian of Christ Church, for these details of Walton's university career, hitherto a matter of some obscurity. He comments:

Walton's preparations for examinations are not easy to understand, for . . . he did not have to pass both the Music Prelim and Responsions but only one or the other in order to be eligible for a B.Mus. yet he seems to have been sent down for failure to pass Responsions. But he would in any case have had to pass one of the four groups Greek or Latin, French, German, or English Literature . . . We have no means of knowing for which particular one he

were asked what they knew of, among other works, Parry's *Job*, Schoenberg's *Gurrelieder*, Strauss's *Ariadne auf Naxos*, Stravinsky's *The Firebird*, Holst's *The Hymn of Jesus*, and Ravel's *Daphnis et Chloé*.) Walton's sixth and last term of residence was Michaelmas 1920, when he did not live in college. On 20 October the Christ Church governors postponed renewal of his Exhibition 'for the dean and Censors to make a definite proposal as to the amount'. On 10 November they agreed to pay him a sum not exceeding £150 to 'clear him from his reasonable liabilities'. Walton's name was removed from Christ Church books at Christmas 1934 but was replaced in 1941 when he received the honorary degree of Doctor of Music. He was elected an Honorary Student in 1947.

Walton said in November 1977, in an interview with Arthur Jacobs, that he 'didn't want a degree. But Sir Hugh Allen and all those people who were looking after my education thought I should be a Mus. Bac. and I should be a schoolmaster. That was my career as mapped out by them, and I saw that coming and withdrew, I think.' A permanent residue of the uncertainty over finance at Oxford was Walton's abomination of poverty. Although some years were to pass before he was self-supporting, he grew ever more conscious of the value of and need for money, and was always to be acutely aware of the worth of his work in the market-place.

When one remembers that *Façade* was, in 1918, but three years away, it is obvious that Walton read, heard, and absorbed as much music as he could at Christ Church. Over sherry at lunchtime on Sundays, Dr Strong played avant-garde music of the day, and Walton and his friends laughed uproariously at the first of Schoenberg's *Six Little Pieces* for piano. Musical experiences were not confined to Oxford and visits to London: they were available in Lancashire too. His father took him to Hallé concerts in Manchester, where (as Walton told Stephen Williams in the *Evening Standard* of 5 November 1935) the first orchestral work he heard was Tchaikovsky's *Casse-noisette* suite ('I have never lost my affection for it'). An even stronger impact

prepared, and his name does not appear in any list of candidates. Walton did not submit a musical exercise (the third part of B.Mus.) because he never entered upon his ninth term of residence.

was made by Beecham's Manchester opera seasons in 1916 and 1917, especially by the colourful productions of *Boris Godunov* and *Le Coq d'or*. When he heard the Carl Rosa in *Madam Butterfly*, *Faust*, and *Mignon* in November 1918 he told his mother that 'they were done very well for a travelling company'.

At Christ Church in 1918 Walton began to compose his first large-scale work, the Pianoforte Quartet, which is dedicated to Dr Strong. ('I have a most lovely Bechstein upright in my rooms,' he told his mother.) This ambitious piece is of interest still,[9] partly because of its inherent romanticism and largely because of the clues it provides to the influences at work on the sixteen-year-old composer. Here, for the first time, the Elgarian flavour can be detected and parts of the slow movement sound like Vaughan Williams—a composer with whom Walton has little affinity. Or is this because the real influence in that movement was Vaughan Williams's teacher Ravel, since the theme of the Andante tranquillo is 'cribbed' directly from 'Le Martin Pêcheur', the fourth of the *Histoires naturelles*?[10]

In spite of its immaturities, the Piano Quartet was a significant milestone in Walton's development. Its unashamed romanticism is in a French rather than a German manner. It has four movements, the second being a Scherzo with a fugal section, based on the opening theme of the first movement (a procedure he followed in his A minor String Quartet over twenty-five years later). The work's opening theme, first played by the violin, in fact recurs throughout the work in various guises, another pointer to a structural feature of the concertos. The harmony and rhythm of the Scherzo have the pungency which was later to become familiar from *Portsmouth Point* and the Scherzos of the concertos. The manuscript score was 'lost in the post' between Italy and England for two years. When it resurfaced in 1921, Walton rewrote some of it, particularly the Finale. It was published in 1924. Thereafter there were intermittent performances and it was recorded on 78 r.p.m. discs in 1938 and on LP in 1954. The LP was made by the

[9] The Pianoforte Quartet was published in 1924 on the recommendation of the Carnegie United Kingdom Trust. The adjudicators were Allen, Vaughan Williams, and Sir Henry Hadow.

[10] I am indebted to Mr Angus Morrison for drawing my attention to this similarity.

Robert Masters Quartet, whose pianist, the late Ronald Kinloch Anderson, found that some passages were almost unplayable. He sought Walton's advice and was surprised to discover
with what accuracy the composer remembered details of his
early, then thirty-year-old, composition. Walton agreed to
some small changes in the piano part. After an Aldeburgh
Festival performance in June 1973, Walton decided to revise
the score now that the original publishers had relinquished the
copyright and Oxford University Press was proposing to bring
out a new edition. Anderson was invited to edit the score,
incorporating the earlier alterations. This was published in
1976 and is the edition used in all subsequent performances
and recordings.

While at Oxford, Walton made friendships which were to
endure for many years: with the poets Siegfried Sassoon and
Roy Campbell and the novelist Ronald Firbank. (His meeting
with Sassoon can be exactly dated from the letter he wrote
home on 12 February 1919: 'I went to the Musical Club last
night. We had a fine concert. I met John Masefield and
Siegfried Sassoon the poets. They are great men.') Most
important and lucky of all, certainly most significant, was his
meeting Sacheverell Sitwell, at twenty-one the youngest of the
three children of Sir George Sitwell, Bart. Sacheverell thought
Walton 'had a very intelligently shaped head. I felt he needed
support and help.'[11] As we know from his brother Osbert's
autobiography, Sacheverell designated Walton as a musical
genius and fetched Osbert to Oxford in February 1919 to meet
him and to hear this apparently tongue-tied genius play the
slow movement of his Pianoforte Quartet. 'Talk was desultory
. . . the atmosphere was not easy,' Osbert wrote.

Osbert's attention was principally reserved at this time for
'an undoubted and more mature musical genius, Bernard van
Dieren', and this enthusiasm was evidently soon communi-
cated to Walton, whose letters to his mother in 1919 convey a
vivid sense of the excitement of this new friendship:

[15 May] Sitwell came back the other day from Spain. I went to dine
with him last night. We are busily arranging about the van Dieren
concert which we hope to come off about June the 10th.

[11] Interview in BBC *Music Weekly*, 28 Mar. 1982.

[1 June] Sitwell and myself have arranged a terrific concert for the 13th of this month. We are doing a lot of van Dieren, and new songs by Delius and myself. If it is a success we ought to make about £20. But that is a mere detail.

[29 June] I am going to London on Wednesday to stay with the Sitwells . . . Our concert never came off. We had everything ready, bills out and everything and at the last moment Helen Rootham who was going to sing was taken ill so there was nothing else to do but to postpone it . . . However as a consolation some of my songs are going to be sung at Lady Glenconner's concert in London on Friday. Also I shall be conducting two pieces for orchestra at the Russian Ballet. [This surprising statement cannot be verified.]

[17 November] I went to hear Cortot give a recital last Monday. He is simply magnificent . . . Lady Ottoline Morrell asked me over to her house last Sunday. It was very entertaining. I went to London yesterday for the afternoon and saw the new ballet *Parade*. It was very marvellous, especially the scenery. The music was by Erik Satie, a Frenchman. I am to meet Stravinsky next month or perhaps before, so that will be too exciting for words.

Osbert Sitwell's first impression of Walton was of 'a rather tall, slight figure, of typically northern colouring, with pale skin, straight fair hair, like that of a young Dane or Norwegian . . . Sensitiveness rather than toughness was the quality at first apparent in him. He appeared to be excessively shy . . .'[12] Roy Campbell described him as

one of the greatest men who have been there [Oxford] this century, a real genius, one of the very finest fellows I ever met in my life . . . We walked out with two young ladies, who were also good friends, and who were employed as waitresses. Of course, needless to say, Willie's one eventually became a countess. Something magical seems to happen to everything he touches . . .[13]

If he had not met the Sitwells, how would Walton's career have shaped itself? Musically, he would probably have developed in the same way, but it might have been a slower and more laborious development and it certainly would have not been initiated by the unique and astounding *Façade*. Nor, perhaps, would he have visited Italy so soon. He loved Italy from his first visit with the Sitwells in the spring of 1920 and it undoubtedly influenced the sound of his music. The visit

[12] O. Sitwell, *Laughter in the Next Room*, pp. 169–72.
[13] R. Campbell, *Light on a Dark Horse* (London, 1951), p. 181.

'changed my whole attitude about life and music', he wrote.[14] It was made possible by a gift of £50 from Dr Strong. 'I somehow suspect that he knew it was much more worthwhile to me than passing examinations.' In old age, Walton's face lit up as he described the train journey across Europe ('raining all the way, just like Oldham') until they emerged from the Alps into Italy—'there it was, ablaze with sunlight. I've never forgotten it, a new world.'

Walton's sojourn with the Sitwells began as a result of his exam failures.

I said to Sachie 'What the hell am I going to do?' So he said 'Why not come to stay with us?' I went for a few weeks and stayed about fifteen years. They seemed to me very nice, extraordinary people. I'd never met anyone like them. I never knew why there was such a fuss about them. I didn't understand about intellectual circles, seeing I'd 'coom oop from Lankysheer', you know.

Even if the Sitwells had not materialized, Walton would have gone to London. Dr Strong, Hugh Allen, and Henry Ley were determined on that course. Adrian Boult

got a passionate letter from Henry Ley about this wonderful Lancashire boy who had been kept at Ch. Ch. by the Dean and was now to come to London and MUST find a job—could I do anything? . . . Anyhow I fixed up an interview (with Goodwin and Tabb, the publishers) and I thought W.W. nicely started proof-reading for a go-ahead publisher etc. etc. A few days later a letter arrived: 'Thank you for the introduction but I'm afraid I have decided to starve in a garret and compose all day rather than enjoy a nice job.' Three weeks later he took up residence with the three Sitwells!! Some starving! And the 'garret' was a jolly little house just off Chelsea Embankment.[15]

Walton became 'adopted, or elected, brother' to Edith, Osbert, and Sacheverell, living with them first at Swan Walk, Chelsea, and later in Osbert's house, 2 Carlyle Square. He had a physical resemblance to them, especially to Edith, and there was a rumour, which Sacheverell rather enjoyed, that he was 'the offspring of my father and Dame Ethyl Smyth!' Asked in 1977 how the Sitwells influenced his artistic and cultural

---

[14] H. Anson, *T. B. Strong: Bishop, Musician, Dean, Vice-Chancellor* (SPCK, 1949), p. 117.

[15] Letter to the author from Sir Adrian Boult, 1 Sept. 1977.

development, Walton replied: 'It was just general life, they weren't there all that much. My great pal in the house was the cook, Mrs Powell. When they were away she looked after me and they were away a lot.' Yet assimilation into this milieu did not of itself make Walton a great composer; had he been anyone else, indeed, it might have damaged him. But Walton's taciturnity was an outward sign of strength of character, and there is no reason to suppose that the sardonic wit and the ability to prick bubbles of pomposity and pretentiousness which characterized the mature man were any less potent in his teens and twenties—*Façade* being proof of this, in any case. Back in Oldham, not unnaturally, William's new habitat was regarded with suspicion and Charles Walton journeyed to London to investigate. Failing to find Osbert or Edith—or indeed his son—at home, he found his way to the Sitwells' octogenarian aunt, who charmed and reassured him.

2
London and Salzburg, 1920–3

THE primary result of Walton's 'adoption' was practical and financial: not only the Sitwells but Dr Strong, the composer Lord Berners,[1] and the poet Siegfried Sassoon combined to guarantee him an income of £250 a year, sufficient in those days to enable him to devote his life to composition without the need of a daily job. A secondary result was that the Sitwells provided a background, almost comically different from anything he could have expected in Oldham, in which his character and genius could mature and flower, in which art was paramount and war must continually be waged against the Philistines. The Sitwells, through their social position, were a link between the many and varied young talents which they gathered about them, as they had gathered Walton, and the older generation of wealthy patrons of the arts. The latter was personified by Lalla Vandervelde, daughter of Edward Speyer and wife of the Belgian Socialist leader Emile Vandervelde, and by Frank Schuster, friend of Elgar and Fauré and of Sassoon and Adrian Boult, presiding in his homes in Old Queen Street, Westminster, and by the Thames at Bray over gatherings of musicians, painters, and actors, all pursuing new artistic experiences whether they proved to be illusory or seminal. Walton recalled in 1977 that the Bloomsbury group (Lytton Strachey, Virginia Woolf, Roger Fry and others) was 'a circle on its own and the Sitwells were another circle and they

---

[1] Lord Berners (1883–1950) succeeded to the barony in 1918. He studied music in Dresden and Vienna and was a diplomat, mainly in Rome, from 1909 to 1920. His compositions, slightly influenced by Stravinsky, are among the best by an Englishman in the early 1920s. Their importance has been overshadowed by their author's much-written-about eccentricities. He was a friend of the Sitwells.

never met, so to speak. They knew each other, but they weren't on artistic terms, or any terms really.'[2]

Not that the Sitwells meant much in 1920 outside the small and exclusive circles in which they moved. Today their name is practically synonymous with that epoch, but it should be remembered that Osbert's first books of verse did not appear until 1919 and that only Edith (five years Osbert's elder as he was five years Sacheverell's) had achieved some sort of notoriety with her poems published in 1915 and 1917 and, even more, with the anthology of modern poetry, *Wheels*, which she edited from 1916 to 1921. If their aim was to make the English bourgeoisie sit up, their effect on some of their own class was amusingly expressed by Lady Colvin, wife of Sir Sidney, biographer of R. L. Stevenson: 'They are quite nice and amusing young people if only they would not write poetry.' The Sitwells' claim on posterity's interest may depend on their work as propagandists and catalysts as much as on their creative work.

The Sitwells claimed the credit for keeping Walton from entering either the Royal College of Music or the Royal Academy of Music. They brought him into brief contact with Ernest Ansermet, in London as conductor of the Russian Ballet when it first performed *Tricorne*, and with Edward Dent and Ferruccio Busoni. 'I am having lessons from Ansermet and Dent,' he informed his mother in September 1921; and he had earlier shown some of his music to Busoni, who was apparently not impressed. Writing to his wife on 5 July 1920, Busoni said: 'The young man Walton (who was at the Spanish Restaurant) sent me some manuscript music. He has a little gift for counterpoint. In other respects, they all write according to a formula: notes, notes, notes, all "hither and yon", without imagination or feeling.' The Sitwells also introduced him to literary figures, T. S. Eliot, Wyndham Lewis, and Ezra Pound. Ten years earlier, as Walton may have mused, he was playing marbles in the street in Oldham.

The Sitwells enabled Walton to attend concerts of contemporary music and thereby to keep 'in the swim'. All this,

[2] Interview with Arthur Jacobs for BBC World Service, at Savoy Hotel, 1 Nov. 1977.

however, could not have led to *Façade* unless Walton himself
had worked intensely hard. This we know he did. Because
Walton spoke of his music in a detached, Richard Strauss-like
manner, we should not be deluded into supposing that he was
any less obsessive about music than any great composer always
is, making any and every sacrifice and taking every opportunity
in order to create what must be created. Roy Campbell wrote
how Walton gave him, in their undergraduate days, 'a sense of
vocation and how a man can live for his art . . . William was
already equipped for greatness, with a metaphorical self-starter
and internal combustion.'[3] Cecil Gray, in his autobiography,
described

the slow, sure, steady way in which he has built himself up into a
mature, self-reliant personality out of such comparatively unpromis-
ing beginnings, and in his art has acquired a formidable technical
capacity which was lacking in his early days, by dint of sheer
unremitting hard work. If ever anyone has deserved success it is
William Walton; no one has ever worked harder to attain it . . .[4]

Osbert Sitwell's account in his *Laughter in the Next Room* of
Walton's early days among them tells a similar story. On
Walton's first visit to Swan Walk in July 1919 he spent most of
the time shut in his room at the top of the house sitting by the
window for hours on end 'eating black-heart cherries from a
paper bag and throwing the stones out of the window'. Bored
and lonely, obviously; until a piano was hired when 'he
remained in his room for longer periods even than before'.
Osbert stresses that Walton 'had always to be near a piano' and
was 'in despair' on an occasion when it seemed that he would
not be able to have the use of one during a holiday. It was the
same at Christ Church; and if the catalogue of his published
works gives the impression of an enormous leap from the Piano
Quartet of 1918–21 to *Façade* (composed November and
December 1921), we now know there was quite a considerable
completed output, granted the fastidiousness of Walton from
which one may infer many other discarded or abandoned
efforts. Two songs, 'The Winds' and 'Tritons', a setting for tenor
and chamber orchestra of Marlowe's 'The Passionate Shep-

[3] *Light on a Dark Horse*, p. 181.
[4] C. Gray, *Musical Chairs, or Between Two Stools* (London, 1948), pp. 285–6.

herd', and a 'pedagogic overture', *Dr Syntax*, inspired by a Rowlandson etching (as *Portsmouth Point* was to be), cover the years 1918 to 1921. On 24 September 1921 Walton wrote to his mother that 'Goossens has *Syntax* but has not returned from his holiday so the date of performance (?) is still indefinite'. It did not materialize. Another letter home dated 18 December 1921 refers to a performance of 'The Winds' 'last Wednesday'. From this period also dates the String Quartet, two movements of which, Moderato and Fuga, were written in 1919 for Walton's Exhibition at Christ Church. These movements were performed at the London Contemporary Music Centre's first concert on 4 March 1921, when they were played by the Pennington Quartet, formed at the Royal College of Music. Its violist, Bernard Shore, told Guy Warrack years later: 'Sir Hugh Allen asked us to play a quartet by a young lad from Oxford— William Walton. I regret to say that having played it, the quartet considered that it would sound better played backwards! Such was the effect on us youngsters at the time.' After this performance Walton rewrote the two movements and inserted a Scherzo and the work was selected for the festival of the International Society for Contemporary Music at Salzburg in 1923. It was played in London at the Royal College of Music on 5 July 1923 and at Salzburg on 4 August. A *Toccata in A minor* for violin and pianoforte of 1922–3 was performed in London in 1925.

'We were able', Osbert Sitwell wrote, 'to keep him in touch with the vital works of the age.' What were they? What was the English artistic climate in 1920–2? There is much to be said, in retrospect, for the theory that the single most vibratory impact on the arts in England in the first twenty years of this century was made by Diaghilev's Russian Ballet in 1911. Here, in a perfection of ensemble which legend leads one to believe has rarely if ever been approached, was an incomparable amalgam of music, dancing, and painting. The colour, the novelty of design, the sense of youthful artistry liberated, made an impression to which the word sensational can for once be justly applied. (Walton's description of its effect on him, after years of cathedral music, was that it caused 'rather a jolt'.) The music of Stravinsky, particularly his *Petrushka* and *The Rite of Spring* and of Ravel's *Daphnis et Chloé* had a far-reaching effect. There are echoes of *Petrushka* in Vaughan Williams's *A London Symphony* (1914), for instance, and in the music of Arnold Bax. When the Russian Ballet returned after the war, its colour was diminished but Diaghilev, who was as much an artist as an impresario, had discovered the great Falla score *Tricorne* for Nijinsky's successor, Massine, and, always sensitive to changes of public taste, had swung from the pre-war opulence and exoticism of *Scheherazade* to the frivolity of *La Boutique fantasque*. When, in 1921, he staged a marvellous *Sleeping Princess* in London, the theatre was half-empty night after night and he lost a fortune. Sensitive to this barometric reading, he put his faith instead in Satie's *Parade*, Stravinsky's *Les Noces* and *Pulcinella*, and Prokofiev's *Chout*, before turning to what has so perceptively been termed 'relentless pursuit of chic'.[5]

Of the non-Diaghilev composers, the influence of Debussy[6] was particularly strong. In the period just before the outbreak of war there had been a vogue for Skryabin, but this had faded by 1920. The colossus who bestrode the pre–1914 years, Richard Strauss, was *démodé*, too, written off (somewhat prematurely, as is now apparent) by the new generation of critics. Sibelius fever had not yet arrived in epidemic proportions. (His latest symphony, the Fifth of 1915–16, did not reach London until February 1921.) Bartók's impact was still to come. Janáček?—a name known to a handful of English musicians. *The Diary of a Young Man who Disappeared* was performed in England in 1922, otherwise, until 1926, it is doubtful if any of his music had reached these shores. Mengelberg's heroic Mahler Festival in Amsterdam in May 1920 sent no ripples across the North Sea. What of the great Viennese trio Schoenberg, Berg, and Webern? Except perhaps for a song or two—and even this is doubtful—there is no record of any music by Berg and Webern having been performed in England much before 1926. Schoenberg had fared rather better. There had been two performances in London (1912 and 1914) of the *Five Orchestral Pieces*, the second being

---

[5] The phrase is Mr Richard Shead's, whose brilliant biography of Constant Lambert (Lutterworth, 1973) I commend to all students of twentieth-century English music.

[6] True, he composed *Jeux* for Diaghilev, but he was not a Ballets Russes composer in the Stravinsky sense.

conducted by the composer; his First and Second String Quartets were performed in 1913 and 1914 respectively, and the *Kammersymphonie* in 1921; and *Pierrot Lunaire* had its first English performance in Kensington in 1923, conducted by Milhaud. But Webern and Berg had to wait until the 1930s and the formation of the BBC Symphony Orchestra for wider dissemination of their work in Britain.

Stravinsky was the darling of the London musical intelligentsia in the 1918 era. Thus, within a short space of time, came the first English performances of *Pribaoutki* (February 1918), *Three Pieces for String Quartet* (February 1919), *Ragtime* (April 1920, conducted by Arthur Bliss), the tone-poem *Le Chant du rossignol*, the Suite from *L'Histoire du soldat* (conducted by Ansermet) and *Berceuses du chat*, for contralto and three clarinets (all in July 1920), the *Trois histoires pour enfants* (1920), *Symphonies of Wind Instruments* (June 1921), and the *Suite No. 2*, for small orchestra (June 1922). It is likely that the Sitwells and their 'elected brother' sought out these performances, and some of these works were reflected in the music of *Façade* (particularly *L'Histoire du soldat*, as Walton admitted). Other musical echoes went into it, too, from France and Italy. Satie's assembly of Nouveaux Jeunes, of whom Auric, Poulenc, Honegger, Tailleferre, Durey, and Milhaud became known as 'Les Six' and collaborated in a notorious Cocteau farcical ballet, *Les Mariés de la Tour Eiffel*, in June 1921, attracted attention among their English contemporaries, and their music crossed the Channel surprisingly quickly. For example, thanks to the advocacy of André Mangeot and of Eugene Goossens, Honegger's *Sonatine* for two violins and First Violin Sonata were played in London in May 1922 and his *Pastorale d'été* for orchestra in October 1921. Poulenc's Cocteau songs, *Cocardes*, were sung in London in April 1921 and Artur Rubinstein played the twelve *Promenades* in the Wigmore Hall in July 1923. Some of these performances, as can be seen, were pre-*Façade*. As for Italy, its heat and light are in *Façade* even if we did not know (from Sacheverell Sitwell) that the fanfare which opens 'Long Steel Grass' was adapted 'from an itinerant fortune-teller's trumpet-call which we had heard in Syracuse or Catania'.

The final influence on the music of *Façade* was, of course, jazz.

Within a week of the Armistice, the first jazz band from New York appeared in Paris at the Casino, opening the ears of such as Milhaud to new possibilities of rhythm and sound. The craze had gained impetus before and during the war. Now it swept Europe, giving its name to an epoch, the Jazz Age. It is a more apposite title than the 'silly twenties' of which, ironically, *Façade* is often regarded as a frivolous symbol. The 1920s— which in the British Isles included the Irish troubles and the General Strike—were a time of experiment and ferment in all aspects of human endeavour. It is unfortunate that they should be superficially and journalistically remembered as if they had consisted solely of the post-war party habits of a very small segment of 'café society'. Much of what was produced in the 1920s has proved not to be ephemeral—jazz was not ephemeral, music written in the 1920s survives in the repertoire of concert-hall, opera house, and ballet, the 'brittle' work of Noël Coward survives (he mocked *Façade* in 1923, alas, but learned from it, as is shown by *Mad Dogs and Englishmen*), Gershwin and Kern survive, and *Façade* survives. So do other major English musical works of the time, Vaughan Williams's *Pastoral Symphony*, Holst's *Fugal Concerto*, Delius's *Requiem*, the chamber works of Ireland, Bax, Bridge, and Bliss, and the songs of Warlock.

France, with Les Six, was not the only country in which a new enthusiasm for music was being publicized and exploited to the full. In London the British Musical Society gave its inaugural concert in June 1921 with works by Holst, Vaughan Williams, Holbrooke, Eugene Goossens, and Cyril Scott. Seven weeks later the first Donaueschingen festival of contemporary chamber music produced works by Krenek, Berg, and Hindemith. In August 1922, in Salzburg, the first international festival of contemporary music was held, the brainchild of Edward J. Dent, friend of the Sitwells and of Busoni. Its programmes included songs by Strauss, Bliss's *Rout*, Bartók's First Violin Sonata (already heard in London a few months earlier), and works by Milhaud, Honegger, Stravinsky, Ravel, Falla, Pizzetti, Malipiero, Castelnuovo-Tedesco, Poulenc, Nielsen, Busoni, Szymanowski, Webern, Wellesz, Bloch, Hindemith, Kodály, Bax, Holst, Schoenberg, and Percy Grainger.

What a line-up, seems the only possible comment today. From this courageous enterprise grew the International Society for Contemporary Music, with its headquarters in London and Dent as its president. It was at its first festival in Salzburg in the following August of 1923 that Walton's String Quartet was performed at the third concert. Among other works performed that week were Berg's String Quartet, Bartók's Second Violin Sonata, Janáček's Violin Sonata, Hindemith's Clarinet Quintet, and Bliss's *Rhapsody*. The jury picked well, though they received little acclaim for picking the Walton.

The quartet, according to the report of the festival in *The Times* of 14 August 1923, was 'unfortunately placed' at the end of a long programme.

The impression it gave was first, that it might be considerably condensed, and second, that some passages might be rewritten with a view to making the work more grateful for the strings to play. The scherzo is the least effective part of the work. The lengthy treatment of an apparently meaningless figure of two notes lost the attention of the hall, and this was never regained. The result was that by the time the latter part of the fugue was reached, the restiveness of the audience made it almost impossible to listen attentively . . . No one grudges Mr Walton the performance of his quartet, but one may doubt the wisdom, from the composer's point of view, of forcing an immature work on the public notice.

This critic wrote of the all-women McCullagh Quartet having 'bravely and successfully overcome the many difficulties of the work'. Osbert Sitwell, writing to Frank Howes in 1942, said:

You can't conceive how *frightful* that poor ladies' quartet was . . . After all the best string quartets in Europe had been playing, these poor good English girls dressed in turquoise tulle put up an abominable performance, added to which the cellist—this you would hardly believe—got the tip of the prong of the cello into the thing that worked a trapdoor above which she was sitting, and began to go down! The audience *rocked* and even she 'came up smiling'... Poor things.

Alban Berg, however, took an interest in the quartet and took Walton and Lord Berners to meet Schoenberg, who was staying at a village in the Salzkammergut. 'As far as I can recall,'

Walton said, 'part of the conversation turned on the use of the piano for composing. I think he was in favour of its use.'[7] This contradicts another account of this incident by Walton, who said that they arrived early and found Schoenberg composing at the piano. He was very angry and sent them away to kill some time. When they returned, all signs that the piano had been in use for composition had vanished.

With the relative failure of the String Quartet, the year 1923—which also included the first public performance of *Façade*—was not auspicious for the twenty-one-year-old Walton. It is hardly surprising that, in the letter to Foss from which I have already quoted, he said that after the quartet he did not know which way to turn and 'produced some rather bad works in various styles now mercifully in the fire'. That statement is worth closer examination, but before it receives it there is something more to be said about *Façade*.

[7] R. M. Schafer, *British Composers in Interview*, p. 75.

3
Façade, 1921–79

EDITH SITWELL wrote her *Façade* poems[1] as studies in word-rhythms and onomatopoeia. They may appear to be non-sensical, but a continuous thread of allusions and images runs through them and evokes the bourgeois culture of turn-of-the-century England—references to Queen Victoria, Tennyson, the Greek goddesses, flowers, trees, the music-halls, Spanish lovers, Negroes, English girls and nursemaids. Satire and parody alternate with nostalgia and a haunting melancholy. The poet and her brothers decided they would be suitable for presentation as a drawing-room entertainment, a highbrow extension of country-house charades, and ought therefore to be accompanied by music. They had a tame composer on the premises, so Walton must write it. But he was uncertain what kind of music was required and declined the request. They had an answer to this: they would ask another of their *protégés*, the seventeen-year-old Constant Lambert. The prospect of rivalry was always to act as a spur to Walton and he began work. Edith was to recite the poems, with music as a commentary. 'I remember so well', Osbert wrote, 'the long sessions that my sister and William had in the rather small room he occupied upstairs and her going over and over the words with him, to show their rhythm and exactly how they went.'[2]

Walton himself said that *Façade* was Edith's idea. 'She began writing things like the Hornpipe as a deliberate kind of exercise. Then when she read it to us, Sachie said: "This would be much better if you had music with it." If you do a kind of whistle at the Hornpipe, then you see it fits in. It developed

---

[1] In the 1920s the poems appeared in collections of poems by Edith Sitwell entitled *Bucolic Comedies, The Sleeping Beauty, The Wooden Pegasus*, and *Troy Park*.

[2] Letter to Frank Howes, music critic of *The Times* and author of *The Music of William Walton* (London, 1965, 2nd edn., London 1974), dated 21 Nov. 1942.

from that.' The idea of having an independent musical score
against which the words were recited 'just grew'.

Osbert claimed to have thought of the manner of presenting
Façade, with a drop curtain, through which a megaphone
(actually a Sengerphone)[3] was thrust. The latter, he said, was
'to get rid of that embarrassment which the physical presence
of reciter seems to bring with it'. He has described the rehearsal
for the first performance in his drawing-room on a January
night in 1922.

The instrumentalists were puzzled and rather angry with the score, the
snow lay thick on the ground and threw up that tremulous vibration
of white light into the room through three windows, full of highly
coloured objects; the lights, too, were on, and it was gay, but the
wretched players were so cold they could hardly use their lips or
fingers. And this seemed to make them more cross and puzzled.
Fortunately my cook Mrs Powell had made some sloe gin the previous
autumn and I brought up an old Chianti flask full of it and dispersed it
in generous measure. They then cheered up. I remember that it was
the first time I had realised William's quality of wit and tunefulness,
as apart from his lyrical and elegiac.[4]

The first performance was on Sunday, 24 January 1922. The
Carlyle Square drawing-room, L-shaped and on the first floor,
could accommodate only about twenty people. They sat on thin
gold chairs. 'Across the narrow opening where had been the
conventional double doors now stretched the Dobson curtain,'
Osbert wrote in his autobiography.[5] The audience comprised
poets, musicians, and painters. They 'were naturally enthusiastic in their reception of *Façade*, for it was essentially an
entertainment for artists and people of imagination'. However,
in the letter to Frank Howes which has already been quoted,
Osbert said that Mrs Robert Mathias 'was one of the few
people not knocked out by it and who completely saw its
point'; and Walton himself said that the audience talked all the
way through and regarded Edith and him as mad. Walton
conducted the four instrumentalists, Paul Draper (clarinet),
Ambrose Gauntlett (cello), Herbert Barr (trumpet), and Charles

---

[3] A device invented by a Herr Senger to magnify the voice of the bass
singing the role of the dragon Fafner in Wagner's *Siegfried*. The statement that
he himself sang Fafner at Bayreuth is incorrect.

[4] Letter to Frank Howes, 21 Nov. 1942.

[5] O. Sitwell, *Laughter in the Next Room*, p. 190.

Bender (percussion). Mrs Mathias asked if a repeat performance could be given in her house in Montagu Square and this
was given a fortnight later, on 6 February.

It becomes ever clearer that before reaching the form in
which it is known today, *Façade* underwent a long gestatory
period of evolution stretching from 1921–2 to 1928. How much
would we recognize of the original version as it was performed
privately in Carlyle Square and Montagu Square? Not very
much, for of the eighteen items performed, only six have
survived into the definitive version published in 1951. These
are: 'Lullaby for Jumbo', 'Hornpipe', 'Long Steel Grass', 'Sir
Beelzebub', 'En famille', and 'Mariner Man'. Gone are 'Switch-
back', 'Springing Jack', 'The Octogenarian', 'Small Talk', 'The
Wind's Bastinado', 'Rose Castles', 'Madam Mouse Trots', 'Bank
Holiday', and 'Aubade'. By the time *Façade* was first publicly
performed, on 12 June 1923, ten more poems had been added
and the audience that afternoon sat through twenty-eight
items, among them 'Herodiade's Flea', 'The Owl', and 'Ass
Face'. 'Long Steel Grass' was temporarily retitled 'Trio for Two
Cats and a Trombone'. Among the additions which have
survived were 'By the Lake', 'A Man from a Far Countree',
'Through Gilded Trellises', 'Valse', 'Jodelling Song', and
'Something Lies beyond the Scene'. At this time 'Daphne' was
included. Walton may have had little thought of putting *Façade*
into publishable form, for he planned another Edith Sitwell
work in 1924, *Bucolic Comedies*, for voice and six instruments.
This was to comprise five songs, all items from *Façade*. Three of
them—'Daphne', 'Through Gilded Trellises', and 'Old Sir
Faulk',—became the *Three Songs* of 1932 written for Dora
Stevens (Mrs Hubert Foss).

The same four players gave the 1923 performance with the
addition of Robert Murchie (flute and piccolo) and F. Moss
(alto saxophone). Walton again conducted and always said that
the performance was musically 'a shambles'. The Sitwells
themselves recalled hostile demonstrations from the Aeolian
Hall audience during the performance, but these seem to have
been their own invention. Walton even persuaded himself that
there had been trouble. But there is no mention of any 'booing
and hissing'[6] in the diaries of Evelyn Waugh and Virginia

[6] S. Walton, *William Walton: Behind the Façade* (Oxford, 1988) p. 58.

Woolf, both of whom were present. Mrs Woolf merely wrote of
'listening, in a dazed way, to Edith Sitwell vociferating
through the megaphone'. The hostility, almost unanimous, was
in the Press next day: 'Drivel They Paid To Hear' was a typical
headline. Most of the comment was fatuous. Only the critic of
Vogue perceived that the function of the music 'seemed to be to
sprinkle jewels and flowers with an apt hand on the pathway
of Miss Sitwell's poetry'. Probably therefore Osbert Sitwell
was exaggerating when he remembered the 'hostile' atmosphere which obliged him and his associates to feel as if they
had committed murder when they resumed their social round.
'When we entered a room, there would fall a sudden unpleasing hush. Even friends avoided catching one's eye'. Herbert
Howells deeply offended Walton by saying to him not long
after the first performance: 'Hullo, William, still fooling
around?' Walton retaliated by quoting to his friends the first
two lines of a quatrain he had learned at Oxford:

> O Howells, Howells,
> He's good for the bowels.

In the audience on 12 June was Noël Coward, then twenty-
three and already a celebrity after his *The Young Idea*. Three
months later, for a revue *London Calling!*, at the Duke of York's,
he wrote a sketch called 'The Swiss Family Whittlebot' in
which the poet Hernia Whittlebot recites poems with her
brothers Gob and Sago Whittlebot. *Façade* was rather effectively
parodied:

> Thank God for the Coldstream Guards,
> Guts and Dahlias and billiard balls—
> Swirling along with spurious velocity,
> Ending what and where and when
> In the hearts of little birds,
> But never Tom Tits.

Walton, as one would have expected, rather enjoyed this
sketch, but Osbert Sitwell was furious and wrote a savage
letter to Coward. Their feud continued for years—writing to
Frank Howes in 1942 Osbert still referred to 'Mr N. Coward's
silly and spiteful music-hall skit'.

The next *Façade* performance, containing twenty-six items,
was at the New Chenil Galleries, Chelsea, on 27 April 1926,

when it became the talk of the town. Dora Stevens, who later married Walton's publisher Hubert Foss, was there with her mother, brother, and sisters and never forgot 'the excitement of that evening . . . William came on to receive the wild applause. Pale, willowy and shy, apparently, but rather appealing and charming, he made his bows . . . At last the clapping and shouting stopped and we wound our way through the crowd down the hall, seeing many famous faces, including those of Arnold Bennett and Augustus John.' Lambert helped Walton as copyist for this performance and, according to Sacheverell Sitwell, was 'dazzled and amazed at the speed and unerring understanding of the settings'.[7] Ernest Newman, in his *Sunday Times* review dated 2 May, described Walton as 'a jewel of the first water' as a musical joker. 'Mr Walton ought to seek out a librettist after his own heart and give us a little musical comedy in the jazz style.' Direct evidence of Lambert's connection with *Façade* was the inclusion at this performance of a new item, 'Four in the Morning', of which he composed the first eleven bars. Other additions were 'Country Dance', 'Scotch Rhapsody', 'Trams', 'Polka', 'Old Sir Faulk', and 'Tango' (listed as 'I do like to be beside the seaside'). The actor Neil Porter was the reciter at the April performance, but when *Façade* was repeated at the New Chenil Galleries two months later, on 29 June, he was replaced by Lambert, then still two months short of his twenty-first birthday. 'Tarantella', 'March', and 'Mazurka' were first included on this occasion. Walton regarded Lambert as the best reciter of *Façade*, Peter Pears the next best.

The friendship with Lambert began in the Sitwells' house, where Walton also met Philip Heseltine (Peter Warlock), whom he did not much like, Bernard van Dieren, Spike (Patrick) Hughes, Cecil Gray, and the pianist Angus Morrison. Heseltine and van Dieren were enthusiasts for Delius's music. Replying to some comments by his publisher, Alan Frank, in 1967, Walton wrote:

I always disliked D. and his works so it doesn't really surprise me that he behaved like that on his deathbed! . . . On thinking over about V.D. [van Dieren] and the Warlock lot, I don't think they were a very savoury lot or really produced anything in particular. W. is either

---

[7] S. Sitwell, *Façade: An Entertainment* (Oxford, 1972), p. xv.

'Elizabethan hearty' or nondescript wanderings in the V.D. and Delian styles. As for V.D.'s music it is as far as I remember so invertebrate that it would hardly stand up.

Walton first met Delius, through Heseltine, while the composer and his wife Jelka were living in Hampstead in March and April 1921. Some years later Jelka's memory of the encounter was recalled in a letter by the soprano Cecily Arnold: 'She told me . . . he sat for 2½ hours & never said a word in spite of all her efforts—he was so shy!'[8]

Walton and Morrison, exact contemporaries, met briefly at a Wigmore Hall recital in 1920. Morrison, as he recalled later, was impressed even then by 'a sense of latent power, an inner fire, still hidden below the surface but which one felt would inevitably declare itself in the fulness of time'.[9] Walton cultivated his friendship with these musicians, perhaps because he was aware that, for all the advantages of his life with the Sitwells, he was not as well versed in the craft of music, in its practicalities, as he would have been if he had been allowed to enter a music college. He grew fondest of Lambert, brilliant as composer, conductor, and critic, dissipating his marvellous talents on all three—and other—occupations. Lambert was his closest musical ally, though he himself lacked that ferocious determination to exist solely to create music which Walton's contemporaries recognized in Walton—for example, Heseltine: 'Walton is specifically musical or nothing.' K. Goldsmith (violin) and Morrison gave the first performance, in Queen Square, Bloomsbury, on 12 May 1925 of the *Toccata* for violin and piano which Walton had written in 1923. Walton later withdrew it, though Lambert preferred it to the String Quartet. It had 'greater and more genuine vitality'[10] and showed 'traces of Bartók and even Sorabji'. Walton was introduced to the composer and critic Cecil Gray by Philip Heseltine. In his autobiography, Gray wrote: 'He was a shy, diffident, awkward, inarticulate, rather devitalized young man, and I frankly admit that I was not greatly impressed either by

[8] Lionel Carley, *Delius, A Life in Letters, ii, 1909–1934* (London, 1988), p. 244.

[9] A. Morrison, 'Willie: The Young Walton and his Four Masterpieces', talk given at British Festival of Recorded Sound, 31 Jan. 1984 reprinted in *RCM* magazine 80 3 (1984), pp. 119–27.

[10] *Boston Evening Transcript*, 27 Nov. 1926.

his personality or by the work of his which I saw. But Philip thought otherwise. "You will see", he said to me one day, "that youth will go a long way." '[11]

Walton and Lambert shared an enthusiasm for jazz which embraced Duke Ellington and the coloured singer Florence Mills (who first appeared in London in a Cochran revue in May 1923). It was about this time that Walton worked with Debroy Somers's Savoy Orpheans but never had anything played 'except at rehearsals . . . They were more occupied with the current "hits" to bother very much about my somewhat clumsy efforts . . . I wasn't quick enough really to be of any use. I used to be allowed a free tea. Quite a help in those days'.[12] This work was, I suspect, quite a help to *Façade* but Lambert's influence was probably even more significant. It was Lambert who provided an important clue to the transformation of *Façade*. Writing in *Radio Times* of 7 August 1936 he said:

The same period [i.e. 1925–6] also produced the second version of *Façade* . . . In the original version, which dates from his Central European period, the instruments were mainly occupied by complicated arabesques and the melodic interest was slight—the second version, however, is one good tune after another and each number is a gem of stylisation and parody.

This helps to explain why Paul Draper at the first 1922 performance asked: 'Mr Walton, has a clarinet player ever done you an injury?'

The next *Façade* performances were the two during the morning of 14 September 1928 at the ISCM festival in Siena, when Lambert again recited and Walton conducted. Twenty-two items were performed. What has become the best-known number, 'Popular Song', was first included at Siena.[13] The poem was specially written for Lambert. 'Black Mrs Behemoth' was another newcomer. The 'March' was added to the twenty-one of the definitive version, with 'Daphne' substituted for the much too English 'En famille'. Walton described the Siena performance in a broadcast he made in 1951:

[11] *Musical Chairs*, p. 285.
[12] Letter from W.W. to Alan Frank.
[13] Walton told Edward Greenfield that he wrote the music for 'Popular Song' in Oldham.

It seemed to be going quite nicely, in fact the audience liked the 'Popular Song' so much that we had to play it again and I had to come in front of the drop-curtain to take a call in my shirt-sleeves. But a couple of numbers later and the whole theatre was in an uproar. Things were being thrown on to the stage, people whistled and shouted and protested in the name of Mussolini and the entire Italian people. I was just a little bewildered because all they'd heard was the 'Tarantella'. The trouble was that they considered they'd been mortally insulted. I had parodied Rossini and poked fun at a composer whom, heaven knows, I love above nearly all others. As a result, I'm always a little apprehensive whenever I conduct that 'Tarantella'.[14]

Spike Hughes recalled another amusing incident in Siena. Twenty thousand people had taken their places in the piazza

to watch the Palio, the famous inter-commune, bareback horse race which has taken place in the old Tuscan hill town of Siena every year for the past five or six hundred years. The course round the square was cleared; trumpeters in mediaeval costume played fanfares; the expectant crowds turned their eyes towards a side street from which the traditional procession preceding the race was to emerge. As the final flourish came to an end, there appeared in the piazza not the blaze and pomp of the *folkloristico* procession, but the frail and rather unsteady figures of William Walton and Constant Lambert, who had stopped off at a dark little wine shop and, emerging into dazzling sunlight, had lost their way to their seats. 'I must say' (William Walton has a habit of starting his sentences with 'I must say'), 'I've never had an ovation like it since.'[15]

The twenty-one items of the definitive version were performed on 29 May 1942 in London, but not in what we now regard as the canonical (published) order. In fact, for this performance no full score was extant and the work had to be reconstructed from the parts. Walton took back the score in 1947–8 and made various amendments. At this time, probably, he decided on the present order and, when the score was at long last published in 1951, it was dedicated to Lambert, who had suggested the arrangement of the poems into seven groups of three as a parody of Schoenberg's 'three times seven' in *Pierrot Lunaire*. Incidentally, Walton's statement to Edward Greenfield that he did not hear *Pierrot Lunaire* until he bought a record of it long after the Second World War was incorrect. He

[14] 'Records I Like', BBC Light Programme, 14 Aug. 1951. Walton was introduced by Spike Hughes.
[15] P. Hughes, 'Nobody Calls him Willie now', *High Fidelity*, Sept. 1960.

later said that 'tho' of course I knew *Pierrot Lunaire* from the small score', the first time he *heard* a performance was in 1942 when it was done in London in a double bill with *Façade*, with Hedli Anderson as the singer ('rather a dim performance, not inspiring').[16] Nevertheless, Angus Morrison remembers seeing *Pierrot Lunaire* among a number of modern scores Walton had in his room in Carlyle Square. Lambert used to tell an anecdote, which he swore was true, that Walton found a passage for clarinet in the score which he was sure was unplayable. So, to discover if this was so, he incorporated it into one of the first of the *Façade* items (perhaps it was this that provoked Paul Draper's protest!). Morrison thought that the original *Façade* items were more obviously influenced by *Pierrot Lunaire* than those added later. As the work progressed, burlesque and parody became predominant and Edith Sitwell's poems were based on dance rhythms (tango, polka, waltz, etc).

The complicated history of *Façade* did not end with publication of the score. In 1977 some previously unpublished numbers were performed to mark Walton's seventy-fifth birthday. Under the title *Façade Revived*, these were 'The Octogenarian', 'Aubade', and 'Said King Pompey', all from the original 1922 version, plus 'Herodiade's Flea', 'The Owl', 'The Last Galop', 'Mazurka', 'Daphne', which had dropped out after 1928, and 'March', which had been included only in the 1926 and 1928 performances. For the 1979 Aldeburgh Festival, the composer reworked some of these and other numbers as *Façade 2*, which he dedicated to Cathy Berberian. She, with Robert Tear, later recorded them. They are: 'Herodiade's Flea' (under the title 'Came the Great Popinjay'), 'Aubade', 'March', 'Madam Mouse Trots' (dropped after 1926), 'The Octogenarian', 'Gardener Janus Catches a Naiad' ('Baskets of Ripe Fruit'), 'Water Party' (dropped after 1922 and 1923), and 'Said King Pompey'. One can hear from these items that *Façade* as originally conceived was a sparer, more 'modern' work than it became. Yet one may doubt how Schoenbergian it was, even at the very first performance. One will never know, for the reworking of *Façade 2* rendered the items nearer to the style of the definitive *Façade*. The clarinet and cello solos which open and close the 'Aubade'

[16] Letter to S. Craggs, 22 Nov. 1977.

are perhaps more 'advanced' than some other passages, but are
scarcely atonal. *Façade 2* has not so far 'caught on', but it should,
in due course, because the hornpipe-ish 'March' with its
piccolo start, its castanets, and its jolly trumpet tune, is a kind
of miniature *Portsmouth Point*, the jungle drums in 'Madam
Mouse Trots' are a bizarre touch, and the saxophone and
clarinet at the start of 'The Octogenarian' convey a mysteriously
chilling suggestion of old age. The guitar-like accompaniment
to 'Water Party' is a burlesque touch and the trumpet and
tambourine in 'Said King Pompey' bring this appendix to a
hectic finale. It is perhaps necessary to remind some readers
today that the popularity of *Façade* in its original form is a post–
1942 development. Before then it was known mainly as a ballet
(1929) and in the two orchestral suites (1926 and 1938).

It is misleading to regard *Façade* only as a 'false start',
frivolous and unrepresentative. It is unique and unrepeatable,
as Walton wisely realized, but it contains the essence of him—
even to the 'conservatism' which earned him so much dis-
repute in advanced circles, for what is one to say of an *enfant
terrible* who was so obviously poking fun, witty and elegant
fun, at the modish Parisian-type up-to-dateness in which he
had been immersed up to the neck? Musically, the rhythmical
'snap', the tango-like languor, the genius for a pregnant
melodic phrase are all present. Yet most remarkable of all—in
retrospect—is the work's poetic vein of nostalgia which is not
something that has been added by the passing of years. For all
that *Façade* is rightly called an 'entertainment', for all its
hornpiping, its fanfares, its music-hall parodies, its mock-
Rossini and jodelling, for all its wit and skittishness, beneath
and through these there runs an awareness of true romance—
magical like a spell—evoking warmth of feeling as well as
sharpness of satire:

> And all the ropes let down from the cloud
> Ring the hard cold bell-buds upon the trees—coda
> Of overtones, ecstasies, grown for love's shroud.

4

With the Sitwells, 1923–9

THE poignant numbers of *Façade* recall that it is a work dating from a time when the world was in a period of violent and sudden change, an old order giving place to new. This was the age, in musical England, that rejected Elgar, not yet being able to hear in his music the sound of the ebbing sea. For a symbolic image of a relevant aspect of this period one may return to those soirées presided over by Frank Schuster which tried after 1918 to pretend that only four years by the calendar separated them from 1914. Elgar himself knew that the party was over: his Cello Concerto is musical proof of that, and when he returned in the early 1920s to the ambience of Schuster and of his other patron, Sir Edgar Speyer, he felt like an uninvited guest: 'I could not enjoy it . . . the atmosphere has gone never to return,' he wrote to a confidante. But a much younger man and another poet, though with words—Siegfried Sassoon—captured a memorable glimpse of Schuster at one of his parties peering unsmilingly down from a gallery at his dancing guests 'with the mask of social sprightliness discarded, and as though he were comparing this latest evocation of festivity with those many others which had gone their way of impermanence with the caravan that starts for the Dawn of Nothing when the door is closed behind the last departing guest'.[1] In *Façade*, words and music, this also found expression:

> Why did the cock crow,
> Why am I lost
> Down the endless road to Infinity toss'd?

Osbert Sitwell, from a considerably more sardonic viewpoint, has also used the Elgar-Schuster symbol for the passing of a way of life. Though antipathetic to Elgar's music, he went,

---

[1] In *Siegfried's Journey* (London, 1945), pp. 122–6.

with Walton and Sassoon, to Schuster's seventieth birthday party for Elgar at Bray in June 1927 when the three chamber works were played and 'one could almost hear, through the music, the whirr of the wings of the Angel of Death' (no acknowledgement from Osbert to John Bright for that phrase). When, in later years, I wonder, did Osbert Sitwell realize the irony of that day: that the twenty-five-year-old composer whom he had befriended and patronized, as Schuster had patronized Elgar, was created in the mould of Elgar, both as man and musician, that the youth from the humble Lancashire home and the old man from the humble Worcester shop had more in common than had Walton and Sitwell? And what were Walton's thoughts that day? He had attended one of the earliest performances of Elgar's Violin Sonata at Oxford in May 1919, and he was sufficiently 'committed' in 1930 to be one of the eighteen signatories of the letter sent to editors of musical journals in England and on the Continent protesting against E. J. Dent's cavalier treatment of Elgar in a German encyclopaedia. Spike Hughes, in an article in an American magazine in 1938, described his first meeting with Walton in the early 1920s in the gallery of Queen's Hall. 'We were listening to Elgar's Second Symphony. I had a score; we got talking . . . we found that we both happened to like this Elgar symphony and that we daren't tell anybody because they would undoubtedly sneer at us.' Walton's attitude to Elgar was made explicit in an interview he gave to the *Yorkshire Observer* of 21 December 1942 while he was rehearsing the Huddersfield Choral Society in *Belshazzar's Feast* preparatory to recording it a fortnight later in Liverpool. 'I have unbounded admiration for Elgar,' he said, 'I like even *Salut d'amour*. There's no other English composer to touch him. He's bigger than Delius, bigger than Vaughan Williams. He's becoming bigger all the time. Consider his *Falstaff*. It's much finer than any of Strauss's tone-poems. Some day that will be generally accepted.' It is perhaps necessary to remind present-day readers that these were very unfashionable opinions in 1942. In due course Walton came to be regarded by many English musicians as Elgar's successor; and in 1947 Elgar's daughter, Carice Elgar Blake, gave him her father's Savile Club tie.

In the years from 1923 to 1927 Walton was casting about to

find his true idiom and sampling every new experience, striking out at random and taking something here, rejecting something there, as the self-taught man must. If the String Quartet dabbled in a kind of atonality—a strange mixture of late Beethoven and Schoenberg—was it a sign that Walton veered for a time towards the views expressed in Cecil Gray's *Survey of Contemporary Music*, published in 1924 but mainly written in 1920? This is a fascinating compendium of the fashionable views of mid–1920 'advanced' English criticism. Were the parodies in *Façade* aimed at Les Six as 'a group of charlatans and mountebanks'? Gray savaged Strauss, gave Elgar a cautiously respectful nod, dismissed Debussy, misjudged Ravel and Bartók, and scorned Stravinsky as a clown, already *démodé* and barren. The saviours were Van Dieren, Schoenberg, and Sibelius. Against such a background it can scarcely be wondered that Walton did not know which way to turn.

Walton was dismissive about his Savoy Hotel experiences, but they deserve closer scrutiny. Debroy Somers (1890–1952) was in 1920 appointed arranger and adviser to the Savoy Havana Band formed by the American band leader Bert Ralton and later directed by Cyril Ramon Newton. The better-known Savoy Orpheans were not formed until October 1923, making their first broadcast—conducted by Somers—on 30 October. Walton was introduced to Somers by Richmond Temple, a director of the hotel. Invited to make arrangements of foxtrots for them, Walton spent a year in fruitless endeavours. His interest in jazz had led him to improvise tunes at two pianos with Angus Morrison, but he never committed any to paper. However, in 1923 he completed a large-scale project, a *Fantasia Concertante* for two pianos, jazz band, and orchestra. 'I have to see what can be done with my concerto with these Savoy people,' he wrote to his mother from 2 Carlyle Square on 4 May 1925, 'though I am afraid that there is only a remote chance of anything satisfactory coming of it.' Nothing did. According to Constant Lambert, the concerto—'monumentally planned'— was finished and about to be performed but Walton 'suddenly abandoned the jazz style in a fit of disgust, rightly realizing that the virtues of ragtime, its pungent timbres and intriguing syncopation, are more than handicapped by the deadly mono- tony of the four-square phrases, the inevitable harmonic clichés stolen from Debussy and Delius, the trite nostalgia of the whole atmosphere'.[2] William de Mornys, director of the Savoy dance bands, recalled that in 1924 Gershwin visited London to write his musical *Primrose* (which opened at the Winter Garden on 11 September). 'He gave a "first" of the *Rhapsody in Blue*. At the Savoy it had a snob success. On the concert platform at the Queen's Hall, it had no success and worse on the provincial tours. There was no interest then in that type of music and I feared for the *Fantasia*.'[3]

Walton met Gershwin once. He was friendly with Vladimir Dukelsky, the Russian-born composer who wrote the song 'April in Paris' under the pseudonym Vernon Duke. Diaghilev staged his ballet *Zéphyr et Flore* in 1925. Dukelsky took Walton to meet Gershwin, who was living in a flat in Pall Mall. 'I had been an admirer of his brilliant and captivating tunes,' Walton said. 'He was in the middle of writing his Piano Concerto in F, and I was hypnotized by his fabulous piano-playing and his melodic gift.'[4] (This meeting must have been in May 1925, when Gershwin began to sketch the concerto while in London for the opening of his revue *Tell me more*. At this meeting, Walton said, Gershwin played *Fascinating Rhythm* to him.)

According to Lambert, 'some unmistakable touches of Gershwin' entered the score of the incidental music Walton wrote for Lytton Strachey's play about the Boxer Rising, *A Son of Heaven*, produced in July 1925. 'Easily the dullest play we have walked out of since Arnold Bennett's *The Bright Island*,' the critic of *The Curtain* wrote. Walton told Strachey's biographer Michael Holroyd that the music had 'long ago disappeared. If I remember rightly, he [Strachey] was not very interested in *A Son of Heaven*, in fact I suspect he was against it being put on at all.'

In the spring of 1925 Walton and the Sitwells went to Spain, where Walton was much taken with the rhythm of the Catalonian national dance, the *sardana*. Its influence can be detected in the overture he composed while there, *Portsmouth Point*, although he had begun it before leaving London: the main theme, he said in a BBC interview in 1971, came to him

[2] *Boston Evening Transcript*, 27 Nov. 1926, p. 5.
[3] Letter to Stewart Craggs, 9 Feb. 1976.
[4] *Sunday Telegraph*, 25 Mar. 1962, p. 8.

while he was travelling on the top of a No. 22 bus. The title was taken from the etching by Rowlandson. Lord Berners had begun a ballet of the same name in 1920 but abandoned it. (Walton did not know of this until many years later.) Walton's overture was chosen for performance in Zurich at the ISCM festival. This involved him in copying out all the orchestral parts, a task that occupied six weeks. But it was the work that led to his lifelong association with the Oxford University Press Music Department. This had been founded in 1923 with Hubert Foss at its head. Walton was drawn to its attention by Siegfried Sassoon and he eventually signed a contract giving him a royalty on every copy of *Portsmouth Point* sold and half the Performing Right Society's fee. Writing to Sassoon on 5 October 1926, Walton said: 'I am letting you know about the fruits of your labours. Foss has taken not only those songs but also P.P. Also he has made a contract for five years to publish my works. Thank you so much for the trouble you've taken.' When the full score of *Portsmouth Point* was published in June 1928 it carried a dedication to Sassoon. The piano duet version was published in 1925, the first Walton score under the Oxford University Press imprint.

The overture was first performed in Zurich on 22 June 1926, conducted by Volkmar Andreae. Walton thought it sounded 'ghastly'. Its first London performance was six days later, when Eugene Goossens conducted it as an interlude in an evening of Diaghilev ballets. *Portsmouth Point* is the prototype Walton scherzo. There is more of the *enfant terrible* about *Portsmouth Point* than about *Façade*, which sounds the more mature work (perhaps because of the rewriting mentioned above). Rowlandson's print depicts a lusty, rowdy, and eighteenth-century quayside scene and the music is lusty, rowdy, and twentieth century in its Stravinskyan harmonic acerbity, its cocking of snooks at academic forms, its hornpipe irreverence, constant changes of time signature, irregular rhythms, and brash and brassy orchestral sound. Walton and Lambert had done their share of roistering with Warlock in Kent; the effects went into Lambert's whole mode of life, eventually to destroy him. They went into Walton's *Portsmouth Point*. The fact that a work by a twenty-four-year-old English composer had been selected for an international festival was in itself sufficient to reinforce the general view of him as belonging to the avant-garde. But there must have been some, even then, who noticed that *Portsmouth Point* signified a retreat towards insularity. With its nautical tunes and its added-note diatonicism, it belongs to an English tradition of honourable vintage. The sailors of HMS Pinafore had had a night on the tiles. It was Walton's sole (and characteristically wry) gesture at this period in the direction of the picturesque English past which inspired Warlock's Elizabethanisms and works by Lambert with movements entitled Sarabande, Coranto, and Brawles.

The dedication of Walton's next work, *Siesta*, a short piece for small orchestra, was to another member of the Sitwell entourage, Stephen Tennant, fourth son of Lord and Lady Glenconner, who was nineteen when he and Walton first met in 1925. Walton wrote *Siesta* at Haus Hirth, the house near Munich where the Glenconners stayed.

He, with Siegfried Sassoon, Rex Whistler, and Beryl de Zoete came to stay with me [Tennant recalled]. I was very jealous of his genius. Like all men of true genius, he was a remote, mysterious young man, very quiet and enigmatic. Christabel Aberconway adored Willie. But nobody understood him, really. I often went to the Russian Ballet with him. On the whole he was not very easy to talk to and had no conversational virtuosity. He was an enigma, a sphinx. He was devoted to me and I to him.[5]

Tennant, effeminate and with a tortoise for a pet, was the 'original' of Sebastian Flyte in Evelyn Waugh's novel *Brideshead Revisited*.

Siesta showed even more clearly what kind of composer Walton was—lyrical, romantic, and 'conservative', tapping springs of individual melody without which such a composer becomes merely a carbon copy. Walton is in a tradition, but he is an original. The shapes of his tunes, the sound of his orchestra, whether it is crackling or caressing, are unmistakably his own. Why is *Siesta* so neglected? It is a wistful, shapely miniature suffused by the 'Italian' atmosphere which Walton absorbed, and orchestrated with complete sureness. When *Façade* was performed at Siena, Walton went there with his publisher, Hubert Foss, who in a letter to his wife described an

[5] Letter to Stewart Craggs, 10 June 1975.

incident which vividly suggests the kind of experience evoked by Walton in Italian mood in many of his works from *Siesta* onwards. They had left a reception in order to walk round Siena

in pitch black starlight. Up one little street we stopped on hearing music. We were at the top of the steps to a lower level, and at the bottom was a tiny open space lit by one lamp. Four people were playing tangos on mandolines and whistling the tune with a flexatone to help, and one or two couples were dancing. It was such a beautiful sight, so simple and romantic and peasant-like and such a change from the idiotic reception.

In June 1961 Walton heard Sir Adrian Boult's recording and wrote to Alan Frank: 'It was interesting to me to hear *Siesta* which I'd almost completely forgotten about—a charming piece!'

Siesta had its first performance on 24 November 1926 at the second of the Guy Warrack series of chamber orchestra concerts in the Aeolian Hall, London. At this time Walton orchestrated some numbers from *Façade* as a suite (No. 1) to be performed as an interlude between acts of Berners's ballet *The Triumph of Neptune*, which was to have its première on 3 December. Although Berners himself later completely scored the ballet, he did not have time to do it all in time for the première and Walton helped him. In a letter to his mother in Oldham, he wrote:

I have been so terribly busy this last month. I have had to orchestrate four large numbers for Berners's ballet, and it has kept me hard at it. Friday was a great night—the production of the new ballet ... I conducted my *Façade* suite with good success and again on Sat. Quite a surprise my being asked to do it. I also conducted my piece *Siesta* at the Aeolian Hall about a fortnight ago.[6] Unfortunately I received nothing for it, but it is well worth doing for the experience. We had arranged to go to Rome next Sunday. Owing to this unforeseen event of my conducting (I am doing every night this week), it seems improbable that I shall be able to come and see you as I had intended to do, as the tickets are already booked and we are unable to change them without a great deal of expense, which I can't bear being in very low water, not having received very much for my orchestration. And Sachie is in

[6] This disposes of any doubt about the identity of the conductor of the first performance.

straits too, so I have to keep myself as much as possible. However I enclose you £2 which I am sorry to say is all I can send you at the moment. I am also sending Alec another suit.

Walton did not like conducting and said in later life that he only did it because record companies asked him, which was an over-simplification of the facts. 'At the same time,' he said, 'I think I did and certainly could conduct my music better than most conductors. For instance, *Portsmouth Point* was neglected because of conductors not being able to conduct it. So I had to conduct it myself because it was, to me, child's play and still is.'

In 1926 Constant Lambert received much publicity when he became the first English composer to have a ballet score chosen for performance by Diaghilev's company. This was *Romeo and Juliet*, performed in Monte Carlo on 4 May. This success had its usual effect on Walton—not surprisingly, since Walton himself had orchestrated the work for Lambert. He was strongly jealous of other people's successes—not of the person who had the success, but of the fact of the success itself. Given the chance, he believed, he could do better. In this instance, he was supported by Osbert and Sacheverell, who knew Diaghilev well. They arranged for Diaghilev and his entourage (Boris Kochno, Serge Lifar, Henri Sauguet, Georges Auric, and others) to go to lunch at Carlyle Square to hear three movements of a ballet which Walton had composed in the winter of 1925–6. Unfortunately there was only an upright piano in the house, so after lunch they all walked to Angus Morrison's home at 9 Oakley Street, where he and Walton played the music on two pianos. Diaghilev listened attentively, but rejected the ballet on the spot, saying Walton would write better things. For his second English ballet score he chose *The Triumph of Neptune*. At Lambert's suggestion Walton recast his score during 1927 as a *Sinfonia Concertante* for piano and orchestra, each of its three movements being dedicated to one of the Sitwells. (The dedications were omitted when the revised score was published in 1953.)

The *Sinfonia Concertante* had a very successful first performance at a Royal Philharmonic Society concert on 5 January 1928. York Bowen was the solo pianist and Ernest Ansermet conducted. The bill for the copying of the parts, always an expensive consideration for a young composer, was paid by Siegfried Sassoon (it was about £18). Writing in the *Sunday Times* after a Promenade Concert performance in 1929, Ernest Newman said of it: 'I have heard few new works lately that have given me so much pleasure as this, or filled me with such hope for the authors of them. Here is a young mind strong enough to throw over all the facile composition dodges of the last few years yet show itself unmistakably original . . . The slow movement . . . probes deeper than anything of its kind that I have heard for some time.' Although not on the level of the later concertos, the *Sinfonia Concertante* is a much finer work than its general underrating would suggest. Walton did not help, of course, by his honesty in not calling it a piano concerto; he made matters worse when he revised it in 1943 and retitled it *Sinfonia Concertante for Orchestra with Piano Obbligato*—not the way to entice virtuoso soloists to take it up. (He told Stewart Craggs in 1978 that he thought the original version was 'better and more interesting'.) Superficially one can easily detect the influences which went into its composition, not so much Stravinsky, as is usually said, but the Russians like Borodin and Rimsky-Korsakov, the Poulenc of *Les Biches*, and, most prominently, Ravel (for many clues to Walton's orchestration at this period, listen to *Alborado del gracioso*).

More significant is the foreshadowing of what was to come from Walton himself. The introductory Maestoso, for example, is the first mining of the seam that was to provide the Finale of the First Symphony and is followed by a passage of drooping sequences which are practically Elgar's copyright. Osbert is saluted in this movement by a flowing, pensive lyricism alternating with music in allegro spiritoso vein, similar to *Portsmouth Point* not only in its irregular rhythm but also in its ambiguous harmonic modulations and its avoidance of whole-hogging dissonance: Walton had chosen the middle of the road. The 'Edith' slow movement is the most beautiful and the most Ravelian. There are thematic relationships between each of the three movements—the family nose?—but the second theme of this Andante comodo also has (I think) an uncanny and perhaps intentional resemblance to the melodic fragment which accompanies 'Lullaby for Jumbo' in *Façade*. Again, ambiguity of key, use of 'false relations', and scoring of considerable delicacy and economy contribute to a more serene mood than Walton had hitherto explored, so that the opening movement of the Viola Concerto becomes in retrospect less unexpected. Notable also is his dexterity in building a strong emotional climax from no more than a one-or-two-bar melodic phrase by constantly varying repetition and the addition of a new 'head' or 'tail' to the original idea. The Finale (Sacheverell's movement) is one of the best of Walton's displays of jazzy humour, as good as any by Prokofiev and equalled in English music only by Rawsthorne and Williamson. Where this movement is disappointing is in the contrived coda after the cadenza, when themes from the earlier movements are summoned to form an imposing but not entirely convincing or satisfying conclusion.

After an accomplishment of this order, an even finer work was to be expected; nevertheless few would have prophesied not only the superb technical achievement of the Viola Concerto but its astonishing emotional depths, as though, through some profoundly shaking personal experience, the composer had aged more than his years almost overnight.

5
Viola Concerto, 1928–9

In spite of its twentyish trappings—echoes of Gershwin here and there—the Viola Concerto belongs musically and spiritually to the late romantic tradition: it is essentially within the orbit of Bax, Elgar, and Ireland, a mature, not to say middle-aged, outpouring. (To test this, listen to even the early works of the later-developing Michael Tippett: one cannot believe that he is of Walton's generation, being only three years younger. The gulf seems more like two decades.) Not the least of the extraordinary virtues of this concerto is the choice of solo instrument: the dark and huskily passionate sound of the viola became the perfect medium for the music's prevailing mood of plaintive melancholy. (Readers of Osbert Sitwell's *Laughter in the Next Room* will know the magnificent Rex Whistler drawing of Walton in andante comodo mood, made in April 1929 when this concerto was being written, mostly at Amalfi, Italy.)

Walton left for Amalfi in November 1928. A letter to Sassoon, postmarked 5 December, includes this:

I have been working hard at a Viola Concerto suggested by Beecham and designed for Lionel Tertis. It may be finished by Xmas and is I think by far my best effort up to now. Hoping to be here till April. I imagine I may get two other works finished as well. How long are you remaining in Paris? *Façade* entertainment is being done there sometime in January by the 'Pro Musica' society and the *Sinfonia* is being done on Feb. 22 by Ansermet with the new Orchestre Symphonique de Paris . . . I enjoyed your book immensely [*Memoirs of a Fox-Hunting Man*] and look forward to the second instalment.

He wrote to Sassoon again on 2 February 1929, seeking financial help. He went on:

Both Osbert and myself have been ill with a nasty attack of flu caught from the American tourists who descend on this place in their thousands . . . It has been a great nuisance as it has robbed us both of about three weeks valuable working time. Nevertheless, I finished yesterday the second movement of my Viola Concerto. At the moment, I think it will be my best work, better than the *Sinfonia*, if only the third and last movement works out well—at present I am in the painful position of starting it, which is always full of trials and disappointments, however I hope to be well away with it in a day or two . . . *Façade* has been put off in Paris till April . . . I have really got into this mess with the bank through the fault of the O.P. [Oxford Press] as they hadn't informed me that they didn't pay out royalties till the work was in print and I was expecting about £20 or £30 last October from the many performances [of the *Sinfonia Concertante*] here and America, now I find I can't get it till next October, and I had overdrawn on that expectancy . . . I am sorry to trouble you, but there is this bill also for my piano . . . It will be most kind of you if you can do anything about all this . . .

Ten days later he wrote: 'Thank you so much for your letter and a thousand thanks for the glad news—it is really most generous of you'. Walton also wrote to Angus Morrison from Amalfi, reporting progress on the concerto and saying: 'My style is changing. It is becoming more melodious and mature.'

On return to London in the spring of 1929 Walton sent the completed work to Tertis, whose courageous one-man campaign had restored the viola to the ranks of solo instruments and had elicited major works from Bax, Vaughan Williams, and others. Tertis rejected it by return of post, because of its modernity ('God knows why,' Walton commented in later years):

With shame and contrition [Tertis wrote in his autobiography] I admit that when the composer offered me the first performance, I declined it. I was unwell at the time; but what is also true is that I had not learnt to appreciate Walton's style. The innovations in his musical language which now seem so logical and so truly in the mainstream of music then struck me as far-fetched. It took me time to realize what a tower of strength in the literature of the viola is this concerto . . . [1]

Walton was hurt and bewildered and contemplated converting it into a violin concerto. He knew that solo violists were rare beings and, after all, he had only written the piece for Tertis—he used to say that he knew little about the viola when

[1] L. Tertis, *My Viola and I* (London, 1974), p. 36.

he started except that it made 'a rather awful sound'. This was a typically Waltonian over-simplification. He knew and admired Berlioz's *Harold in Italy* and admitted that in writing the concerto he had been much influenced by Hindemith's viola concerto, *Kammermusik No. 5*. He told Hindemith's biographer Geoffrey Skelton: 'I was surprised he played it. One or two bars are almost identical.'[2] According to Tertis, it was he who suggested Hindemith as a substitute—Hindemith had been violist of the Amar Quartet from 1923 to 1929. According to Walton, Edward Clark of the BBC musical staff sent the concerto to Hindemith, who agreed to perform it at the Henry Wood Promenade Concerts in the autumn. 'There is very good news about my concerto,' Walton wrote to Sassoon on 3 July 1929.

Hindemith is playing it on Oct. 3rd, myself with the 'bâton'. Also I wield it in the *Sinfonia* on Sept. 15th . . . I have more or less got to go to Germany to see Hindemith myself about the V.C. but how I am going to go, I can't think. He is going to be at the Baden-Baden Festival on the 25th and I hope to meet him there as it will be better than going to Berlin . . . I went to the new ballet *The Prodigal Son*. Scenery by Rouault, music by Prokofiev. Lovely the former and mediocre the latter, except for the end which is I think better than anything he has done. The dancing is very good, very eccentric and perhaps dull in patches. 'Jaggers' [Diaghilev] has got a new friend, a composer aged *16* called Igor Markevitch.

Walton had a new friend, too. The postscript to this letter begins: 'Imma has just rung in distress . . .' Imma was the widow of Baron Hans-Karl Doernberg and the daughter of Prince and Princess Alexander of Erbach-Schoenberg. She was twenty-eight. Her husband, twenty-six years her senior, had died ten months after their marriage in 1923. According to Lady Walton's biography, Walton and Imma first met at Edith Olivier's house at Quidhampton in February 1931, but this is erroneous. She was already on the scene during the composition of the Viola Concerto, which is dedicated to Christabel McLaren, wife of Henry McLaren, later Lord Aberconway. Although physically attracted to one another, Christabel and Walton never became lovers but remained lifelong friends.

[2] G. Skelton, *Paul Hindemith: The Man behind the Music* (London, 1975), p. 98.

Walton fell in love with the baroness at first sight. Sassoon described her as 'pretty, sweet, lively, and courageous . . . with a tall, graceful figure'.

Needless to say Walton went to Baden-Baden. A postcard to Sassoon on 28 July says: 'Arrived here safely last night. Thank you ever so much for enabling me to get here. I see Hindemith in half an hour's time.' Hindemith was there for a performance of *Lindberghflug* and he and Walton went through the score of the concerto. The first performance, at the Proms, was a success, even though the rehearsal was a shambles. The orchestral parts were full of mistakes, with bars missing and wrong notes, and Walton sat up all night correcting them. He was deeply grateful to Hindemith, though he did not delude himself that he was the ideal soloist. 'His technique was marvellous, but he was rough—no nonsense about it. He just stood up and played.'[3] Tertis was at the performance and his reaction was: 'The notes were all there, but the tone was cold and unpleasing and the instrument he played did not deserve to be called a viola, it was far too small.'[4] He realized what a mistake he had made and played the concerto on 4 September 1930 at the ISCM festival in Liège and on many subsequent occasions. Since then the work has never lacked performances from a variety of brilliant executants, the best of them, perhaps, being the late William Primrose. Some violinists have also transferred to the viola in order to play the concerto, including Yehudi Menuhin and Nigel Kennedy, who have recorded it, and Pinchas Zukerman.

In its design the Viola Concerto resembles Elgar's Cello Concerto in beginning with a slow or at any rate ruminative movement followed by a fleet Scherzo, and concentrating most weight into the Finale, which ends in a mood of pathos by recalling the principal theme of the first movement. There are many other Elgarian features, for example the devices whereby the solo instrument is never obscured by the large orchestra, and the effective use of tremolando strings to accompany the viola's cadenza-like coda to the first movement. But, of course, the music is thoroughly individual. All Walton's technical fingerprints are there: the melodic exploitation of conjunct motion and wide intervals (sevenths and ninths), sustained

[3] *Paul Hindemith*, p. 98.
[4] *My Viola and I*, pp. 36–7.

notes and looping arabesques, falling cadences; the added-note minor–major diatonic harmony which gives the music its bitter–sweet flavour (the description is almost a cliché but cannot be bettered); the syncopated and irregular rhythmic patterns, jerking the music forward in convulsive leaps. By these means, totally subjugated to his artistic aim, Walton produced this highly original work. Its structure is intensely satisfying and satisfactory: the two faster dramatic outbursts in the opening slow movement; the epigrammatic wit of the Scherzo; and the perfectly balanced Finale, its fugal elements hinted in the bassoons' hesitant tiptoeing initial theme, its climax a great central episode in which the soloist is silent, returning with the concerto's opening theme, to which the

Ex. 1

main tune of the Finale now becomes accompaniment. This device may well have been borrowed from Prokofiev, a composer whose music Walton admired. In his First Violin Concerto the opening phrase of the soloist's G minor theme in the Finale later becomes the background to the return of the D major theme from the first movement. This eloquent epilogue remains the single most beautiful and moving passage in all Walton's music, sensuous yet full of uncertainty, with additional poignancy added by the oboes. There is, underlying the music, a sense of frustrated longing as the predominant emotion. Whatever the hidden personal programme of the concerto, it was something Walton could never repeat.

In 1961 Walton revised the scoring, using double instead of triple woodwind, omitting one trumpet and the tuba, and adding a harp. Why he decided to do this is difficult to understand. For example, in the coda of the Finale, it is not an improvement to have the harp playing the accompanying figure instead of the mixed bowed and pizzicato cellos of the original version. Walton did not, however, withdraw the original score. When he sent the completed revision to his publisher on 16 October 1961, he wrote: 'It is, I think, an improvement on the old version, particularly as regards clarity and definition. The music is the same and the solo part unaltered save for an odd 8ve higher here and there—mostly culled from W.P.'s [William Primrose's] performance.'

An important historical note is that the only occasion when Walton met Elgar (for they did not meet at Schuster's on that day in 1927) was at Worcester in September 1932 when Tertis played the Viola Concerto at a Three Choirs Festival concert at which Elgar also conducted *For the Fallen*. Elgar was in a mood when he would talk only about racing; it is saddening to learn from both Lionel Tertis and Basil Maine that he did not like the concerto. Maine describes in his book *Twang with our Music* (London, 1957) how Elgar paced up and down behind the orchestral gallery during the performance of the concerto 'deploring that such music should be thought fit for a stringed instrument'. How then shall I, one of Elgar's biographers, placate his ghost because I have dared to use the adjective Elgarian so often during this survey? And what did Elgar think of *Portsmouth Point*, which was performed at this festival in the same programme as his *Severn Suite*? Walton told Foss that, if the Three Choirs authorities had known what the concerto was about, it would not have been considered suitable for a cathedral! ('It's dedicated to Christabel Aberconway, so I suppose it's about her,' Walton said in his usual evasive way in 1977.) Such are the advantages of 'absolute' music (and surely Elgar must have smiled to himself for similar reasons when conducting his own orchestral works?).

One of the most perceptive comments made about the concerto was contained in an analysis of it by Donald Tovey, written in the mid–1930s: 'Walton's dramatic power has asserted itself in oratorio; but its unobtrusive presence in this thoughtful piece of purely instrumental music is more significant than any success in an oratorio on the subject of Belshazzar's feast . . . It is . . . obvious that he ought to write an opera . . .[5]

[5] *Essays in Musical Analysis*, iii. *Concertos* (London, 1936), pp. 220–6.

6
Belshazzar's Feast, 1929–31

Belshazzar's Feast began as a BBC commission some time in the second half of 1929. The corporation was emerging as a major and influential patron of music. In 1929 alone fifty works received first performances, or first performances in Britain, under BBC auspices, and plans were going ahead for the formation of the corporation's own symphony orchestra. Adrian Boult was brought from Birmingham, where he had been conductor of the city's orchestra since 1924, to be Director of Music. Walton had decided that his next composition would be choral. Again it was a friend's success that spurred him to try to emulate and outdo it. Constant Lambert's *The Rio Grande*, for chorus, piano, and orchestra, to a poem by Sacheverell Sitwell, had been first performed at a BBC broadcast concert in February 1928, with Angus Morrison as pianist. Its first public performance was at a Hallé concert in Manchester on 12 December 1929, with Sir Hamilton Harty as the solo pianist and Lambert conducting. The critics hailed the work for its poetic fusion of jazz and romanticism. Walton publicly declared it to be 'much better than I have ever written'. When Edward Clark approached him in August 1929 about writing a work for the BBC to broadcast, Walton saw his chance.

On 17 December the *Daily Express* reported: 'Walton is going abroad to execute the first commission ever given by the BBC to a British composer. He has chosen Belshazzar's Feast as a subject, but has uncertain ideas as to how his work will form itself, save that it will be built around a Sitwell narrative.' Accurate reporting except that it was one of three commissions. Two other composers, Lambert and Victor Hely-Hutchinson, had been approached. All three, it was stated, would be willing to accept 'the limitations of apparatus we have in mind, i.e. small chorus, small orchestra of not exceeding 15, and soloist'. Their subjects would be: 'Walton, Nebuchadnezzar, or

The Writing on the Wall, text by Osbert Sitwell; Lambert, Black Majesty (the Emperor of Haiti), text by Lambert from the book of the same name; Hely-Hutchinson, The Town, to a text by Cecil Lewis.'[1] The fees for Walton and Lambert were 50 guineas.

Osbert Sitwell had suggested Belshazzar's feast in Babylon as a subject everybody knew about (Walton said *he* for one did not) and worked on the libretto in Venice in December 1929, although Christabel McLaren claimed that she did most of the research.[2] No matter who helped him, Sitwell produced a magnificent piece of work: the section of the *Book of Daniel* dealing with the feast is given extra impact by the excision of the episode concerning Daniel's interpretation of the Writing on the Wall, and the central drama is 'topped and tailed' by choruses taken from Psalms 137 and 81 and preceded by an adaptation from *Revelation* of the description of Babylon in all its affluence. The whole is introduced by Isaiah's prophecy of doom for the inhabitants of Babylon: 'Howl ye, howl ye, therefore, for the day of the Lord is at hand.' In an interview he gave to the *Yorkshire Evening News* of 7 October 1931, Sitwell said: 'When the libretto was, as thought, complete, I handed it over to Mr Walton. At his suggestion various passages were altered and other passages were substituted.' According to Angus Morrison, Sitwell originally ended the libretto with the nursery-rhyme:

> How many miles to Babylon?
> Threescore miles and ten.
> Can I get there by candlelight?
> Yes, and back again.[3]

Walton scuppered this literary conceit, which, as Morrison said, might have been possible for Mahler to bring off but not Walton, and ended the work as he intended, with a pagan shout of triumph.

Writing to his mother from 2 Carlyle Square in December 1929, Walton said: 'I am off tonight to Amalfi to join Osbert. I

---

[1] BBC internal memorandum from Edward Clark and Julian Herbage, 12 Jan. 1930.
[2] Letter to Stewart Craggs, 20 Oct. 1972.
[3] A. Morrison, 'Willie', op.cit.

am sorry not to be with you all over Xmas but I must get away to begin work on this thing for the BBC. As any how I don't much care for Xmas, I am spending it in the train, where one doesn't notice it. I am sending a small present for my nephew and niece.' At Amalfi every winter the Sitwells and Walton stayed in the remote Albergo Cappuccini, a former monastery built on a cliff-side. Walton worked in what had been a monk's cell. He had the use of an upright piano in a back laundry-room. Progress on *Belshazzar's Feast* was slow at first, as Osbert Sitwell told Sassoon on 17 January 1930 in a mischief-making letter:

The Maestro sends you much love. He is rather depressed, sits long at the piano, and does little work. But it is a stage through which he always passes, if he could only recognize it. Also Constant Lambert's success with his *Rio Grande* has, I am sorry to say, tinged a little their carefree friendship with a certain acerbity. In fact, I thought the other day that I distinctly heard him referred to as 'that little beast'. But do not mention it.

Four days later: 'Willie is writing to you shortly. He has been working very slowly, and is most pathetic, like a dumb white calf, bless him.' But on what was he working? After his return to England in May, Clark wrote a memorandum on 30 May to Boult:

I saw William Walton on Wednesday who had just returned from abroad where he has completed the composition of *Belshazzar*. It is for two soloists, small chorus and small orchestra. Whilst abroad he has shown this work to various people whom it has evidently much impressed, and has been told by Berlin that they wish to broadcast the performance of this work, and also by Volkmar Andreae[4] that he proposes to give a public performance during the course of the next season at Zürich. Part of the arrangement being that the first performance should, of course, be given by the BBC, Walton is asking us to let him know when he may expect this to take place in order that he may make arrangements for subsequent performances abroad.

Was Clark in error in saying the work was 'completed' or had Walton merely spoken of a completed section? A chamber cantata *Belshazzar's Feast* is difficult to contemplate, in any case,

---

[4] Volkmar Andreae (1879–1962) was conductor of the Tonhalle Orchestra, Zurich. He conducted the first performance of *Portsmouth Point* and also conducted the Viola Concerto, with Tertis, in Zurich in 1932.

although Walton wrote to me in 1976: 'I recollect vaguely that I did start on a version for 2 soloists, but it went down the drain.' The visit to Berlin mentioned by Clark had occurred in April, when Walton conducted the Viola Concerto for Hindemith. We find Walton writing to Sassoon from Amalfi on 2 March in familiar terms:

The enclosed bill will show you what straits I am in, as regards my piano, but that is not my only trouble. The question is how am I to exist between April 17th when I leave here for Berlin, and April 24th when I get paid (Mks. 500) for my concert, and incidentally how to get there even . . . You are enough in touch with Osbert's state of finance to know that to expect help from him is not easy or possible: so I am more or less forced to come to you again for help . . . My hopes for the autumn are comparatively high—I had a letter from C. B. Cochran hinting that he would like my help in his next revue!! both as a ballet and as a "jazz" merchant. I cordially accepted and hope that something will come of it, but I suppose that one shouldn't be too sanguine about it.[5] My royalties also ought to be more this year, as I have been having many performances—that hardy annual P.P. having been toured by the Chicago Symphony and played in New York by Koussevitzky,[6] in Paris on the 'radio', in Birmingham by Boult! and even it has penetrated the forbidding portals of the Royal College—the *Sinfonia* has done less well but that is to be expected, only having been played in Paris, Geneva and London . . . Work goes tolerably well, but I always am uncertain of its merit anyhow at this comparatively early stage of its creation.

Sassoon paid for the trip to the Berlin concert.

I am in a bit of a funk about it [Walton wrote to his mother], but I think I shall manage alright as I have four rehearsals. I had a nice letter from Sir Ham. Har. about the *Façade* records[7] , and I think I shall get more of a look-in at the Hallé next season, in fact I believe he is contemplating playing the piano part of the *Sinfonia* as he did in the *Rio Grande* which I see is having a huge and very well-deserved success,

[5] Cochran later met Walton and asked him to write the musical comedy *The Cat and the Fiddle* to a book by Otto Harbach. Eventually, however, the impresario switched to Jerome Kern and it was to be five years before Walton wrote for Cochran.

[6] The effect of Walton performances in the United States was seen in the *Musical Times* of September 1931. It contained a long letter from Mr Charles Mitchell of Illinois in which he averred that 'more bad music has been written in England than in any other country in the world', but Walton was the 'most surprising and inexplicable phenomenon of early twentieth-century music . . . How William Walton, an irrepressibly buoyant, healthy youngster could be born of such a sickly, stupid mother as England is beyond divining.'

[7] The Decca issue of some items from *Façade*.

which makes me happy for Constant as he has not had much of a good life till now and this ought to put him on his feet, both financially and otherwise . . . Ever your Willie.

On 4 September 1930 R. J. F. Howgill of the BBC music staff wrote to Foss at Oxford University Press: 'It seems to have been agreed between Edward Clark and William Walton that *Belshazzar's Feast* has grown to such proportions that it cannot be considered as a work specially written for broadcasting, and the latter therefore proposes to write something else as his commissioned work.' (In fact he never did.) Early in 1931 it was announced that *Belshazzar's Feast* would have its first performance at the Leeds Festival the following October. According to a letter Walton wrote to Foss early in 1933, he made little progress on the work during the rest of 1930: 'In *Belshazzar* I got landed on the word "gold" - I was there from May to December 1930 perched, unable to move either to right or left or up or down.' There is also a pleasing story that he developed a mental block after reading a joke by the humorist 'Beachcomber' in the *Daily Express* that the Writing on the Wall was not 'Mene, mene, tekel Upharsin' but 'Aimée, Aimée, Semple McPherson', who was an American evangelist much in the news at the time. Another possible delaying factor was the intensification of his affair with Imma von Doernberg. By the early part of 1931 they were living together at the Casa Angelo in Ascona, Switzerland. From there, on 8 March, Walton wrote to Sassoon:

It is heavenly here and I am enjoying it very much and am immensely happy. Also I am doing a vast amount of work, as you prophesied I would. I am now on the last chorus. Unfortunately at the moment it doesn't progress too well, but I hope to complete it or practically do so before I leave, which I do a fortnight tomorrow. It is a most frightful nuisance having to come back just at the critical moment, as otherwise if I could stay, I have do doubt that I could complete the whole work for my birthday (29th) as I promised Foss. Also at the rate I am living here, I could stay till practically the end of May and complete the orchestration . . . I feel bound to return otherwise Tertis will take such offence and he has spent so much time and labour on the Viola Concerto that I feel it would be ungrateful of me not to be there[8] especially as I promised him I would . . . Incidentally I wish also to

---

[8] Tertis played the concerto at a Royal Philharmonic Society concert, conducted by Ernest Ansermet, on 26 March 1931.

hear it myself, never having done so properly as yet. Also there will be some business, the Leeds people, Sir Thomas and Foss to see too, but it is all a matter of a few days.

In an interview he gave to Peter Lewis of the *Daily Mail* which was published on 28 March 1972, the eve of his seventieth birthday, Walton said: 'I do get moments of great exhilaration when things are going well. I remember the excitement of getting towards the end of *Belshazzar's Feast* and I still feel it when I hear it now.' The exhilaration and excitement are incomparably conveyed in Walton's two recordings of the work.

By March 1931 the Leeds Festival chorus was being sent its parts so that they could begin to learn the work. Beecham was director of the festival, but allocated the new work to Malcolm Sargent. According to Walton, in his interview with Lewis, Beecham was convinced that the work was doomed and 'declared in his best seigneurial manner "As you'll never hear the thing again, my boy, why not throw in a couple of brass bands?" So thrown in they were, and there they remain.' Brass bands in this context meant two extra brass sections comprising three trumpets, three trombones and tuba. Walton persisted to the end of his life in the belief that the Leeds chorus began by rebelling against the difficulties of the choral writing and that Beecham sent Sargent to pacify them. But Sargent's biographer, Charles Reid, interviewed survivors of the 1931 chorus, who declared that this story was false and that they enjoyed learning *Belshazzar's Feast*, while conceding that 'it is likely that some of the older singers muttered resentfully about the score's irregular metres and other hazards'.[9] Not at all unlikely, one would say! A report of a rehearsal in the *Yorkshire Post* of 14 September 1931 stated: 'The singers, now that the toil of learning the notes is over, are beginning to feel pleased with themselves and actually to like the music.' But that was after nearly six months.

The first performance, in Leeds Town Hall on Thursday, 8 October 1931, with Dennis Noble the baritone soloist, was a tumultuous success, although Walton was dissatisfied with it and calculated that it lasted ten minutes longer than it should

    [9] C. Reid, *Malcolm Sargent: A Biography* (London, 1968), pp. 200–1.

have done. With scarcely a dissenting voice, the critics acclaimed it as a landmark in British choral music, perhaps the greatest work in its genre since Elgar's *The Dream of Gerontius* in 1900 and the biggest choral success at Leeds since Vaughan Williams's *A Sea Symphony* at the 1910 festival. By now *Belshazzar's Feast* is both a classic and a repertoire piece. Ernest Newman, writing after the first London performance, said that, compared with it, Stravinsky's *Symphony of Psalms* 'is very anaemic stuff indeed. Mr Walton works constantly at a voltage that takes our breath away . . . It is difficult to realize that so young a man has so complete a command of his subject, of his craftsmanship, and of himself . . . After this, I should not care to place any theoretical bounds to Mr Walton's possible develop-ment.' With his customary flair for the apt summing-up, Neville Cardus after the Leeds performance described it as 'a clear case of red-hot conception instinctively finding the right and equally red-hot means of expression'. That was an indirect tribute to the choice of subject. There is no more suggestion of a religious or even mystical impulse in Walton's music than there is in Richard Strauss's, but his humanity was well able— in the compassionate 'By the Waters of Babylon'—to encom-pass the feelings of the captive Jews (so well that the pianist Harriet Cohen once told him that he must have Jewish blood!) and the colder, more brittle side of his musical character was apt for expression of the barbaric elements in the tale. 'Stark Judaism from first to last', said the critic of *The Times* on 10 October 1931. 'It culminates in ecstatic gloating over the fallen enemy, the utter negation of Christianity'—words that made their mark on the ecclesiastical powers of the Three Choirs Festival, who refused to admit *Belshazzar's Feast* into their cathedrals until 1957.[10] Perhaps it was not for clerics to salute Walton's musical achievement in eroding the largely artificial barrier between sacred and secular.

[10] Ivor Atkins, organist of Worcester Cathedral, excluded *Belshazzar's Feast* from the 1932 festival because of the unsuitability of the libretto for cathedral performance. Presumably it would not have mattered if the relevant passages had formed the lessons in one of the cathedral services. One of his choristers wrote to the *Daily Telegraph* in February 1932 asking 'in what particular this libretto of Mr Sitwell's is more unsuitable than, say, the text of the Demons' Chorus from *Gerontius* or, for the matter of that, many other works hitherto given by the Three Choirs such as the Wagner selections at Hereford'.

What Cardus did not say was how remarkable it was that Walton should have displayed such mastery of choral technique in his first major choral work. He had written nothing for voice since his boyhood at Christ Church, Oxford, fourteen years earlier. He had seemed to be developing into a purely instrumental composer yet suddenly he produced a work in which all the training of his choir-school days and all his North Country background of massed voices and sounding brass fused to bring into existence a choral symphonic poem of a unique kind. It is in many ways a direct descendant of *Gerontius* in its imaginative use of voices and, of course, in its dramatic intensity. Neither work can be classified as an oratorio in the accepted sense but both achieve revolutionary effects within an evolutionary framework, though, of the two, *Gerontius* is far and away the more revolutionary. To quote Tovey again: 'One of the first essentials of creative art is the habit of imagining the most familiar things as vividly as the most surprising. The most revolutionary art and the most conservative will, if they are both to live, have this in common, that the artist's imagination shall have penetrated every part of his work.'

The foreign listeners to *Belshazzar's Feast* at the ISCM performance at Amsterdam on 10 June 1933, conducted by Lambert, found it a 'conventional' work. They were astute, though they probably did not mean to be complimentary. So exciting is the impression made in thirty-five minutes that one is bamboozled by the delightful confidence-trick that the composer has played. One does not notice that much of the harmony of this 'modern' work is of nineteenth-century vintage, that diatonicism is at the root of the matter, and that string tremolandi, brass fanfares, and masterly use of unaccompanied declamation work their customary spell. The chilling orchestral sounds which introduce the Writing on the Wall derive from Strauss's *Salome*, when Salome is peering into the cistern before Jokanaan is killed. The frenzied excitement and dramatic power of the music are created by the overwhelmingly rhythmical vitality and impetus—and by penetrating imagination. The great work is also outstanding for its economy of method. True, it requires large orchestral forces, with the two extra brass sections of seven instruments apiece; but it is short in the manner of its near-contemporaries

Honegger's *King David*, Stravinsky's *Oedipus Rex*, Kodály's *Psalmus Hungaricus*, and Vaughan Williams's *Sancta Civitas*, it uses only one soloist, and the huge forces are strictly reserved for those moments when their effect will be greatest. Walton's sense of the dramatic in this work is unerring, and it is all done with Handelian choral splendour and Purcellian declamation. There is no elaboration, no repetition, and the only ornamentation is in the pungent orchestral descriptions of gold, silver, iron, wood, stone, and brass. Walton's solution to being 'stuck on "gold"' was one of those splendid march tunes in his own style of pomp and circumstance which served him not only for the barbaric glitter of Babylon but for the Crown Imperial of a coronation and the flight of the Spitfire.

Ex. 2

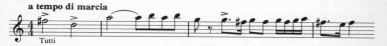

For the conductor, the primary task is to judge the tempos correctly; many performances fail to be completely satisfying because the fast passages are too fast and the slow too slow, with a tendency to pile everything into the work's final chorus of jubilation, which is, from the viewpoint of invention, its most vulnerable point. No one surpasses Walton as a conductor of his music, and his two recordings of *Belshazzar's Feast* will remain as examples to other conductors in the matter not only of tempos but of dramatic emphasis and under-emphasis.

In the seven weeks before the first London performance of *Belshazzar's Feast*, conducted by Adrian Boult on 25 November 1931, Walton rescored several passages and again himself copied the revised orchestral parts. He dedicated the work to Lord Berners, who paid him £50. Sassoon reported in a letter to his friend Sir Henry Head that the ovation Walton received from the Queen's Hall audience was 'really remarkable'. He noted that 'Willie's baroness' shared his 'victory'.

7

In Switzerland, 1931–3

THE year 1931 was also notable for Walton because it saw the start of Frederick Ashton's ballet to the music of *Façade*. This, however, was not the first *Façade* ballet: Günter Hess had choreographed it for the German Chamber Dance Theatre in Hagen, Westphalia, in September 1929. Hess visited London the next year and is said to have exchanged ideas with Ashton. The music for the ballet was taken from the orchestral suite. Edith Sitwell had refused to give Ashton permission to use the original entertainment as a ballet, though Walton was willing. Later, when she saw it, she regretted her refusal. The 1931 ballet comprised seven *divertissements* which satirised various kinds of folk, social, and theatrical dance. 'Country Dance' was added for the Vic–Wells Ballet in 1935 and 'Noche espagnole' and 'Foxtrot' ('Old Sir Faulk') in July 1940.

A curiosity of the period from 1924 to 1936 which irritated Walton and came home to roost in 1977 was the appearance in musical periodicals—*The Sackbut*, *The School Music Review*, *Monthly Musical Record*, and *Musical Opinion*—of articles on various musical topics by a William Walton. (Some of the titles were 'The Status of Jazz', 'Critical Listening', 'The Modern Movement in Music', etc.) The writer was an individual with the same name and it is odd that none of the editors of the periodicals concerned made any effort to inform their readers that he should not be confused with the composer—perhaps they were content to let the misapprehension work to their advantage. It would appear that the editor of *The Sackbut* was, because Walton wrote to him to disclaim authorship and received no reply. Publication of the Craggs catalogue in 1977 attributed these articles to the composer, who protested from Ischia that 'I have never written an article of any kind ever . . . I should perhaps be ashamed to admit it, but I was (for that matter still am) incapable of putting pen to paper in a coherent way'. Amusingly, the seventy-five-year-old Walton wondered if 'his' article about Frank Bridge had been 'disagreeable and malicious. I suspect it was, because when I met him later on, it was a difficult meeting, in fact we never hit it off at all and it could be that he had read this false article and was not best pleased about it and I, being utterly innocent about having done anything to offend him, was quite mystified by his attitude.'

In her biography of Walton, Susana Walton describes him as a pessimist who was sure that 'the end of the world was alarmingly near'.[1] Deep concern over the international financial situation was the theme of letters he wrote to Sassoon in 1931. In August, while he and Imma were in Ascona, he wrote:

The world crisis . . . seems to be having a lull before an even greater storm. The bankers, whom I have been more or less avoiding, seem to think that it will be worse in England than in Germany . . . Imma has recovered what money she had in Damat bank and it seems that she will be alright [Walton always misspelt it thus] to round about Xmas . . . She heard yesterday from her brother-in-law that the Doernberg estate can't pay anything for some time to come and her father also will have to discontinue the small allowance he has been giving her. Her aunt doesn't seem to be going to do anything about it. All of which is very depressing for her and her nerves are consequently in rather a bad state . . .

Imma added a postscript, as she often did, for she was obviously very fond of Sassoon:

Willie looks very well indeed . . . We shall have another week together before he has to go back to conduct at the 1931 Promenade Concerts. This time the parting will be rather sad, as one doesn't know when and if there will be another meeting again for some long time. But I am very grateful for all these peaceful months we could spend together, which is entirely due to you! God bless you for it!

Walton had already written in more detail to Sassoon:

I want to outline the general horror of the coming winter & urge you to do something about it. Even if the Conference in London does put Germany on her feet, the position not only for Germany but for all countries is going to be worse. I have heard that unemployment will in England reach 5 or 6 million & that there will be a general

---

[1] *Behind the Façade*, p. 181.

breakdown about February 1932. Also that the value of the £ is bound to drop (inflation is anyhow proposed in the Macmillan report). I've been informed of this by several Swiss bankers, who say there is no confidence in England & that the only currencies to keep their value will be the American, Swiss and Dutch. Consequently I beg of you to consult with Lassande (who is sure to urge you that everything is all right) and move some money—a great deal—out here while there is time, for a 'law' is sure to be made preventing capital from leaving England . . . I am afraid this will be the last time for long ahead that Imma and I are together, which is too sad for me to dwell on. There is, I feel, no doubt that when I return to England I also should try and get a job of some sort, though what heaven alone knows, as I think that it is more than possible that however much my friends may want to help me, they will not be in a position to do so much sooner than any of them expect . . .

When he did return to England, he spent the evening of 18 October with Hubert and Dora Foss. It was an evening Dora never forgot. 'Willie was (languidly!) extemporizing at the piano with Hubert sitting on the double stool with him, while I stood behind them. We were absolutely enchanted with what he was playing . . . Hubert or both of us said something to the effect that he *must* use that heavenly tune for something, but I cannot recollect the word "symphony" being used.'[2] But they had in fact heard the first stirrings of a symphony.

Imma confided her personal worries to Sassoon. In a letter from Zurich in December 1931 she wrote: 'Willie and I had two more nice days in London, but the parting is always the most painful distress to us both, especially now, when the circumstances are more and more against us and it might be such a long time before we see each other again—I wish I could do more for Willie, but I have the feeling that fate is at the end against the fact that our two lives settle down together.'

The year 1932 began well for Walton. On 28 January he wrote to Sassoon from Faringdon House in Berkshire, the home of Lord Berners:

Last week I was in Manchester & heard a marvellous (the first good orchestral part) performance of the Viola Concerto.[3] Harty has asked me to write a symphony for him. So I shall

    [2] Letter to the author, 15 Jan. 1974.
    [3] Hallé concert in the Free Trade Hall on 14 Jan. 1931. Harty conducted and the soloist was Lionel Tertis.

start on that when I come to Edith.[4] A rather portentous undertaking, but the Hallé is such a good orchestra and Harty such a magnificent conductor, besides being very encouraging, that I may be able to knock Bax off the map. Anyhow it is a good thing to have something definite suggested and a date to work for. I've been writing some songs for Mrs Foss during my stay here. They'll teach her! I come to Edith about Feb. 18th. We leave here on Sat. I shall stay in Carlyle Sq. for the following ten days as I want to see Furtwängler and Klemperer.

The three songs for Dora Foss (Dora Stevens) were 'Daphne', 'Through Gilded Trellises', and 'Old Sir Faulk'. All were from *Façade* and two had been part of the abandoned *Bucolic Comedies* (1923–4). The 1932 versions were revisions and were first performed on 10 October 1932 by Dora and Hubert Foss, to whom they were dedicated. Walton sent Dora the first completed song, 'Through Gilded Trellises', on 21 December 1931. 'It is not so difficult as it first looks,' he wrote, 'especially the piano part, which sounds difficult but it is in reality as easy as Sidney Smith and I may say not unlike . . . I've now to write an Xmas carol by tomorrow morning for the *Daily Dispatch*.[5] It is bad to be tempted by filthy lucre!' The second song to be composed, 'Old Sir Faulk', followed from Faringdon House in early January 1932. 'Here is another song. Alas! not the "Lydian" one, which is proving a little obstreperous. I am not sure that you will approve of this one, but it is, I think, fairly good of its type and at any rate ought to evoke a touch of lunacy in any programme.' The last song, 'Daphne', was sent from Quidhampton in February. 'I am sorry to have been such a time with it but I could not get it satisfactory—this is the 3rd version—all I can say about it is that it is better than the other two. I am here for some weeks, trying to start on a symphony. What a fool I am, treading where so many angels have come a "cropper".' A month later he wrote: 'About the order of the songs. I had conceived it as being 1. *Daphne*, 2. *Through gilded trellises*. 3. *Old Sir Faulk* but perhaps Hubert is right and 1 and 2 should be reversed. Though I think my order is better as I think

[4] Edith Olivier, a society hostess, whose country home was The Daye House, Quidhampton, in Wilton Park, near Salisbury.
[5] 'Make we joy now in this fest', for unaccompanied choir. Published in the *Daily Dispatch*, Manchester, on 24 Dec. 1931.

Old Sir Faulk won't come as such a shock after 2 as after 1.' The songs were published in Walton's preferred order.

Hubert Foss and Dora Stevens had been married in 1927 and Walton frequently stayed with them after 1930 at their home 'Nightingale Corner' in Rickmansworth, Herts. Dora Foss's recollections of those days describe Walton—he was 'Willie' to them—as 'rather shy and diffident, and at that time desperately hard up'.

He had borrowed Osbert Sitwell's pyjamas, a most striking orange and black pair, to come to us, but his hairbrushes and other dressing-table accoutrements appeared to date from his choirboy days. However, he always appeared to be immaculate and extremely elegant in what he informed me were Moss Bros. misfits . . . We sat long over meals while Willie regaled us with gossip and comments on friends and foes alike. How we laughed! He was most appreciative of good food, and as Hubert and I were too, we always took the greatest pains to give him the best within our means—and how (comparatively) easy that was then! He loved simple things to eat too, like home-made pickled shallots, which he could not stop eating. He was allergic to physical exercise in any form . . . He was very anxious to further a very superficial acquaintance with Arnold Bax, so we invited them both to dine with us [on 28 October 1931] . . . I doubt if I spoke at all, other than uttering the necessary 'hostess' remarks at dinner . . . By the time we returned to the drawing-room, the two composers were more or less oblivious of us and we just sat back and enjoyed ourselves. Part of the discussion and general talk was about other composers, both past and contemporary, and the contemporary ones were wholeheartedly and jovially pulled to pieces. One incident I remember—for some reason or other they turned on the wireless for a programme of Russian music and 'guessed' the composers—nearly always incorrectly—which caused them much amusement . . . It was while Willie was staying with us that he first met Sir Henry Wood. We took him up to the Woods' at Appletree Farm House one Sunday and on that occasion Hindemith was there too.

Walton worked on the symphony in the early part of 1932, first at Edith Olivier's home at Quidhampton in February and March, then in April at Weston Hall, Towcester, home of Sacheverell and Georgia Sitwell. At Weston Hall he was banished to a stable, where a piano had been installed—'a senile, disintegrating upright', as Dora Foss described it, 'which Willie kept together and attempted to keep in tune with the aid of a spanner'. The Fosses stayed to lunch at Weston Hall when Roger Fry and Arthur Waley were also guests.

Lunch was very late. Sacheverell kept on retiring back stage to see if it was ever coming. There was a butler and a young boy waiting. We started the meal with baked beans which appeared (and tasted) as if they had been an hour in the oven. This course was followed by slices of meat (species unidentifiable) covered with white sauce. The butler, when this was finished, solemnly dealt plates to all and, equally solemnly, removed them and replaced them with small pudding plates. A very small apple pie and a minute jug of cream were then rather gingerly partitioned.

Not exactly high life in the 1930s as one imagined it.

The letter from Walton to Sassoon telling him about the request from Harty also asked him to guarantee a bank overdraft. However, it added some significant news: 'I deposited my 120 Decca shares with the bank and the manager seems quite pleased with them! Incidentally they are now mine and dear Lil, as I saw in her will, forgave all paltry sums that she had lent the various "down and outs". Also I've 200 Swiss francs in my bank at Ascona . . .' 'Dear Lil' was Mrs Samuel Courtauld, wife of the financier, textile manufacturer, and patron of music, who had died in London on Christmas Day 1931. A few days later Walton was informed of more details of her will. He wrote to Sassoon from Carlyle Square on 6 February:

The miracle has happened . . . Dear angel Lil Courtauld has left me in her will the magnificent sum of £500 per annum for life [the equivalent in 1988 of £13,500] and with the disposal of the capital in case of my death. It seems quite unbelievable and I am feeling more than a little hysterical about it. It is too marvellous for words. I can hardly realize what it means to my life and really to everyone else to whom I have been a willing burden. I suppose it will take a little time before I begin to receive anything and if you can put up with me till then, I shall be more than grateful, as I am for all you have done for me in the past.

Sassoon evidently wrote to Samuel Courtauld. His letter has not survived, but Courtauld replied from 20 Portman Square on 25 February:

I know that you are an old friend of Willie Walton, and I am very pleased if you really think that my wife's bequest will help him to produce the best that is in him. We both thought that what he has already done is far ahead of anything written in this country for a very long time—and in the very front rank anywhere. I don't object to your phrase (though it is irksome to be coupled with riches) though, as a constitutional contradictor, I always find myself defending whatever is attacked. I rather expect that our real views on this subject are practically identical. At any rate I never succeeded in making any friends among this scarcely redeemable class.

Walton's financial euphoria soon evaporated. In late May he was back in Ascona with Imma, whose health was bad. He wrote to Sassoon on 29 May: 'Life has been very worrying and such things as symphonies have fallen far into the background. However as it is already decided that a performance of it is to be given here [Zurich] on next March 27 I suppose I shall be able to get away with it soon.'

He then asked Sassoon to guarantee his overdraft of £300, because he feared a drop in the rate of exchange:

Therefore as I hope and expect to be here most of this year and of the next, I should feel very much safer about it if I had between £200 and £300 here always in a Swiss bank. If I have to be in England more than I expect, it can easily be moved back . . . I may say that even with the expense involved by coming here, my financial position is absolutely sound, so please don't think that I have any other motives than I have stated.

Sassoon, of course, obliged, and in his letter of thanks Walton reported that 'her highness' (Imma) was 'beginning to be her usual sweet self'. She was leaving to stay with Neil McEacharn at Pallanza for a few days, 'so she can lead a life of luxury . . . And I must get down to this symphony.'[6] The next news of the work comes in a letter to Sassoon dated 10 August 1932: 'As regards that symphony, I must confess that I've been also doing a good deal more ruminating than actual work— nevertheless I've collected a number of symphonious bars which promise well.'

Still from Ascona in October 1932, Walton wrote to Hubert

---

[6] Imma went to stay at the Villa Taranto, Pallanza, with its magnificent gardens. This belonged to a Scotsman, Captain Neil Boyd Watson McEacharn, whom she was eventually to marry on 1 July 1940, when he was fifty-five. This was a marriage of convenience, for he was a homosexual. She died from leukaemia in London on 14 March 1947. Shortly before her death she telephoned Walton at Ashby St Ledgers, where he was snowed up and no trains were running, so he was unable to travel to see her, as he had hoped. McEacharn died in Italy on 18 April 1964 and the gardens were later presented to the Italian Government for the formation of a national botanic garden.

Foss that the symphony 'shows definite signs of being on the move, a little spasmodic perhaps, but I've managed to get down about 40 bars which for me is really something. What hopes for it being completed for April I should hardly like to say.' This was the first mention of a possible first-performance date. On 20 December Walton wrote to Sassoon:

The Casa Angelo being in the process of demolition for a new house to be built, we are now installed all by ourselves in a nice flat with Finni looking after us. I have my piano here as well, as it became unbearably cold in my room at the Monte, which is quite empty [Montegufoni, the Sitwells' home in Italy]. Though the symphony begins to progress a little, I've no hopes of finishing it in time for the April performance so I've written to Harty cancelling the date. He has been awfully nice about it and I feel much relieved and now being fired not to be done in by old Elgar's No. 3 I hope I shall produce something good.[7] 'Belshazzar' in the meantime I hope will keep me afloat both by name and financially, for there are something like 15 performances in England and America this season and something near 3,000 copies sold. There is also a vague prospect of it being recorded by Sir Hamilton Harty, but I have heard nothing definite as yet . . .[8] I hope you now see how right I was to change as much money as possible into francs.

Harty—a composer as well as a conductor—meanwhile had written (5 December) to Foss after conducting *Belshazzar's Feast* at Nottingham. He was troubled that Walton

may have been annoyed by a letter I wrote to him recently in which I rather questioned whether his work needed all the complexity he has given it in order to produce his desired effect. Anyway, he has not replied to it. The work is certainly one of genius and it may be no other way was possible to him. If you could get it into his head sometime that I am a terrific admirer of his . . . it would be good of you.

At this same date Walton also wrote to Foss to say 'the symphony is not getting on in the way I feel it should do'. He recalled how he had been 'stuck' while working on *Belshazzar's Feast*

---

[7] It had been announced in December 1932 that the BBC had commissioned a symphony from Elgar, who died in February 1934 leaving only some sketches.
[8] This did not materialize. The first recording, conducted by Walton, was made in January 1943 in Liverpool.

and I'm now in a similar distressing position, but it is not such a nice chord (in fact it is only an octave on A) . . . I should like to hear your considered opinion on Bax's 4th and the new Bliss work [probably *Morning Heroes*]. Instinct tells me that with the Bax, we have heard it all before at perhaps even greater length. Harriet Cohen told me it was all so gay, just like Beethoven, but perhaps better rather than that master, but my instinct (or is it prejudice) tells me otherwise.

In his letter to Harty, Walton apologized for his slowness but explained that

I don't think [the symphony] is any the worse for it, in fact I hope and think that it promises to be better than any work I've written hitherto, but that may be only an optimistic reaction to the months of despair I've been through when I thought I should never be able to write another note. However, the 1st movement is finished and the 2nd ought to be in another 10 days or so. But having disappointed you once, I feel wary about fixing any date to its ultimate completion, but it ought to be ready some time for next season [1933–4]. I must say I think it almost hopeless for anyone to produce anything in any of the arts these days. It is practically impossible to get away from the general feeling of hopelessness and chaos which exists everywhere, however one may try—so you mustn't think I'm an exception and one capable of encompassing all difficulties and produce a masterpiece. But I'm trying my best . . . I really hope to produce something worthy of your genius as a conductor.

Harty wrote to Foss in February 1933: 'Why don't you go over to Switzerland and wrest poor W.W.'s Baroness away from him so that he can stop making overtures to her and do a symphony for me instead! (Rather a good joke!)'

8

Unfinished Symphony, 1933–4

WHEN Walton returned to London in the spring of 1933, he took the completed first movement and Scherzo of the symphony with him and played them to Angus Morrison. At this time, too, he learned that Harty had resigned as conductor of the Hallé and become conductor of the London Symphony Orchestra. In June, while in Amsterdam for the ISCM performance of *Belshazzar's Feast*, he showed the two movements to Foss, who wrote to his wife: 'Willie's symphony is most exciting—really on the big scale and in the purest symphonic manner: just like Beethoven and Sibelius and yet very personal. Rather tragic and the second movement scherzo really sinister. I think I have persuaded him to use for the slow movement that ravishing idea he played to us ages ago on 18 October 1931, when he was starting the work . . .'

The 'ravishing idea' was an allegro version, intended as the beginning of the symphony, of what is now the opening theme of the slow (third) movement. Walton told Edward Greenfield[1] in 1972 that after beginning the allegro with the theme which is now given to the flute in the slow movement (Ex. 3), he 'wrote on

Ex. 3

[1] Interview in *The Guardian*, 29 Feb. 1972.

for twenty-odd bars and found himself stuck over the entry of the clarinet theme with its fluttering figure and strong rising interval. When he was stuck there he lighted on the idea which became the haunting opening theme of the first movement':

Ex. 4

Walton respected Foss's opinions and regarded him as a 'constructive' influence. He told Arthur Jacobs in 1977: 'He was a very intelligent musician and could criticise quite well. If he thought something was wrong, he was probably right. He was very good to talk to about one's difficulties.' There was also Angus Morrison, who described the hiatus in the symphony in the talk from which quotation has already been made. Referring to the slow movement, he said:

It seemed to me a strong indication of how desperate he felt about getting it started at all that he came to me more than once just to discuss and try out nebulous ideas long before anything had been put down on paper. Various ideas were tried out, discussed and as quickly rejected, but always he kept recurring to a little phrase—sounding extraordinarily vague and amorphous as he played it—which finally, after some months of uncertainty, he decided to use as the opening of the movement. It has since transpired that he originally thought of this phrase as much faster and as the opening subject for the first movement. It was only the gradual substitution of a much slower tempo that brought to the surface the latent sadness . . .

Once a start had been made, the movement was completed in short score fairly quickly, and soon he asked if he could come and play it to me. At first it was a much longer movement and contained an extended middle section recalling the malevolent mood of the scherzo, but which was not to my mind nearly so effective, partly because he had said it all before. Willie himself obviously had considerable misgivings about this section and had already discussed removing it with Cecil Gray . . . By the time he came to me I think he had almost made up his mind, and his playing of it then and our discussion afterwards finally clinched it. There and then he marked the cut in his copy, tore out the intervening two pages and left them behind on my piano.

Walton told Morrison during the months when he was 'stuck' that he would like to 'write something like the slow movement of Schubert's string quintet'. Morrison is interesting on a comparison between Lambert and Walton when they played their works to him:

Constant was inclined to be touchy about criticism and one had to choose one's words very carefully when making a suggestion. With Willie it was quite different. One could say exactly what one thought without the slightest danger of his being hurt or upset. 'Willie, that's absolutely terrible, you *must* cut it out and think of something better' would be taken in exactly the same spirit as 'Oh, Willie, that's marvellous, you mustn't change a note'.

This view of Walton was shared by Cecil Gray who, in his autobiography, expressed admiration for Walton's 'dogged tenacity' in realizing his artistic potentialities and 'his infinite capacity for accepting criticism and learning from it—a very rare quality indeed, and one that he possesses to a greater degree than anyone I have ever known'. Walton 'positively invites adverse criticism—that is what he wants. He is not interested in receiving facile compliments, and would not thank you for them if you were to proffer them ... Such humility and readiness, even eagerness, in soliciting criticism is a very sure sign of an artist of great integrity, who puts his work before his own personal feelings and pride.'[2]

While Walton was working on the slow movement, in the late summer of 1933, he also sketched the opening and the coda of the Finale. He told the late Hugh Ottaway many years later that, while the slow movement was being written 'ideas were jotted down for the last movement, only to be discarded and eventually re-constituted'.[3] He visited Dora Foss in September, who noted that 'he had been working on the last movement of the symphony and was very depressed with its progress'.

Meanwhile, interest in the first performance of the symphony was intense: after the success of *Belshazzar's Feast*, Walton's every move was watched with interest. Harty included it in the London Symphony Orchestra's 1933–4 prospectus, advertising it for performance on 19 March 1934 at Queen's Hall. Adrian Boult wanted it for the London Music Festival of May 1934

[2] *Musical Chairs*, pp. 286–7.
[3] *Musical Times*, Oct. 1973, vol. 114, pp. 998–1001.

with the BBC Symphony Orchestra. From Weston Hall, Towcester, Sacheverell Sitwell's home, Walton wrote in January 1934 to Kenneth Wright of the BBC music planning staff:

Having been ill, I wrote to Sir Hamilton Harty telling him that it had put me so much behind with my symphony that it would be best not to announce it for performance on March 19th. In his reply he says amongst other things: 'As for me, I must look forward to the first performance of this work whenever it is finished and I take it for granted that you will reserve it for me. If not ready in time for this season, it could be produced early in the next, etc.' So you will see that the May Festival performance hangs on whether I can finish in time for March 19th. I'm not in complete despair about doing so, but I just want to warn you in time, in case I don't.

A month later, on 16 February 1934, Walton wrote to Boult that there was 'no hope' of his completing the work for 19 March, so the May festival performance would have to be abandoned too. 'I am sure you will realize that it is a rather painful situation for me to live down.'

At this juncture, Walton scored the completed three movements and left the Finale sketches as they were. For a considerable time there was no further progress. He was not well. From Amsterdam in June 1933 Foss wrote to his wife: 'Willie looking thin'. Nevertheless, Foss, Walton, and Lambert enjoyed themselves in Holland, visiting the Rijksmuseum to see the Van Goghs. 'Also', Foss wrote to his wife, 'there is an exhibition of old instruments which is thrilling and Willie behaved abominably of course: beating all the drums in ecstasy and turning a plaque of Richard Strauss to the wall.' [4] And he was in emotional turmoil because his relationship with Imma had ended—'jealousy and hatred all mixed up with love', he said many years later.[5] She left him for a Hungarian doctor, Tibor Csato, who later said Walton had been impotent at this time, a condition that can temporarily afflict creative artists when all their energy is concentrated on, in this case, composing a symphony. A friend, explaining the delay in completing the symphony, is reported to have said: 'The

[4] In Lucerne in 1947 he and Yehudi Menuhin visited a music shop which had a display about a forthcoming Britten opera production there. Walton turned the large photograph of Britten to the wall.

[5] BBC *Music Weekly*, 28 Mar. 1982.

trouble was that Willie changed girl friends between movements.' As Walton wrote to Alan Frank: 'I changed from Imma
D. (once a Baroness always a Baroness!) to Alice W.'[6] Alice
Wimborne was Viscountess Wimborne, whose husband Viscount Wimborne, formerly Ivor Guest, was an extremely
wealthy steel magnate and had been Lord Lieutenant of Ireland
in 1915–18. Alice Katherine Sibell Wimborne was twenty-two
years older than Walton, being fifty-four in 1934.[7] She looked
twenty years younger than she was and her family knew that
she enjoyed the company of men younger than herself. She and
her husband led independent lives, teaming up again to
entertain either in London at Wimborne House, in Arlington
Street, near the Ritz Hotel—the private concerts organized by
Lady Wimborne at Wimborne House were a feature of London
social life in the 1930s—or at their country home at Ashby St
Ledgers, near Rugby. This manor house, standing in 31 acres,
belonged to the Catesby family from the fourteenth to the
seventeenth centuries. Robert Catesby and the Gunpowder
plotters, including Guy Fawkes, fled there after the failure of
their plot to blow up Parliament in 1605. Sir Edwin Lutyens
made additions and alterations for the 1st Viscount Wimborne
and a half-timbered Suffolk house was re-created and linked to
it. Ashby belonged to the Wimbornes from 1903 to 1976.
Walton, quoted by his widow, described Alice Wimborne as
'beautiful, intelligent Alice. She was very kind, full of all the
virtues. Moreover I even got on very well with her husband . . .
The family seemed to like me, I don't know why.'[8] It was not
until Lord Wimborne died, in the summer of 1939, that Alice
bought a small London house where Walton frequently stayed
with her.

But there was also a practical reason for the delay over the

[6] Letter dated 3 Nov. 1981.

[7] Alice, born on 26 September 1880, was a member of the Grosvenor
family, the younger daughter of the 2nd Lord Ebury and a cousin of the Duke of
Westminster. In 1902, the year Walton was born, she married Ivor Guest, heir
to the Guest steel fortune and son of the 1st Lord Wimborne. They had a son
and two daughters. Guest became Liberal MP for a Cardiff constituency in 1910
and was created a peer in his own right, as Lord Ashby St Ledgers, in the same
year, while his father was alive. He succeeded to the Wimborne title in 1914
and became a viscount in 1919. A doggerel verse about him at the time was
'One must suppose that God knew best, When he created Ivor Guest'.

[8] *Behind the Façade*, p. 79.

symphony. In the summer of 1934 Walton was asked to compose the music for the film version of Margaret Kennedy's successful play *Escape Me Never*, to be directed by Paul Czinner, with the Austrian actress Elisabeth Bergner as the star. A ballet sequence was an important feature of the film and Czinner asked his assistant, Dallas Bower, who should compose the music for it. Bower pressed for Walton and was supported by Richard Norton (later Lord Grantley), the executive producer. Herbert Wilcox, the producer, wanted Vivian Ellis. Bower's views prevailed. Walton was paid £300 for the score, a princely sum at the time and negotiated by Foss. Walton thus became the first 'serious' British composer to write film music, although Holst and Elgar had previously been asked.[9] Writing the score 'nearly drove me to a lunatic asylum', Walton said, but he wrote twelve minutes of music overnight on one occasion and, to his surprise, 'soon found myself writing five to ten minutes' music a day without too much difficulty'.[10] The payment Walton was made, like his bequest from Mrs Courtauld, was a turning-point in his life in regard to his financial independence. 'I'd never seen such a sum in all my life,' he told Arthur Jacobs in 1977. 'From then on the finances were no real question. I could always do a film.'

Walton's score for *Escape Me Never* used an exotic percussion section, particularly for the scenes in the Dolomites, where atmosphere was created by celeste, cowbells, glockenspiel, and vibraphone. The ballet sequences were choreographed by Frederick Ashton and Ninette de Valois and filmed at Drury Lane. They are brief in the finished film, but are notable for glimpses (in long-shot) of the young Margot Fonteyn. Spike Hughes remembered Walton's claiming that the tango he first wrote for this film was

a natural hit. He was quite right: it most certainly was a natural hit and had been, in fact, for some months—as the theme song of a film starring Lawrence Tibbett. Accepting expert opinion, Walton scrubbed his world-beating pop from his score, but protested rather

[9] Holst wrote music in 1931 for a film of *The Bells*, an adaptation of the play in which Sir Henry Irving had had such success. No score survives. Elgar was approached in September 1933 to write or arrange music for the film *Colonel Blood* and in 1927 had provided music for a silent film, *Land of Hope and Glory*.
[10] *Sunday Telegraph*, 25 Mar. 1962, p. 8.

sadly that never in his life had he even heard tell of Mr Tibbett's song. He had, however, long known and admired Isham Jones's classic of the 1920s called *Spain*. As also, it was clear, had the later composer who had liberally based his Lawrence Tibbett number on it.[11]

On 13 September 1934 Harty wrote to Foss:

'Ecco! Il Duomo! as the little Fiesole tailor said to me after the 12th fitting of the suit he was trying to make for me (. . . he was as sardonic as he dared) . . . I shall not be sardonic, but very thankful, when I say these words to Willie W.—but he, poor boy, is right to wait until he is quite satisfied. I do so hope (and think) it is going to be all right this time. But what a difficult accouchement! No matter—pass the chloroform, Nurse—these are the authentic pains!

These words suggest that Harty was prepared to wait for the completion of the symphony. Nevertheless, a few weeks later it was announced that the three completed movements of the work would be performed at Queen's Hall on 3 December 1934 by the London Symphony Orchestra conducted by Harty. In his late years, Walton said on several occasions that he was persuaded into this course by Harty and Foss against his will—'sort of pushed into it' was one of his phrases—and that he had always regretted it. There are good reasons for scepticism about that. Foss wrote the sleeve-note in 1952 for Walton's LP recording with the Philharmonia Orchestra in which he said: 'The rare decision was taken *by the composer* [my italics] and his advisers to perform the symphony in public incomplete . . .' Roy Douglas, who was librarian of the LSO at the time of the first performance, recalls that the directors of the LSO were annoyed by the 19 March cancellation in 1933 and were determined to play the work in some form in 1934. It was a feather in their cap to give the first performance and good publicity for them, but it was also a box-office risk to give a symphony by a British composer. The directors were playing-members of the LSO. They engaged the conductors, who were not allowed their own way in all matters.[12]

The performance was a success, even if some of the woodwind players complained that parts of their music were 'unplayable'. The audience was overwhelmed by the impact of

---

[11] 'Nobody Calls him Willie now'.
[12] In letters to the author, 1983.

the symphony and eager for the composer to finish it. After the concert, Lady Wimborne gave a sumptuous supper party at Wimborne House, described by Siegfried Sassoon to Dora Foss as 'Rome before the Fall'. Mrs Foss remembered that in the supper room seventeen footmen looked after thirty-five guests. 'After supper we moved into the vast music room, with its fantastically lovely crystal chandeliers lit by hundreds of candles.' Walton himself looked ashen and ill at the performance. Shortly afterwards he went into hospital for tests and had his head x-rayed. Undoubtedly the emotional strain of the break with Imma was intense. The late Gerald Jackson, in his autobiography, recalled a wartime performance of the symphony conducted by Walton: 'Throughout, his face was marked by intense suffering . . . My impression was, and remains, that the composer must have endured a considerable amount of spiritual turmoil during its composition and this he re-lived.'[13] It is worth noting that, despite the end of their relationship, Walton still dedicated the symphony to Imma. Harty's reaction to the work was ambivalent and, as will be seen, astonishingly percipient. He wrote to Foss on 4 December 1934: 'Some day I should like to talk to you about the young man. Enormously gifted—something further has to happen to his soul. Did you ever notice that nothing great in art has lived that does not contain a certain goodness of soul and a large compassionate kindness? Perhaps he has not noticed it either! . . . All is not well there, I feel'.

Walton's improved financial situation allowed him to buy a London house, 56A South Eaton Place, a necessity in any case because it had been made clear to him that he was no longer welcome at Carlyle Square. His estrangement from the Sitwells continued until Lady Wimborne's death in 1948. 'They all joined in', Walton wrote many years later, 'and it was very unpleasant indeed at times, as it involved all their considerable number of friends, with notable exceptions like Christabel Aberconway. But the girls were an early issue only coming to a real head over A.W.'[14] Alice Wimborne disliked and disapproved of the Sitwells and was proud to have 'rescued'

---

[13] G. Jackson, *First Flute* (London, 1968), p. 62.
[14] Letter to Alan Frank, 2 Apr. 1979.

Walton from their influence. Osbert strongly disapproved of her relationship with Walton, not on moral grounds, it would seem, but out of jealousy and resentment at Walton's no longer needing to feel beholden to him. In any case, David Horner had now moved in with Osbert in a homosexual relationship and he did not want Walton there. It was to Horner that Sitwell spitefully wrote that he had seen Walton and Alice walking in King's Road, Chelsea, and that she 'looked old and footsore and slummy'.[15]

At this period Walton was a fairly regular *habitué* of the George public house on the corner of Great Portland Street and Mortimer Street, nicknamed 'The Gluepot' by Sir Henry Wood because his players so often became stuck in it. There, until too frequent visits were discouraged by Alice, he would meet Lambert, Hyam Greenbaum, Alan Rawsthorne, Moeran, Leslie Heward, Elisabeth Lutyens, Ralph Hill, and Spike Hughes, among others. It was on one of these occasions, according to Dallas Bower, that Lambert said: 'I've written an essay which I hope will be published after my death entitled "Sir William Walton O.M. by the late Constant Lambert, D.T." ' A bitterly exact prophecy.

[15] J. Pearson, *Façades: Edith, Osbert and Sacheverell Sitwell* (London, 1978), p. 311.

9

Finished Symphony, 1934–5

THE symphony had two further performances in its incomplete form, at Courtauld–Sargent concerts on 1 and 2 April 1935. Walton resumed work on the Finale around this time. He still did not know what to do about the middle of the movement. According to Walton himself, he rang up Lambert, who suggested a fugue. 'But I don't know how to write one.' 'There are a couple of rather good pages on the subject in Grove's Dictionary.' So he read the Grove entry and wrote the fugue. But using it still worried him, perhaps because Vaughan Williams had also resorted to a fugue in the Finale of his Fourth Symphony, which had its first performance on 10 April 1935. Writing to Foss from Sacheverell Sitwell's home at Towcester on 9 July, Walton sent him ten bars of the main fugue subject, saying he had produced it within the previous ten days and describing it as 'the third subject'.

Ex. 5

Twelve days later he was writing to Dora Foss about starting the movement all over again:

Only a chance remark of the gardener's wife as she brought in the famous ham saved me from destroying it already. 'How pretty your music is getting—it sounds just like a great big band.' Considering it is more than likely she knows what she likes than I know what I like and perhaps it may be a piece for the mob, at any rate I thought I'd better keep it to see what Hubert thinks on Monday ... I hope you enjoyed your holiday in spite of the adjacency of the Blisses. I love your description of his studio and of course he is just like a moustachioed cod-fish so he will be in the right environment. I hope it won't make his music even more watery.

On 11 August Foss went to Towcester and reported to his wife:

The last movement is nearly complete, only 1½ or 2 minutes to do yet. Apart from the fugue subject, he's actually thought of another idea altogether and is working them all together. I'm delighted with it and think he really will complete in a week or ten days. The next thing is a ballet to a libretto about Bath by Osbert: I read it and thought it very good. It's for the de Basil Ballet.[1] His royalties this year are £96— really more than that but there are some deductions for arrangers' fees. Quite good. Far better than Lambert's.

This letter also includes one of the few descriptions of Walton together with the Sitwells:

Sachie was dressed in a (dirty) bright yellow open-necked shirt, an old coat and red trousers! Willie in grey flannels and a blue tennis vest, and Edith in a black hat like a mortar-board without the stiffening, and a fawn (veal) dress, with enormous jade bracelets and a vast jade ornament on her chest ... Oh, Anthony Eden told Osbert Sitwell that there is imminent risk of war between England and Italy over the Sudan and Abyssinia and that war is expected to be declared early in September. Cheerful!

On 30 August Foss told his wife: 'I've just heard that Willie rang up to say he'd actually finished his symphony!' Walton played the whole work to five friends on 13 October. The first performance of the completed symphony was given in Queen's Hall on 6 November 1935 by the BBC Symphony Orchestra conducted by Harty. Inevitably attention was concentrated on the Finale. It was the only part of the work new to the critics, and it was assumed that it was to some extent an 'afterthought' added *in toto* after the 1934 performances. This assumption

---

[1] Walton himself mentioned this project, which came to nothing, in a letter to Foss from Towcester on 9 July 1935: 'I may be in London on the 18th for a day or two and then on to Renishaw for the weekend to discuss the ballet question with these Russian people who are coming also. I think it is almost as good as settled that I do it.'

annoyed Walton, but he had only himself to blame. Interviewed on the morning of the first performance, he said: 'All
went well until I came to the last movement. I did not touch it
for eight months—I had to wait for the mood—I could not
think of the right thing to do. Then it came'.[2] That does not
imply that none of the Finale had been written, but one could
be forgiven for drawing that inference—and most people did.
Critical opinion on the Finale was divided almost equally
between those who considered that it was too 'easy' a solution
to the problem posed by the other movements and those who
thought it was the best movement. Walton, in his last surviving
letter to Sassoon, written on 21 October 1935 inviting him to
the first complete performance, said: 'In some ways I think the
last movement to be the best of the lot, at any rate it will be the
most popular, I think. There has been a slight chilliness
between me and Carlyle Square, so I've settled down here
[South Eaton Place] and I must say I much appreciate being on
my own.'

The only genuine response to this remarkable symphony
should be emotional. One may be intellectually intrigued by its
stylistic affinities with Sibelius and others and by its formal
devices such as the virtuosic use of pedal-points, but in the
long run it is the music's eloquent emotional outpouring which
keeps it fresh and exciting—'the climax of my youth,' Walton
described it. Who that loves it can forget first acquaintance with
its opening: the horn-call rising over the throbbing of the
timpani and strings (a jazz influence here, surely), the oboe's
repeated-note theme sinuously curling its way into one's mind
forever, then the convulsions and catastrophes which are
thundered out for the next ten minutes? It was Richard Capell,
in *The Listener* of 4 October 1935, who first applied the adjective
'bardic' to this great movement. At the same time he called it
'less self-contained' than the first movements of other composers, making the effect of the opening act of a drama but
' "prelude" would be an inadequate word for it'. It is therefore
fascinating to learn from Greenfield's interview that Walton at
one point wondered whether to make it a one-movement
symphony, leaving the first movement as a separate entity.

[2] *Evening News*, 7 Nov. 1935, p. 1.

Walton's problems seem to have stemmed from a conflict between his personal emotion and his declared intention, according to Foss, of avoiding the nineteenth-century romantic symphony. Walton was, it seems, well aware of the problems which his excess of emotion created for his interpreters. Bernard Shore recalls Walton's telling the BBC Symphony Orchestra in the first movement: 'I don't like this movement any slower—it's too emotional as it is and it gets unbearable if that side of the picture is drawn out.'[3]

The influence of Sibelius on the symphony is often cited, but today one is also conscious of Prokofiev; indeed, one is tempted to call Walton the English Prokofiev. The symphony belongs as much to the world of Shostakovich as to that of Sibelius. Walton's own comment on the latter influence, made in a letter to Alan Frank written on 3 November 1981, was:

On the occasional time I listen to a record of S.1 I'm always struck by how unlike it is to Sib. in sound and content. There is admittedly more than a slight likeness in the 1st announcement of the 1st theme, especially at the end of the phrase. His music, as you say, was not much in the Sitwellian world and I got to know it chiefly thro' Constant and Cecil Gray and became a great enthusiast for some time, still am for that matter.

But why speak of others when Walton himself is writ large in every bar? Only he could have composed the 'malice' Scherzo, which rushes headlong to its final climax as if the devil was on its heels. Its chords on the wrong beats, jagged metres, angular rhythms, displaced accents, and furious energy are a sophistic- ated development of the Stravinskyan methods which Walton first used in *Portsmouth Point*. The music sounds as if some kind of fission process is at work in it. As in the first movement, a flattened seventh is prominent in the harmony, but just what makes this Scherzo malicious is not easy to define. That Walton succeeded is borne out by the statement of Sir Adrian Boult in old age that he did not want to conduct this symphony any more because he could not face 'all that malice'.

The slow movement, originally marked 'un poco lento con malincolia' which was changed to 'andante con malincolia'

---

[3] B. Shore, 'Walton's Symphony', in *Sixteen Symphonies* (London, 1949), p. 374.

when the score was published in 1936,[4] reminds us by its cool, limpid flute melody that Walton had thoroughly studied the scores of Ravel and Debussy. The movement rises to a tragic climax and falls back from it with what seems like inevitability: it is all in one sustained line and it is hard to believe that the composer had had trouble with its form and that a section was removed. As an expression of mood-indigo anguish and of dark warring forces within him, it is unsurpassed in his whole output. The Finale begins in Walton's ceremonial style, his *Crown Imperial* manner. If the fugue section does sound as if it has been stitched in, it also conveys immense exhilaration and joy. There are poignant episodes too, none more effective than the trumpet's ethereal 'last post' before the drums and brass launch the majestic and triumphant coda. Only in this movement, incidentally, is the orchestra larger than Brahms used: Walton requires an extra set of timpani, cymbals, side-drum, and gong, and how thrillingly he employs them.

There is a constant danger, if the symphony is not extremely well conducted and performed, that the bellowing brass and screaming woodwind, the physical impact of the salvoes of fortissimo, can give an impression of uncontrolled vehemence, of a noisy and overcrowded score lacking the balance between emotion and its expression which distinguished the Viola Concerto. Alan Frank recalled a rehearsal of the symphony when Walton turned to him and said: 'You know, Alan, those eight bars mean absolutely nothing at all—just sheer passage-work. Mean nothing!'[5] The tension is almost too strongly sustained. Walton was right to balance the symphony with a Finale of lighter substance, but there was so much to balance that the Finale has to gird its loins for a culminating onslaught on our feelings.

Still, one would not have it otherwise. The symphony is masterly and its reception in 1935 was ecstatic, both from musicians and public. (Curiously, when Ormandy conducted it in Philadelphia in 1936, two hundred of the audience walked out and there were noisy demonstrations against its modernity.) It was regarded in England as a liberating masterpiece.

---

[4] There is no such word in Italian as *malincolia*. The Italian for melancholy is *malinconia*.

[5] BBC *Music Weekly*, 28 Mar. 1982.

'Historic night for British music,' said one headline, summing up what was generally believed. Sir Henry Wood wrote to Foss after the first complete performance: 'What a work, truly *marvellous*, it was like the world coming to an end, its dramatic power was superb; what orchestration, what vitality and rhythmic invention—no orchestral work has ever carried me away so much.'

The second performance of the complete symphony was given in Birmingham by the City of Birmingham Orchestra conducted by Leslie Heward, on 22 November. Foss reported to his wife:

It really was a magnificent performance—the band is not really full of good players, but they went all out for this and pulled it off. The tears were rolling down my cheeks during the Epilogue, so they were down many others! In some ways it was better than Hay's[6] but not in breadth of experience. The thing that pleased me was the way it gripped the public: Willie had an ovation, took his bows splendidly and was noble all through.

It was recorded by Decca on 10 and 11 December 1935, only five weeks after its first performance. Foss described the first day as 'rather rushed and fussed and I had to have a few words (which Willie meant but would not say) with Harty. He became like a lamb after losing his temper. We got 5 sides down which means 7 tomorrow.' Walton rescored the wood-wind parts in the Finale's coda. After listening to the records, John Ireland wrote to Walton:

This is the work of a true Master—unlike any other English symphony, this is in the real line of symphonic tradition. It is simply colossal, grand, original, and moving to the emotions to the most extreme degree . . . It has established you as the most vital and original genius in Europe. No one but a bloody fool could possibly fail to see this . . .

I quote these letters to show not only the enthusiasm which this work created at its birth, but also the burden which it placed on its creator. Royalty attended the Wimborne House party after the complete performance. There was a full-page feature on Walton in the *Evening Standard* that day. 'I felt I had arrived at last,' he told his wife years later. Mengelberg and

6 'Hay' was Sir Hamilton Harty.

Furtwängler sent for scores of the symphony and it was taken up as early as 1938 by one of Walton's finest interpreters, George Szell, who conducted it in Scotland and Australia. Walton wrote to Harty the day after the performance to thank him for 'the infinite trouble, care and energy you put into the performance . . . There is no way of describing it except by that well-worn word "inspired".' For a new symphony to be recorded so soon was in itself testimony to Walton's standing in English music at that time. The occasion also had its lighter side. Walton discovered that the bawdy words sung to the march *Colonel Bogey* fitted the coda of the symphony's Finale. 'The first performance', he wrote, 'took place in the recording room during Harty's recording it for Decca. Those taking part (alas, all deceased) were Constant, Hyam Greenbaum, Hubert Foss and Walter Yeomans of Decca, led with great gusto by the composer. I remember the recording staff being much bewildered, not to say shocked, by this frivolity.'[7]

[7] Letter to John Warrack, 11 Aug. 1958.

PART II

William

10

Crown Imperial, 1936–7

THE First Symphony crowned an astonishing decade for Walton, ten years in which he had produced the three masterpieces on which his fame most securely rests and which had seen the revised *Façade* and *Portsmouth Point*. All this in a no less fruitful decade for English music as a whole, for within the same period came Vaughan Williams's *Flos Campi*, *Job*, and Fourth Symphony, Bax's Third and Fourth Symphonies, Bridge's *Oration*, Lambert's *Rio Grande*, and a host of lesser works. Not content with this, the critics were greedy for more, especially from Walton. Only Newman, taking the hint from Walton's music for *Escape Me Never*, advised him to 'keep his hand in with trifles' for two or three years. This Walton did.

Walton's letter to Sassoon on 21 October 1935 contained news of some of these 'trifles':

I've several things, chiefly commercial, on foot. Firstly a ballet for Cochran, with Osbert doing the libretto (in spite of the coldness), then music for the film *As You Like It* with Bergner and lastly music for Sir J. M. Barrie's new play with Bergner and Augustus John doing the scenery. So I shall be able financially to keep my head, I hope, well above water for the time being.

A Cochran revue was one of the highlights of the theatrical season in the inter-war years. The chorus consisted of 'Mr Cochran's Young Ladies', a well-drilled and beautiful line-up from whose ranks many an eventual star emerged (for example, Anna Neagle and Sarah Churchill). Décor and choreography were of the highest class. The 1935 revue was designed to be particularly good, as this was the year of King George V's Silver Jubilee. It was entitled *Follow the Sun* and the Sitwell–Walton ballet, *The First Shoot*, was the nineteenth scene in Part II. Sitwell's libretto was a tragic story of how the ex-chorus girl wife of an Edwardian peer is accidentally shot by her admirer at a country-house shooting party. Although it lasted only ten minutes, it was said to be 'very amusing', in spite of the story, and Walton's music 'of the same general character and genre as *Façade*'.[1] The choreography was by Frederick Ashton, scenery and dresses by Cecil Beaton. The revue opened in Manchester on 23 December 1935, but its London première was postponed until 4 February 1936 because of the King's death during January. (In a telephone call to Foss on 1 January 1936 Walton described the ballet as a 'wow', but the rest of the show 'has not got a funny line in it'.) A waltz sequence was twice more used by Walton—in the films *Went the Day Well?* (1942) and *Three Sisters* (1969)—and in 1980 Walton rescored the ballet for brass band for the Grimethorpe Colliery Band and its conductor Elgar Howarth. Walton later said that Cochran 'was either planning great things for me to do with him or else, as he did so subtly with everybody he knew, he was just quietly picking my brains. Did I know of a young conductor? Could I find him a new orchestrator? And so on—it was a ceaselessly active mind and it influenced the taste of a whole generation as surely and rewardingly as the Russian Ballet had done.'[2]

Cochran was also the producer of *The Boy David*, Barrie's last play, which dealt with the story of David and Jonathan and the slaying of Goliath. It opened in Edinburgh in November 1936 (having been postponed from the previous February because of Bergner's illness) and in London a month later and was a disastrous flop. Walton's score was pre-recorded. A few of the discs still survive, otherwise nothing is known about the music beyond a report in the *Daily Mail* for 21 November 1936 that the pre-recording had been necessary 'owing to the difficulty and expense of finding expert exponents of the rare instruments, for which the music is scored, who would be free to play every night'. Replying in 1950 to a query about it from Alan Frank, Walton said he had no idea where the score might be and cared even less.

Bergner and Czinner were also involved in the next film for which Walton wrote the music, an adaptation (suggested by Barrie) of Shakespeare's *As You Like It*. This was a considerable score and was played on the sound-track by the London

---

[1] Letter to S. Craggs, 6 May 1972, from Frank Collinson, who conducted.
[2] 'Records I Like'.

Philharmonic Orchestra, conducted by Efrem Kurtz. The song 'Tell me where is fancy bred', from *The Merchant of Venice*, was incorporated into the film (the setting was not, as has sometimes been said, the same as that written by Walton on 2 July 1916 for the Christ Church choir school's production of *The Merchant of Venice*), but the best-known song in *As You Like It*—'Under the Greenwood Tree'—was not used. Walton, however, had set it and it was published separately by Foss in 1937. Some of Walton's score was orchestrated by Hyam Greenbaum, a conductor whom Walton regarded as an 'unrecognized genius'. Greenbaum sometimes assisted him and other composers in this way when they were pressed for time. The title music of *As You Like It* is an incongruous piece of pastiche Sibelius, one of the cleverest ever written. There is no attempt elsewhere in the score to concoct mock-Tudor music; indeed, Delian picturesqueness and nostalgia pervade the Waterfall Scene.

It was on the set of *As You Like It* that Walton first met Laurence Olivier, who was playing Orlando. Olivier later recalled the occasion:

Everybody was extolling him to the skies as England's musical genius. As I looked at him, there was something about him that made me believe it. He was pale; pale-eyed, pale-skinned, even pale personality plus pale green hair. He was cold to somebody younger than he at first meeting ... the paleness and the coldness made the passionate blaze in all of his music a thing of wonder and amazement.[3]

The 'coldness' with a new acquaintance was a sign of Walton's innate shyness and insecurity.

There is a curious puritanical snobbery in English musical criticism which decrees that a composer's 'off-duty' music is of necessity of little value and that the major and the lighter works are mutually exclusive. For a long time it was rarely suggested that Elgar's lighter works, even the exquisite *Wand of Youth*, could have any important artistic connection with his large-scale masterpieces. Walton has suffered in a similar way, with the added opprobrium that film music—horror of horrors! —is financially lucrative. Even allowing that the composer

[3] In an eightieth birthday greeting in 1982, published in an Oxford University Press catalogue.

himself may regard such tasks as relaxation (though they often require more discipline than the comparatively leisurely progress of a symphonic undertaking), the keepers of our composers' consciences rarely concede that film and occasional music can be and often is a fructifying agent for larger works. It also can scale its own heights. Walton's 'Touch her soft lips and part' from *Henry V* is as tender as any miniature in English music.

The year 1936 brought other offers, some of them more significant. Benny Goodman and Joseph Szigeti asked him for a work for clarinet and violin, but another commission intervened and they applied to Bartók instead, hence *Contrasts* (1938). The other commission arose when Jascha Heifetz asked Spike Hughes if he knew 'a young man by the name of Walton' with whom he wanted to discuss a violin concerto. Hughes introduced Walton to Heifetz at lunch at the Berkeley Hotel, where the violinist commissioned a concerto. (He had been urged to approach Walton by the violist William Primrose.) Walton's version of what happened at the Berkeley was given in a letter to Alan Frank on 10 October 1973: 'He'd already been disappointed by the Bax and was v. wary. I wasn't all that keen, knowing how difficult it could be. However, we got as far as terms—1,500 dollars or £300 (the pound was worth more than the dollar then).' During 1936, Walton made some revisions in the scoring of his Viola Concerto.[4]

At this time Walton was also asked to write the music for another Czinner–Bergner film, *Dreaming Lips*. In addition, plans were being made for the Coronation of King Edward VIII in May 1937. A BBC memorandum dated 27 November 1936, a few days before the Abdication crisis startled the nation, stated that

Walton . . . would love to be commissioned by the corporation to write a really fine symphonic Coronation march. No one will doubt that his immense technical ability should produce a march of equal value to the existing Elgar marches . . . Walton is at the moment engaged on a violin concerto for Heifetz, a choral work for next year's Leeds Festival and music for a film, so that he feels that some sort of immediate

---

[4] A BBC memorandum from Kenneth Wright, Assistant Director of Music, to Julian Herbage, dated 27 January, 1937, mentions 'Walton's new version of his Viola Concerto'.

financial encouragement from the corporation would be justified in order to enable him to put some of his work on one side . . .

Terms were agreed by March 1937, the sum paid for *Crown Imperial* being 40 guineas. By this time the Coronation was to be of King George VI. Walton wrote the march at Ashby St Ledgers as this undated letter to his mother shows:

I hope you received my wire and have settled about taking your new house. I am down here for some days trying to get going on my Coronation March for the Abbey, so I'm pretty busy and have been for some time what with one thing and another. When you next write to Alec tell him that Lady Wimborne has very kindly written to the wife of the Canadian Governor-General (she is her cousin) and if he's ever in Montreal and near the Governor's special train, he is to write his name down and they will be delighted to see him. Here is a cheque to be going on with. I expect you will be having a busy and expensive time.

The first performance of *Crown Imperial*, a magnificent example of Walton's 'pomp and circumstance' manner which, as the BBC official rightly forecast, can rank alongside Elgar's marches, was on 16 April 1937, when Boult and the BBC Symphony Orchestra recorded it for HMV. It was broadcast on 9 May conducted by Clarence Raybould. In Westminster Abbey on 12 May, Boult conducting, it was played to accompany the entry of the Queen Mother, Queen Mary. The march—its opening marked 'allegro reale' (royally)—has a superscription 'In beautie beryng the crone imperiall' from William Dunbar's 'In Honour of the City of London', the poem Walton had decided to set for the Leeds Festival which, in Coronation Year, naturally hoped for a successor to *Belshazzar's Feast*. Yet Walton claimed that he took the title from a passage in Shakespeare's *King Henry V*, Act IV, Scene i, which, he said, had 'a whole list of titles for Coronation marches':

I am a king that find thee, and I know
'Tis not the balm, the sceptre and the ball,
The sword, the mace, the crown imperial,
The intertissued robe of gold and pearl,
The farcèd title running 'fore the King,
The throne he sits on, nor the tide of pomp
That beats upon the high shore of this world—
No, not all these, thrice-gorgeous ceremony,

Not all these, laid in bed majestical,
Can sleep so soundly as the wretched slave . . .

For the Coronation of Queen Elizabeth II in 1953 he jibbed at
Ball and Sceptre but made it *Orb and Sceptre* and said he had
reserved *Bed Majestical* for King Charles III.

Sargent conducted the first performance of *In Honour of the City
of London* at Leeds on 6 October. The first London performance
followed on 1 December, when Walton conducted. Walter
Legge, writing in the *Manchester Guardian*, said that it 'at once
showed up the shortcomings of the first performance and the
intrinsic worth which the first performance partly obscured'.
This short, laboured, and somewhat hectic cantata has never
become a particular favourite with choral societies. Walton
himself rather liked it, perhaps because it was neglected,
though he wrily remarked that it sounded as if it ought to have
been written before *Belshazzar's Feast*. The choral writing is
harsh and angular, with ungrateful leaps for the sopranos in
particular and the orchestra is continuously busy and energetic.
The setting lacks contrasts and the essential quality of repose.
As a musical interpretation of the poem, Sir George Dyson's
earlier setting (1928) is to be preferred. Walton uses no soloists
in his setting. The six verses are set for chorus in four parts,
expanded when necessary up to eight. Each verse is in a
different rhythm, the simplest being the 4/2 of verse 4 which
is concerned with the Thames. The cantata begins with chordal
cries of 'London' followed by a busy rising figure on the violins
which later accompanies the 'merchants full of substance'. The
'famous prelates' are introduced by plainsong, with a—
mocking?—pizzicato counterpoint. After a characteristically
jaunty orchestral interlude, the second verse ('con agilità
e molto ritmico') introduces two new short motifs—they are
scarcely themes—in the accompaniment to the chorus's four-
part singing. For the third verse Walton resorts to homophony,
building to a powerful climax of thirds. For the fourth
(Thames) verse, the energy at last slackens while women's
voices sing their long melodic lines. The men enter only for the
last two lines and the verse ends in unaccompanied eight-part
harmony. Another breezy orchestral passage precedes the
men's salute to London Bridge and the Tower. In the last verse,

rhythmical brass chords illustrate the 'artillary', the 'massive walls' evoke sturdy octaves, and the river is again depicted in orchestral triplets. The cantata ends joyfully in E major, with the bells ringing.

11

Violin Concerto, 1938–9

WALTON was deeply interested in everything his fellow-composers were doing. He had been at all the rehearsals for Vaughan Williams's new Fourth Symphony in April 1935 and heard Constant Lambert tell Arthur Benjamin on the day of the first performance that it was 'the greatest symphony since Beethoven'. He also noted the arrival on the scene of a new talent, Benjamin Britten, whose *Our Hunting Fathers* had its first London performance on 30 April 1937. Perhaps it was Walton's interest in this astonishingly adventurous and imaginative score that led to their first meeting. Britten's diary for 28 July 1937 (he was then 23) contains this entry:

Haircut—& then lunch with William Walton at Sloane Square. He is charming, but I feel always the school relationship with him—he is so obviously the head-prefect of English music, whereas I'm the promising new boy. Soon of course he'll leave and return as a member of the staff . . . Anyhow apart from a few slight reprimands (as to musical opinions) I am patronised in a very friendly manner. Perhaps the prefect is already regretting the lost freedom, & newly found authority!

Britten was at this date composing his *Variations on a Theme of Frank Bridge* for the Boyd Neel String Orchestra to play at the Salzburg Festival on 27 August. Walton heard the first London performance on 5 October and wrote to Britten to congratulate him on 'really a fine work'. He at once arranged a private performance and the same letter continued: 'I do hope that the rehearsals for the Wimborne House performance will be enough as it sounds rather difficult, and a doubtful perform-ance would do a lot of damage to a work where every note tells.' Walton also attended the BBC concert on 19 November 1937 at which Sophie Wyss sang the first performance of Britten's *On this Island*, his settings of Auden. Britten's diary again: 'Wystan comes & we—with the Gydes [Sophie Wyss's family], W. Walton & Lennox Berkeley have a party after.' A month later Walton ended 1937 by going into hospital for a bilateral hernia operation. From the London Clinic on 23 December he wrote to Britten (now 'My dear Ben'):

I am so very sorry not to have answered your letters before . . . I'm now getting on well, in fact I leave tomorrow for the country, but shall have to be in bed for another 10 days. How are you? Better I hope, and I trust nothing more serious than overwork (which incidentally I couldn't reprimand you too seriously for). I gather the Wimborne House performance was excellent and it went down very well. It was disappointing that you couldn't be there, in fact that we both weren't. And I liked listening-in to *A Boy Was Born* and liked it much more than when I heard it last time . . . Incidentally I saw Ansermet and the *Variations* are chosen for the next I.S.C.M. Festival along with Lennox and Rawsthorne, which is all to the good. Many congratulations.

For his convalescence, Alice Wimborne took Walton to the beautiful Villa Cimbrone, overlooking the sea above Ravello. As always, Italy worked its spell on Walton. His ambition had always been to build a house there. Cecil Gray saw 'a beautifully constructed model dream palace made by his own hands to be built to his own specifications on the Gulf of Naples, near Posilito, his spiritual home, where Gesualdo also had a palace'.[1] Walton went to Italy now, hoping to begin work on Heifetz's concerto and on 26 January 1938 he wrote to Hubert Foss: ' "Morning sickness" is beginning, but otherwise not much progress'. For the next three months he worked at it. 'Alice was very good at making me work', he said, 'and got very cross if I mucked about.' In mid-April he received a letter from Arthur Bliss, writing on behalf of the British Council, asking him to write a violin concerto to be performed at concerts given as part of a music festival associated with the New York World Fair which was to be held from 30 April to 31 October 1939. Bax, Vaughan Williams, and Bliss himself were also being asked for works, the terms being £50 for a first performance only, with no other rights sought except possibly the dedication, and £100 extra for the trip to New York.[2] Walton replied on 28 April:

[1] *Musical Chairs*, p. 288.
[2] The other works all materialized and were performed: Bliss's Piano Concerto (soloist Solomon), Bax's Seventh Symphony, and Vaughan Williams's *Five Variants of Dives and Lazarus*.

The proposal suits me admirably, that is if everything can be arranged. You may or may not know that Heifetz has commissioned me to write a violin concerto for him and I have just got started on it. The only question is if he will agree to the 1st perf. being under the auspices of the British Council. I think he probably will as it seems to me as good a 1st perf. platform as he will ever get. My terms with him are that he pays me a certain sum for the right to be the sole performer for a certain length of time. I don't know if the British Council would insist on a British violinist appearing in New York—if so I am afraid that it would dish the whole thing, or at any rate I should have to decide whether Heifetz or the British Council took it. But it is obvious that the best arrangement would be for the 'world première' to be played by Heifetz under the auspices of the British Council . . . I don't think there is any complication about the dedication.

A fortnight later, on 11 May, Walton poured out his troubles to Foss. He told him about the British Council offer. (Incidentally, it is too much of a coincidence that the Council should have asked him for the type of work he was already writing—it must have been an open secret that he had a concerto for Heifetz on the stocks.) Then, in May, the film producer Gabriel Pascal telegraphed to him offering him 550 guineas to write the music for the film version of Shaw's *Pygmalion*. Walton refused, but the producer persisted in a series of telephone calls to Ravello. Pascal offered two more films at the same or higher price, 'which means refusing £1,600 within the next twelve months. But to accept would mean refusing the American offer, for I should have to return now . . . As I say I've turned it down. Whether it is a wise decision I am in some doubt.' What would happen, he wondered, if Heifetz didn't like the work.

What, however, seems to me the greatest drawback is the nature of the work itself. It seems to be developing in an extremely intimate way, not much show and bravura, and I begin to have doubts (fatal for the work, of course) of this still small voice getting over at all in a vast hall holding 10,000 people . . . It all boils down to this: whether I'm to become a film composer or a real composer. I need hardly say that no one likes refusing the prospect of £1,650, but on the other hand . . . I am fairly alright for the moment. In fact I think I can safely wipe out films, which have served their purpose in enabling me to get my house, etc.

Fascinating as it would have been to contemplate Walton as a forerunner of *My Fair Lady*, he was adamant in his refusal and the *Pygmalion* score was composed by Arthur Honegger. He now concentrated wholly on the concerto. The following month he reported the first movement completed—'not too bad'—and added that, having been bitten by a tarantula, he had celebrated with a kind of tarantella, 'presto capricciosa-mente alla napolitana—quite gaga, I may say, and of doubtful propriety after the 1st movement'. He returned to London with two movements written. These he lent to the Spanish violinist Antonio Brosa, whom he had first met at Siena in 1928 and encountered again during the recording of the music for *Dreaming Lips* in 1937. (Brosa played parts of the Tchaikovsky and Beethoven concertos on the sound-track while the actor Raymond Massey 'played' them on the screen.) Walton went to visit Brosa, who later described the scene at his house:

[He] played it with me and I made a few suggestions and so on, and he wrote to Heifetz telling him about this and sent him as samples the two movements. Heifetz replied that he was not quite sure he liked them as Walton wanted them and he suggested that he went to America and worked it out with him. Walton was very upset about this. 'For tuppence I would give it to you'. I said: 'I am not Heifetz. He can play it anywhere he likes. He can make records—I cannot.'[3]

In October 1938 Walton completed the recording—begun in 1936—of the second suite from *Façade* which he had scored for symphony orchestra. This had had its first performance in March 1938 in New York, when John Barbirolli conducted it, and had been introduced to London by Sir Henry Wood at a Promenade Concert in September. Walton also laid aside the concerto to write the music for *Stolen Life*, last of the Czinner–Bergner films. Walton did not leave Rugby to attend the première because he was busy with the concerto and asked Dora Foss to tell him what she thought of the music. This she did, as she told her husband: 'It is first-class cinema music—and by that I mean that it is streets ahead in quality of any I've heard, but still fills the bill. It is very very Waltonesque and as I said in my "report" to him, "I thoroughly enjoyed meeting some old friends, so to speak!!" Bits of the symphony etc. The

[3] H. Dawkes and J. Tooze, *A Conversation with Antonio Brosa*, RCM magazine 65 1 (Easter 1969), p. 10. Brosa in 1940 gave the first performance, in New York, of Britten's Violin Concerto.

storm movement is fine . . .' At the end of July Walton was visited by Frederick Stock, conductor of the Chicago Symphony Orchestra, who was commissioning works for the orchestra's fiftieth anniversary in the 1940–1 season. This was to result in the comedy overture *Scapino*. Another pleasant interruption was to compose a wedding anthem for the marriage of Alice's son Ivor Guest, in Kensington on 22 November. He chose a text from the *Song of Solomon*, 'Set me as a seal upon thine heart'.

In January 1939 Dora Foss wrote to her husband, who was in the United States, that Walton had telephoned her to say that 'his difficulty is making the last movement elaborate enough for Heifetz to play it. He says he will never write a commissioned work again.' On 4 February she wrote:

William for the best part of an hour on the telephone. I want to try and remember what he said as a lot is significant. He feels Heifetz won't do the Concerto—he's a bit 'off' Szigeti who 'whines'. Kreisler is his latest idea—a sort of repetition of the Elgar Concerto triumph in his later years. William isn't at all pleased with the last movement—says it wants two months' more work—Brosa not very enthusiastic—William repeats that he will never work to commission again. I asked what he was planning next. *He said he was going to learn composition* and start in chamber music, beginning by a duet, a trio, a quartet and so on, but *not* unaccompanied violin! He is telling Heifetz that if he does it, he (William) will go over any time after Ap. 1st and hold his hand while he learns it etc.

He'd been to the Bergner film [*Stolen Life*] last night and said you could hardly hear most of the music (I must say I didn't hear 27 minutes of music). He says he has had two more offers of film music as a result of it. He says he'll be terribly busy now orchestrating the concerto—he says if Kreisler should do it, he'd have to alter it a bit. Some parts, I gather, really only playable by Heifetz. I tried to cheer him up about the concerto and said in farewell 'Anyhow I'm sure it's better than anyone else could write' (or something like that) and he said breezily 'Oh, I've no doubt about that', which was so typically William. He said a great deal you know—Heifetz's fee etc., and he talked politics for hours. He is very anti-Jew *au fond*. Thinks that they *do* want war and that they create the 'jitters' that the ultra-Left intelligentsia get. He says that they have got a German 1924 outlook—completely *démodé*. He says he can't afford to be rude to the Jews as they are so powerful musically (or to that effect). He says he thinks it's a menace all the German Jewish musicians that are coming in the country and taking work from Christians etc. etc. . . Oh, he says it's the first piece of *bad* music he's written. I *do* hope he's mistaken, but I do feel he is so detached that he can really judge his own work.

A month later, on 3 March, Walton telephoned her—'terribly excited'—to say Heifetz had cabled 'Accept enthusiastically'. He then withdrew the concerto from the World Fair festival, 'not', as he wrote to the conductor Leslie Heward on 28 March, 'because it's unfinished but because Heifetz can't play on the date fixed (the B.C. only let him know about 10 days ago!) Heifetz wants the concerto for two years and I would rather stick to him.' In May he and Alice sailed in the *Normandie* to discuss it with Heifetz, who was more interested, Walton claimed, in the garden of his Connecticut farm where they met. 'As I was leaving,' Walton recalled (in his letter to Alan Frank of 10 October 1973), 'the question of being paid arose. Was it £300 or 1,500 dollars? I said £300, not realising by then that the pound was a bit shaky even in those days. So he took out a bit of paper, rang up his bank and gave me 1,493 dollars and some cents! He'd made on the deal!' Heifetz's contribution to the work's final form was minimal—some added accents in the Scherzo and alterations in the first movement cadenza which he found too easy so, to quote Walton, he 'jazzed it up'. The concerto was completed on 2 June 1939, but revised in 1943.

I find it difficult to believe that anyone could fail to have heard at the outset that with this concerto Walton showed that he had maintained the imaginative level of the First Symphony, increased his command of orchestration, and regained the emotional poise of the Viola Concerto. His remark to Foss about its 'intimate' nature is crucial, for it shares more than the key of B minor with Elgar's concerto: both works are the media for a peculiarly personal and introspective expressiveness which paradoxically is conveyed in music on a large and dazzlingly virtuosic scale. For sheer beauty of melody, brilliance of detail, and impassioned eloquence, it ranks among the greatest of modern violin concertos (few begin more magically or plunge more swiftly to the heart of the matter). Its Elgarian qualities include the Italianate atmosphere of the music, as redolent of the South as Elgar's tone-poem. Like Elgar, Walton reserves his strongest effort for the Finale, which is emotionally a summing-up and contains a passage of thematic prestidigitation, when the principal subject of the first movement, the opening theme of the Finale, and the Finale's broad and lovely second subject are combined—again, a Prokofievian procedure.

Walton let it be known in later years that the concerto expressed his feelings for Alice Wimborne. Like Elgar and his concerto, he could have inscribed it 'Here is enshrined the soul of Alice'. The unforgettably romantic theme of the elaborate first movement is introduced 'sognando' (dreaming) by the soloist, with another theme accompanying it on cellos and bassoons. The rising seventh gives the melody its yearning quality (Ex. 6).

The 'official' second subject is a fluent theme for flute and

Ex. 6

strings, in the remote key of E flat minor. Melodic and rhythmic transformations of the first subject are the principal dramatic feature of the movement, after the first of which Walton inserts a cadenza. That Walton was ever concerned that the concerto might lack the bravura to attract Heifetz when he had written a movement as brilliant and difficult as the Scherzo is hard to believe. The movement seems to be poised on a knife-edge— an effect obtained by the precarious tonality, with many flattened notes and a veering between F, B flat, and A—and only in the trio section, its melody announced by the horn and taken up several octaves by the violin, is there a rest from the activity. With the markings 'napolitana' and 'canzonetta' in this movement, we are left in no doubt of the Italian provenance of the music. Even more Italian in its sunlit languor is the marvellous second subject of the Finale, one of Walton's most inspired inventions:

Ex. 7

The whole movement is contrapuntally and thematically Walton at his ingenious best (notwithstanding the uncanny resemblance between the opening theme and that of Prokofiev's Third Piano Concerto); as in the Viola Concerto, the movement moves inevitably towards a spacious recall of the first movement's principal subject from which develops the Elgarian accompanied cadenza—again 'sognando'—in which all the thematic strands are interwoven and interrelated and which is followed by a jovial march-like coda, a last touch of volatility in this volatile concerto.

  Today this brilliant, sensuous, and successful concerto has gone into the repertoire of leading violinists the world over and is as well loved as any of Walton's works. It is piquant to remember it made a rather muted entry on the English scene, because of wartime conditions which prevented its being launched in London by Heifetz, and was seized upon by some commentators as the harbinger of decline. The late Colin Mason, in his article on Walton for the influential paperback *British Music of our Time* written when he was barely twenty-one years old, described it as 'a retrogression to the plane of the Viola Concerto' (*sic*!) and as a technical retrogression because of its chromatic idiom which 'may produce ephemeral, exotic hothouse plants'. Mason was puzzled by its seeming 'to leave something unsaid' in the Finale.[4] A more tenable criticism would be that what is essentially a lyrical work sometimes seems to drag in the virtuosic element by the scruff of its neck, but a really superb performance banishes this reservation. Nevertheless, the concerto does mark a turning-point in Walton's music. Compared with the works which preceded it, especially the First Symphony, it is more relaxed, more indulgent, less spiky, less brittle. There is more compositional sophistication.

There is no doubt that Walton felt himself to be at a critical point in his development when he was writing it. While he was in New York with Heifetz in 1939 he said that the spread of the knowledge of music through radio and the gramophone had reduced music's 'life expectancy' and he added:

Today's white hope is tomorrow's black sheep. These days it is very sad for a composer to grow old—unless, that is, he grows old enough to witness a revival of his work. I seriously advise all sensitive composers to die at the age of 37. I know: I've gone through the first halcyon periods, and am just about ripe for my critical damnation.[5]

These were to be prophetic words, though the damnation was to be delayed in its arrival by the intervention of the war. After his exertions since 1926, Walton's words to Dora Foss about 'going to learn composition' ring true, for he had reached a point in his spiritual and technical maturing where he was ready to take stock and to recharge his batteries.

Walton had intended to return to the United States for the first performance of the concerto. This was originally planned to be with the Boston Symphony Orchestra, with the composer conducting, with several other performances to follow. Heifetz

[4] C. Mason, 'William Walton', in A.L. Bacharach (ed.), *British Music of our Time* (Harmondsworth, 1946), pp. 147–8.
[5] *New York Times*, 4 June 1939.

intended to give the first London performance at a Courtauld–Sargent concert in March 1940. But when Britain declared war on Germany on 3 September 1939 all arrangements were altered. The concerto had its first performance in Cleveland, Ohio, on 7 December 1939, with Artur Rodzinski conducting. Walton wrote to Heifetz from Ashby St Ledgers on 15 October:

Alas, I don't think owing to this something war, there is the slightest likelihood of my being able to get over for it, what with the difficulties of travel and the difficulty, not only of expense, of obtaining any dollars, also I'm of military age and am liable to be called up, but when, the powers that be (there are far too many of them!) will decide. Meanwhile, I have become an ambulance driver for the local A.R.P. though I must admit that it doesn't entail much, as everything, as you will have read in the papers, has been very quiet as regards air raids. In fact this is a peculiar war and not working out as predicted and it looks as if the side that can bore the other most will win . . . Is the concerto going to be broadcast? If so, could you be so kind as to let me know the time and wavelength. Also, if it is, could you have a record taken of it 'over the air' and send it to me as it seems to me to be the only way I shall have, or may ever have, of hearing the work, at any rate for a very long time . . . As for your alteration, I approve of it and send you some alternatives on a separate sheet, which pray heaven, a well-meaning, if unmusical censor won't obliterate under the mis-apprehension that it is a code! Going through the parts, I have come across several mistakes in the full score, also here and there I've made a few alterations in the scoring which will, I trust, be all to the good.

Heifetz must have relaxed his two-year lien on the concerto because the English première was announced for February 1941 in the Sheldonian Theatre, Oxford, to coincide with the conferment of an honorary doctorate of music on Walton by his university. But the set of orchestral parts was delayed on its way from the United States. Moreover, Heifetz's copy of the solo part, with his bowings and fingerings, was lost aboard a ship which was torpedoed in an Atlantic convoy; a photographic copy, which had sensibly been made in New York, was flown over in its place. The performance was eventually given in the Royal Albert Hall on the afternoon of 1 November 1941. The soloist was the Danish violinist Henry Holst, a former leader of the Berlin Philharmonic under Furtwängler, who had settled in England in 1931. Walton conducted. One of the London Philharmonic Orchestra's trumpeters that day was

Malcolm Arnold, who recalled that the parts were so full of mistakes he believed they had been sabotaged.

Holst and Walton had worked on the concerto together at Ashby St Ledgers on 31 August and 1 and 2 September and again in October. Another musician present at those sessions was Roy Douglas, who first worked with Walton in 1940. For the next thirty-two years he helped to prepare almost all Walton's works for performance and publication, correcting mistakes in proofs and parts and, on occasion, orchestrating some of the film music. Most of the sections he scored were comparatively short, seldom more than twenty or thirty bars, sometimes as few as eight or ten. While staying at Ashby St Ledgers, Walton worked in a long studio containing polished boards, a grand piano and stool, some bookcases, and not much else, except for an array of pipe-racks, each holding six or eight pipes, hanging at intervals along the walls. When he had finished smoking one pipe, he would select another from a rack, fill it and light it—he was almost a pipe chain-smoker.

At one of their sessions with Holst on the concerto, Walton asked Douglas why orchestral players disapproved of composers conducting their own music. Douglas replied that the composers frequently arrived at rehearsal unprepared, without having studied the score and without having decided how and what they were going to beat—an amateurish rather than professional approach. After that, while Holst and Douglas rehearsed the concerto, Walton took a baton and practised conducting. Douglas told Holst he would not accompany his playing but would strictly follow Walton's beat. It was from lessons like this that Walton developed into a good conductor of his own music.

There was an unusual final run-through of the concerto at Ashby on the evening of 2 September after a good dinner and generous tots of brandy. Walter Legge was present, sprawled in a chair, and Alice Wimborne reclined on a sofa in a silk gown smoking one of the cigars occasionally sent to her by her late husband's cousin, Winston Churchill. Walton picked up Holst's violin and sawed away at a section of the violin part, accompanied rather more expertly by Henry Holst, while Roy Douglas conducted.

It was at about this period that Walton began to be called

William rather than 'Willie'. Spike Hughes recalled that at twenty-four Walton—'tall, thin and all elbows and knees when seated'—looked like 'Willie'. But he was

allowed to grow out of his diminutive except by overfamiliar acquaintances, who get a reproachful look if they fail to call him "William". The owner's preference for his baptismal name is not due to any sense of self-importance; he just prefers the sound of it, and has always signed even his most intimate letters that way. [Not always.] Pomposity is something entirely foreign to his nature.[6]

[6] 'Nobody Calls him Willie now'.

12
War Films, 1939–42

WALTON intended, if the war had not intervened, to fulfil his plan of writing chamber music. In August 1939 Kenneth Wright of the BBC was in touch with him about 'your long-promised Quartet' for the Blech Quartet. Instead his first wartime task was the transcription of music from J. S. Bach's cantatas selected by Constant Lambert for a Sadler's Wells ballet with choreography by Ashton and décor by Rex Whistler. *The Wise Virgins* had its first performance on 24 April 1940 with the principal roles danced by Margot Fonteyn and Michael Somes. (It is said, accurately I hope, that posters for the ballet season included the words '*The Wise Virgins* (subject to alteration)'.) The ballet, which was slow to attract audiences, is said to have been an enchanting spectacle, with Whistler's designs among his best, though he wrote to Edith Olivier a week before the opening: 'I fear the scenery and dresses are going to be intolerably hideous, but we must all *face the music*, I suppose (or Constant Lambert will be offended—not to mention Willie').[1] There were nine numbers, the eighth a repetition of the third but differently scored. Although transcriptions of Bach are no longer fashionable where they involve four horns, three trombones, timpani, and harp, Walton's are in excellent taste, being neither mock-archaic nor extravagantly 'modern'. That of 'Schafe können sicher weiden' ('Sheep may safely graze') is an exquisite piece of work, its craftsmanship impeccable.

A suite of six numbers was recorded under Walton's direction in July 1940. The first movement is the opening chorus from Cantata 99 'Was Gott tut, das ist wohlgetan'. Strings and woodwind are used until the brass carry the big tune. Next comes the '*Passion*' chorale, 'Herzlich tut mich

---

[1] L. Whistler and R. Fuller, *The Work of Rex Whistler* (London, 1960), p. 64.

verlangen'; although Walton 's transcription is of its form as the organ chorale prelude in B minor (BWV 727). Cor anglais and muted horn lend a dark colouring to the scoring before the strings predominate. The third movement is the tenor aria 'Seht was die Liebe tut' from Cantata 85. After the opening for strings, the solo line is given first to solo flute, then oboe. 'Ah! how ephemeral!', the fourth number, is the opening four-part chorus of Cantata 26, 'Ach wie flüchtig, ach wie nichtig'. Vigorous passages for strings are punctuated by interjections from woodwind and brass, and the choral tune is given to trumpets and trombones. For the famous and well-beloved recitative and aria 'Schafe können sicher weiden' from the secular Cantata 208 'Was mir behagt, ist nur die muntre Jagd'—the melody which suggested the ballet to Ashton— Walton uses the harp to suggest the continuo of Bach's day. A solo violin plays the introductory recitative while flutes and cor anglais have the principal melody. The Finale is a transcription of the choral Finale of Cantata 129, 'Gelobet sei der Herr, mein Gott'. This is the most densely scored movement, with majestic brass—Walton took his cue from Bach himself, who scored it in 1726–7 for three trumpets, timpani, flute, two oboes, oboe d'amore, strings, and continuo.

In the first half of 1940, also, Walton composed a series of nine piano solos, *Tunes for my Niece*, written for Elizabeth Walton, daughter of his brother Noel. But Dora Foss told him they were too difficult, so he rearranged them as *Duets for Children* to include Elizabeth's brother Michael. In this form they were completed on 15 May. A tenth number, the 'Trumpet Tune' finale, was added in June. The duets were later orchestrated as *Music for Children* and first performed at the Queen's Hall by the London Philharmonic Orchestra under Basil Cameron on 16 February 1941. If this could have been Walton's *The Wand of Youth*, it has never attained much popularity in spite of the attractively piquant orchestration.

At the same time Walton was occupied with the score for a film of Shaw's *Major Barbara*. Gabriel Pascal had pinned him down at last, but not as first choice. Apparently Shaw and he had tried to commission music from Toscanini, surely one of the most bizarre and unlikely-to-succeed projects ever to be contemplated even by men of the theatre and cinema. 'No use waiting for Tosca. I cut no ice with him', Shaw said,[2] so Walton was sent for. Pascal took him to lunch with Shaw in Whitehall Mansions, where, as Walton used to tell it, 'my eyes were glued on him, for every time he opened his mouth, his dental plate came down and then went up—very off-putting'. Shaw, reliving his years as a music critic, could not forbear to make suggestions to Walton about his setting of the Utopian ode to the future: 'In bar 37, the major third does not suit the word "lamenting". Make it minor; flatten the B . . . If I were a musician I should not presume to suggest an alteration; but an amateur's comments should always be listened to. Even when they are technically asinine they often draw attention to some hitch or other. Mozart did not disdain the idiocies of Schikaneder. Otherwise extremely O.K.'[3] The score was completed at the end of 1940, some of it being orchestrated by Roy Douglas, who visited Ashby St Ledgers for the first time in November 1940. The house was far from warm, because of wartime shortage of fuel, and they had meals in a long and draughty dining-room. Each of the three had an oil-stove beside his chair. Douglas had taken his rations with him. 'We can have butter with our muffins today, William,' Alice said, 'because Mr Douglas has brought his ration.' After dinner Walton would depart to compose, and Douglas and Lady Wimborne sat on the club fender with their feet in the fireplace trying to keep warm and talking of music, literature, and people. She had a low, husky, attractive voice and was always friendly and relaxed. As the war lasted, the beautiful gardens became progressively more neglected because no gardeners were available, and eventually sheep were encouraged to graze on the lawns. Domestic staff, too, were impossible to find. By 1943 the butler was one of the retired gardeners, by then extremely short-sighted and with thick spectacles. On more than one occasion after he had left the dining-room, a loud crash of crockery was heard.

Walton's other task between July and 28 December 1940 was composition of the overture *Scapino* for Stock and the Chicago Symphony Orchestra. He produced a virtuoso score in the exuberant manner of *Portsmouth Point*, but with a pronounced

[2] V. Pascal, *The Disciple and his Devil* (London, 1970), pp. 94–6.
[3] Letter to Walton from G. B. Shaw, 5 May 1940.

Italianate atmosphere and more poise and elegance than in the
older but younger-spirited work. The original score was for
large orchestra with triple woodwind, two cornets, and a big
array of percussion. In the 1950 revision the orchestra was
reduced to normal proportions except for the percussion.
Walton took the idea for the work from an engraving of
Scapino by Jacques Callot (1592–1635). Scapino was one of the
less familiar *commedia dell'arte* characters, a little like Till
Eulenspiegel and perhaps an ancestor of Figaro and Leporello.
He was a rascally servant who helped his master in various
escapades (the word is derived from Scapino), especially
amorous. So the music is appropriately pert and scintillating,
full of what Elgar would have called 'japes'.

Scapino himself stands before us in two guises in the very
first bars: the lively rogue, with a theme 'molto vivace' on the
trumpet, preceded by a flurry of strings and woodwind and a
crash of percussion. Typical Walton, no one else. Another
theme, oboes and violas, is less obstreperous but still perky.
After these motifs have been tossed about a bit, violas and cor
anglais introduce a more fluent melody which is to become
important later:

Ex. 8

Walton develops this material with a sure touch, for the
delightful detail is one of the overture's chief features. A
'giocoso' rhythm for trumpets is used as a framework for
variation of the other themes. The clatter subsides into a duet
for side-drum and tambourine, and the trio section, in A flat,
begins with a solo cello playing an augmented version of the
third Scapino theme

Ex. 9

The composer's direction for this episode is 'come una serenata' and obviously this is one of the amorous escapades—we can hear the cheeky first Scapino theme chuntering away as a counterpoint. Soon the sentiment is swept aside, the first section is recapitulated after a fashion, and—with considerable wizardry expected from everybody in the orchestra—the work ends uproariously.

The first performance was given in Chicago on 3 April 1941 and the overture was repeated by Stock a fortnight later. Walton conducted the first British performances, at Bedford on 12 November and in London on 13 December. On hearing the 'giocoso' section, the conductor Boyd Neel jocularly accused Walton of quoting Rimsky-Korsakov's *Scheherazade*. Walton's postcard in reply said: 'Ass, don't you know *Tancredi* when you hear it?!!' Neel did not capitulate. The passage does sound like Rimsky, but Rossini is there too; and, as Rossini was the composer Walton admired above all others, his word must be accepted.

Walton received his calling-up papers for the armed forces early in 1941. However, Jack Beddington (1893–1959), director of the films division of the Ministry of Information (1940–6) and a former advertising manager of Shell and British Petroleum as well as a discerning patron of the arts (hence the famous Shell Guides), arranged for Walton's exemption from military service on condition that he would write music for films deemed to be 'of national importance'. Walton would not have resisted call-up on any moral or ethical grounds. He was not a pacifist like his contemporaries Tippett and Britten. Indeed it is impossible to divine Walton's political opinions from anything he said or wrote. His views on this aspect of an artist's life were set out in 1977 when he talked with Arthur Jacobs at the Savoy: 'I don't believe there is any point in a composer "joining in" at all, either as a war hero or as avoiding war or anything. It's nothing to do with us. Tree conservation is much more my line. Very much a dendrologist I am. I've abstained from "causes" because it's not my business.'

The first war film with which Walton was involved was the celebrated *Next of Kin*, written by Thorold Dickinson and Basil Bartlett at the War Office's request as a piece of propaganda about the need for security—the wartime slogan on hoardings in Britain at this time was 'Careless Talk Costs Lives'. Walton wrote the score, thirty-two minutes of music, between 2 and 22 December 1941. He was assisted between 18 and 21 December by Roy Douglas, who scored 345 bars for him. It is not a particularly distinguished score. The opening march might unthinkingly be described as Elgarian but more justly as in the manner of Eric Coates. Other sections lean heavily on Ravel's *Daphnis and Chloë*. *Next of Kin* at first was shown only to members of the Services and civilians with security passes. Churchill asked for the violence to be toned down before he would agree to its being shown to the public on general release.

Walton spent much time now at Ashby St Ledgers. His house in South Eaton Place had been destroyed in the heavy bombing attacks on London in May 1941 which also destroyed Queen's Hall and when in London he lived at Alice Wimborne's Lowndes Cottage in Belgravia. During May 1941 he began a correspondence with his friend Cecil Gray, the critic and composer, about the possibility of collaborating on an opera. They settled on the subject of the composer Carlo Gesualdo, Prince of Venosa, who in 1590 murdered his first wife and her lover. Gray had years earlier written a book about Gesualdo with Philip Heseltine (Peter Warlock). Walton kept up interest in the subject until the end of 1942. He wrote no music for it but wrote several letters about adjustments in the libretto and got as far as wondering whether Michael Ayrton was the right artist to design the scenery. Gray wrote in his autobiography: 'Whether the composer felt that it [the libretto] was a failure or merely unsuited to him, or alternatively that he did not feel equal to the task, I do not know—a bit of each perhaps'.[4]

At about the same time in the summer of 1941 Walton agreed to write incidental music for John Gielgud's new production of *Macbeth*. He went to some early rehearsals and, Gielgud recalled, 'immediately decided to compose background music for the scenes of the witches in accordance with the rhythms of their verses and the rhythms I was trying to achieve with the actresses in those scenes. I gave him the timings for the

---

[4] *Musical Chairs*, p. 243.

interludes and some scene-changes and was amazed to find he had observed them meticulously without having to consult me further.' The play was due to open in Manchester on 16 January 1942. Walton cut it fine. 'I managed *Macbeth* by the skin of my teeth,' he wrote to Roy Douglas in January 1942, 'twenty minutes in eight days.' To Dallas Bower he gave more detail: 'Started on Xmas Day and finished on New Year's Day. And not all of it bad, either! In fact *Macbeth* is pretty good.' Discs of the music were recorded (by HMV) by the London Philharmonic Orchestra conducted by Ernest Irving. I saw the production in Manchester. To a sixteen-year-old's eyes it was magnificent, like the cast, and I was impressed by the music in spite of the dreadful sound-recording system by which the music was relayed in the Opera House. Hearing some of the surviving discs today, one can appreciate Walton's satisfaction with it. It is a substantial score and, though, like so much music for Shakespeare plays, it consists largely of fanfares and a few bars of scene-setting atmosphere, there are longer and more striking passages. With the witches, he achieved exactly what Gielgud described (as well as some intentional and witty references to Wagner's Valkyries); there is some eerie 'bagpipe' music; and Banquo's ghost is conjured up by the use of the flexatone and woodwind over chords from strings and horns. It is unmistakably music of the theatre, some of which could have gone straight into an opera.

Walton entered 1942 with two commissions under his belt for the films division of the Ministry of Information. The first was a score for Michael Balcon's *The Foreman Went to France*, with Tommy Trinder and Clifford Evans, about a foreman (in real life Melbourne Johns) who persuaded his manager to let him go to France to retrieve some special machines before they were captured by the Germans. The script was by J. B. Priestley. The second was music for Leslie Howard's film about R. J. Mitchell, designer of the Spitfire fighter aircraft, to be called *The First of the Few*. On 25 January 1942 Walton wrote to Roy Douglas:

Tomorrow I start on Cavalcanti's [assistant producer] film at Ealing, a rather boring one to fill in the time till the Howard film which I saw yesterday and is very good. That shouldn't be ready for music till the middle of March. However, you never know, so please let me know your movements. After that is over it will be about time I settled to something else. A string quartet? A clarinet quintet? Perhaps neither. *Scapino* has not yet gone to press, but I shall be most grateful if you would help me with the proofs.

Boring or not, *The Foreman Went to France* contained some effective music, as can still be heard when the film turns up occasionally on television. The refugees section, in particular, belongs to the world of the First Symphony's slow movement. Ernest Irving described how, in this sequence, what Walton wrote 'was completely different from the idea given to him by the producer who, by the luckiest of mistakes, failed to make himself understood. The music thus composed under a mis-apprehension really makes your feet sore and your knees sag and excites the exact emotional reaction required'.[5] The parodist in Walton responded to the French setting: two of the instrumental sections are subtitled 'La Ravelse' and 'Moto Poulencuo'.

On 26 January 1942 Dallas Bower, who was now working as a BBC radio producer, put two projects to Walton. The first was music for a BBC programme about Christopher Columbus to mark the 450th anniversary of his discovery of the New World. This was to be written by Louis MacNeice, the BBC Chorus and Symphony Orchestra would be available, Laurence Olivier was to be Columbus, and it had to be ready by mid-March. The second was a 'private effort' to raise money for the Greek Red Cross by producing the Yeats translation of *Oedipus Rex* in the Royal Albert Hall, with Robert Donat as Oedipus. Walton replied next day:

This is most irritating—there is nothing I should like better than to accept both propositions . . . Unless you can get the Corp. to postpone the date to mid-May instead of mid-March, I don't see how I can do it . . . In fact I don't see how else you can do it, as there are precious few composers about nowadays. I don't suppose it would be possible to get Alan Rawsthorne out of the army again . . . He would be best . . . Benjamin Britten is by way of returning—he may be already back, but I gather he joins the RAF music dept. on landing. Who else is there except the old gang of V.W., J. Ireland, A. Bax if you want any of them.

Britten and Peter Pears returned from the United States,

[5] E. Irving, 'Music and the Film Script', in *British Film Yearbook, 1947–8* (London, 1947), pp. 50–1.

where they had been since 1939, on 17 April 1942. When Britten appeared before a conscientious objectors' tribunal, Walton spoke on his behalf. Britten at first said he would serve in a non-fighting unit, hence the idea that he might direct the RAF band, but he changed his mind and said he would not serve in anything which involved him in wearing uniform. Walton was left only with a plea to the tribunal that it would be a crime to prevent such a fine composer from writing music. The chairman thereupon inquired what authority Walton had to assess the importance of Britten's music. Walton replied: 'Well, if you don't know who I am, there is no point in going on.' Britten was required to give concerts for the Council for the Encouragement of Music and the Arts (forerunner of the Arts Council) and in prisons.

On 30 January 1942 Walton wrote again to Bower:

Thank you for the sketch of C.C. I'm a bit terrified of accepting it, since you know what films are like and I'm worried that the Howard film won't be over in time to give this the music it deserves . . . For the music will have to be good and one can't rely on a quick film extemporization technique for it, so it will need more time, trouble and care . . .I should hate to have to deliver it to the tender mercies of V.W. though on the other hand he might do it rather well, but I must admit it is more up my street.

In mid-March he had had a request from Bliss, then Director of Music, BBC, for a suite for brass band. He agreed to write it if there was 'no real hurry'. It never materialized, but apparently it was to be based on a selected list of folk-songs, for on 27 April he thanked Kenneth Wright for 'the list of tunes'. He continued:

I must bore you further and ask if you would be kind enough to let me know what exactly your band consists of, as I must confess I know precisely nothing about brass bands. So if you would let me have a list of instruments, their compass and characteristics etc. I should be grateful . . . Another thing, can I beg, borrow or steal some or all of the volumes from which I can take the tunes. Never having gone in for folk-songs etc. I must again confess almost complete ignorance of most of those in the list. So what I turn out may be a complete disaster, but on the other hand I think I shall enjoy doing it and anyhow it needn't be accepted if you don't like it, and I shan't mind.

Walton was desperately anxious to write the *Columbus* music

and the delays over *The First of the Few* irritated him. In addition he had a bad dose of influenza in March, followed by sciatica. In a letter to Bower from his sickbed he urged postponement of *Christopher Columbus* from its planned broadcast at the end of May—'I don't see that it can be put on during the double summer time at all because nobody is going to sit indoors and listen for 2 hours to something they don't know whether it's good or bad when they can be outside.' (During the war clocks were advanced by two hours on GMT to enable agricultural work to continue late into the evening.) Walton then made some interesting comments on his attitude to film music:

Actually from my point of view, I can't treat C.C. in any way different from a rather superior film. That is, that the music is entirely occasional and is of no use other than what it is meant for and one won't be able to get a suite out of it. Which is just as it should be, otherwise it would probably not fulfil its purpose. That is why I'm against my film music being played by Mr Stanford Robinson or anyone else. Film music is not good film music if it can be used for any other purpose and you've only got to have heard that concert the other night to realize how true that is. For all the music was as bad as it could be, listened to in cold blood, but probably excellent with the film. So I don't care where *Major Barbara* is or any other of my films. The music should never be heard without the film.

Walton began to compose *The First of the Few* on 31 May—two days after he had conducted the definitive (twenty-one items in seven groups of three) version of *Façade* in the Aeolian Hall, London, with Lambert reciting and a new curtain designed by John Piper. Ironically, this was the film music which was to mock his words above, for the film was an outstanding success to which the music made such a remarkable contribution that people wanted to hear it again. The splendid march during the opening titles and the dazzling fugue which accompanied the assembly of the Spitfire's parts in the aircraft munitions factory immediately wooed audiences' ears. The film was first shown in August 1942 and by the end of the year Walton had rescored the *Prelude and Fugue* as a concert-hall work. This was first performed in Liverpool on 2 January 1943, conducted by Walton, and in London seven weeks later conducted by Sargent. The first recording, by the Hallé under Walton, was issued in August 1943. Walton had composed the score throughout June 1942, the month which included the gloomy news of the Fall of Tobruk. From 11 to 18 June he was joined at Ashby by Roy Douglas, who scored some of the music. Douglas was able, on these occasions, to observe the beneficent influence of Alice Wimborne on Walton.

She was [he wrote] far more than merely a society hostess. She was a charming and very cultured woman of much artistic and literary discernment. I stayed at Ashby on many occasions, perhaps a dozen or so times, and was well aware of her discriminating and helpful criticisms of William's music. For instance, after dinner William would often go away and write some film music and then bring it to play to Lady W. and myself and I have known her to say: 'That's not really good enough, William, you can write a better tune than that.' And he would meekly go back to the music room and do so. She listened to many of the rehearsals of the Violin Concerto when I was there playing it through with Henry Holst, and she frequently expressed acute and valuable opinions. I am also strongly of the opinion that she had a very good influence on his character. When he was with her he was a much kinder and more thoughtful man and not so inclined to be bitchy at other people's expense. He was a weaker character than is sometimes realized, though he could be very obstinate.[6]

Alan Frank, of the Oxford University Press, recalled seeing Walton and Alice Wimborne together towards the end of the war. In spite of the gap in age, they looked and acted, he said, 'like a couple of kids enjoying themselves'.

[6] Letter to the author, 1 Sept. 1987.

13

Columbus and Henry V, 1942–5

WITH the Howard film out of the way, Walton was free from
July to compose *Christopher Columbus*. During June someone at
the BBC had at last realized that 12 October 1942 was the exact
anniversary of Columbus's discovery, so the broadcast was
fixed for that evening, much to Walton's relief, though, as he
drily pointed out to Bower, 'I see Oct. 12 falls on a Monday so
you had better get busy in case any "Monday Night at 8"
business has been scheduled. C.C. should have a priority for
this date.' In 1942 there were only two radio channels, the
Home and the Forces. Most serious programmes were on the
Home, but 'Monday Night at 8' was a popular 'magazine' show.
Walton had apparently not realized there was another possible
rival—12 October 1942 was Vaughan Williams's seventieth
birthday.

He composed a very substantial score, completing the music
for another Cavalcanti film, *Went the Day Well?*, at the same time.
Roy Douglas assisted with the orchestration of both scores in
mid-September. Rehearsals for *Columbus* began in September
in London and Bedford (where the BBC Symphony Orchestra
was then stationed). Boult conducted and the BBC Chorus was
augmented by thirty singers. Ten days before the performance
Bower reported to Walton, who was taking a rest in North
Wales, that 'Adrian is a little fussed at the complicated nature
of the whole undertaking', but after the broadcast he wrote:
'Boult worked like a navvy and . . . really got the shape of it.'
Bower was upset by the miscasting of Margaret Rawlings as
Beatriz (the recording shows he was right to be) when he had
been denied Vivien Leigh because of incompetence by the
BBC's casting office. Also in the cast was MacNeice's wife, the
singer Hedli Anderson. Olivier's stirring performance vocally
anticipated his *Othello*. There was much disappointment that
Walton had not been present at the performance. He did not

hear all the broadcast because of poor radio reception in Wales. But unquestionably he had been stimulated by the play into writing some fine music. The choral sections are strikingly good and there are foretastes of the future composer of opera in the scene where a plainsong Kyrie leads into Beatriz's 'When will he return?', an exquisite song which is the only part of the score to have been published (not until 1974). Columbus's procession through Granada is another operatic sequence, there are two lively sea-shanties, and the music which accompanies Columbus's sighting of land conveys the authentic *frisson* of a new world discovered. Altogether it is music of greater worth than most of his film music. The play itself is characteristic of radio-drama of the day, ambitious, high-flown, sometimes banal (as in the Portland Place Mummerset speech of the lower-class Spaniards), yet with a loftiness of aim that compels admiration. Like the adaptation of *The Pilgrim's Progress* in 1942, for which Vaughan Williams wrote the music, and *The Rescue* (1943) and *The Dark Tower* (1946) which had marvellous Britten scores, *Columbus* represents BBC radio-drama at its serious best. It was rebroadcast in 1973.

Walton himself had no high opinion of the work. In December 1966, when he heard that a BBC producer, Michael Pope, had resurrected the score from the BBC archives and was urging Oxford University Press to publish the songs, he wrote to Alan Frank: 'About Chr. Columb. I can't believe that there is anything worth while resuscitating from that vast and boring score. I don't remember a thing, so I had better have a look at those songs, not that I can do much about them as I think the BBC bought the whole thing outright for next to nothing.' On 12 January 1967 he returned to the fray: 'One must draw a line somewhere about the horrors of one's past being allowed to be dragged up and I am for a complete ban on those songs even to destruction of the MSS. I'm not at all sure that MacNeice wouldn't feel the same about his lyrics!' He was no keener in May 1973 when Pope let Gillian Widdicombe hear the original recording and she urged publication of Beatriz's song, saying it was 'really beautiful'. Walton's comment to Alan Frank was: 'I feel if it was as beautiful as all that I should remember it!' But he allowed publication.

During 1942 Walton became civilian music adviser to the

Army Film Unit, there being no room for him in the unit as an officer. 'I can't say I mind,' he later told Bower; 'I can say what I think, speak out of turn etc. in my present status than if I had to remember I was speaking to a general and minding my ps and qs. My first experience of a government department, it stank of red tape, but I found everyone very nice.' In January 1943 he went to Liverpool to conduct the first recording of *Belshazzar's Feast*, with the original soloist Dennis Noble, Huddersfield Choral Society, and the Liverpool Philharmonic Orchestra. This was the second of a series of recordings of British music made during the war under the auspices of the British Council. Walton in this, and in his later recording, obtained a more exciting and convincing performance than anyone else, mainly (as has already been said) because his tempos are right and his sense of contrast so vivid.

One of Walton's next important commissions was first mooted in Bower's letter to Walton of 2 October 1942, written during the *Columbus* rehearsals, although there had obviously been discussions. He wrote: 'The situation regarding *Henry V* has become complicated by another matter quite extraneous to it, and I have now to fight yet another private war in order to smooth things out. We appear to be surrounded in this country by past masters in the art of mismanagement. However, I think we shall be in production eventually . . .' Bower in 1938 had prepared a version of Shakespeare's *Henry V* for BBC television, then in its infancy at Alexandra Palace. It was rejected as too ambitious. During the war, when Bower was in the army, he converted it into a film scenario and, when he was transferred to the Ministry of Information, suggested its production. Again, it was rejected. During work on *Christopher Columbus* Bower, now a BBC producer, mentioned the project to Olivier, who was enthusiastic and interested the Two Cities Films producer Filippo del Giudice in the idea. Del Giudice suggested Walton for the music, and Bower's letter to Walton of 14 October had a PS: 'Larry and Vivien are determined to have you for Henry.' Work on making the film began on 6 October 1943 and finished on 3 January 1944, but the music and sound-effects were not added until the spring of 1944.

Walton's first preoccupation in 1943 was to compose the music for a new patriotic ballet to be choreographed by

Frederick Ashton, who was given special leave from the RAF
for the purpose. There was little time, as the first performance
was planned for 6 April and the ballet had to be rehearsed
while the Sadler's Wells company was touring the provinces.
Walton sent the music a few pages at a time. But the creation of
The Quest is best related by Walton as he described it in a letter
to John Warrack written on 27 August 1957:

About *The Quest*. Not much to be said for it . . . The subject, not very
inspiring, had been concocted from Spenser's *Faerie Queene* by some
woman or other whose name I cannot remember [Doris Langley
Moore], but it was a difficult scenario. The music was composed under
adverse conditions and was written more or less as one writes for the
films, first come first served! so some of the ideas were not too bad and
some better not mentioned, but as it had to be done quickly as Freddy
had only limited leave, it had to be done that way—45 mins' music in
less than 5 weeks. It was not much of a success from anyone's point of
view, with choreography, scenery or music all suffering from the same
causes. I remember having to bribe guards on trains to take a minute
or so's music to wherever Freddy happened to be with the ballet—to
Wolverhampton, Preston or somewhere. Some of the dances were
done before the music arrived! However, considering everything,
there were some good moments, though at the moment I can't recall a
note of the music—the "seven deadly sins" came through at the end, a
passacaglia could have been magnificent if there had been any
orchestra. And Bobby Helpmann looking more like the Dragon than St
George! However there have been quite persistent attempts to revive
it, but both Freddy and I, perhaps wrongly, have fought shy of it.
There was one particularly ghastly moment towards the end when all
the girls (there were no men at the time, unless you include B.H. in
that category!) pulled out their handkerchiefs and began to wave to
what was apparently an invisible train off to the war . . .

The Quest had a distinguished cast. Scenery and costumes
were by John Piper, and among the dancers, besides Helpmann, were Fonteyn, Alexis Rassine, Moira Shearer, Julia
Farron, and the fifteen-year-old Beryl Grey. Lambert conducted. Walton finished the score on 29 March 1943, his forty-first
birthday. In the previous week at Ashby he had called in Roy
Douglas and Ernest Irving to help with the scoring, Douglas
doing about two hundred bars of the 'Seven Deadly Sins'.
Writing to Douglas a few weeks later, in May, Walton said:
'About the ballet—when I've time I shall rescore the "7
deadlies", also make some cuts in all the scenes. Having seen it

three times I think I know just what is needed to brighten it up.' Douglas believes that Walton did rescore it. The ballet has never been revived and, at the time he wrote to Warrack, Walton did not know what had happened to the score. Warrack tracked it down in 1958 to a warehouse in North London where Sadler's Wells kept various items. A suite of four movements was arranged by Vilem Tausky in association with Walton in 1961 and had its first performance at the Royal Festival Hall on 3 June of that year. If not top-class Walton, the score breathes the atmosphere of the theatre with an ease that augured well for an opera. Incidentally, the suite opens with the direction 'allegro malizioso', the malice this time not being directed at a lover who had discarded him but at the magician in the ballet who had laid a spell on St George. There is a lightly scored and lilting siciliana, a long flute solo which has characteristic Waltonian wry charm.

Walton's letter to Roy Douglas about the ballet also indicates that he had begun to think about his music for *Henry V*.

I've been working on the battle of Agincourt [he said] and luckily didn't get very far as all the footages were rearranged two days ago and they forgot to tell me! So I must now start again. I am by way of recording it on the 21st May but doubt if I'm ready. 10 mins. of charging horses, bows and arrows. How does one distinguish between a crossbow and a long bow musically speaking? Will you be available some time round then to play for the guide-track?

Walton encountered Vaughan Williams about this time and mentioned that he was 'in a bit of a dilemma what to do about identifying the French musically speaking (I could hardly use the *Marseillaise*), and he suggested, if I remember rightly, *Réveillez-vous, Piccars* & told me where to find it, not mentioning that he had used it himself. It is in fact, to me, a rather typically English V.W. tune & I nearly used it instead of using the Agincourt Song.'[1]

Early in July nothing had progressed. On 15 July Douglas recorded the guide-track—so that the music and action could be synchronized—on the piano, but it was never used. It was originally intended to cut the film to fit the music, but later the

[1] Walton letter to Roy Douglas, 21 Apr. 1979. 'Réveillez-vous, Piccars' is a fifteenth century French battle-song. Vaughan Williams arranged it for voice and piano in 1903 and used it again in 1933 in his brass band piece *Henry V*.

composer was required to fit the already composed music to the finished film. For the latter half of 1943, therefore, Walton was relatively unoccupied.

I've been taking the opportunity during a lull in *Henry V* [he wrote to Douglas on 23 December] to re-score the Vl. [Violin] Con. I started out to do a little patching here and there, but found it was not a satisfactory way of doing it, so more or less I started from the beginning and I have even gone so far as to introduce a bass clarinet! etc instead of the timp. I sent it to be copied next week in the hope the parts will be ready for a performance in Birmingham on Jan. 17th . . .[2] I think now that I've got it as good as I can ever get it. It is also true that I've revivified the Sin. Con. [*Sinfonia Concertante*] chiefly by eliminating the pfte and making it easy enough even for Harriet Cohen to play.

In November 1943 Walton received two commissions from the BBC. With the victory at El Alamein and the defeats inflicted on the Germans in Russia, it was now obvious that the Allies would win the war and the Corporation asked Walton and Vaughan Williams for a 'victory anthem', the condition being that it should be symbolic of all the fighting freedom nations and not last for more than five minutes. Not surprisingly, both composers declined.[3] More attractive was the invitation to compose a work for first performance at the 1944 Promenade Concerts to mark Sir Henry Wood's fiftieth anniversary as their conductor. In a letter to Wood on 23 December Walton told him he was 'safely launched' on a Te Deum for chorus and orchestra. But on 22 February 1944 Walton wrote to Julian Herbage of the BBC music staff:

Before it is too late, wisdom dictates that I should take some evasive attitude about this projected *Te Deum*. The situation is this—*Henry V* should have been finished six weeks ago and I need hardly add, through no fault of my own, there is no prospect of it being finished before the end of March. Which would give me about 6 to 8 weeks in which to write the *Te Deum* and with another 4 weeks for copying the

[2] In fact the performance on 17 January 1944 was at the Civic Hall, Wolverhampton. It was played by Henry Holst, with the Liverpool Philharmonic conducted by Sargent. Performances followed in Birmingham on the 18th and Liverpool on the 19th. In effect Walton rescored the whole work.

[3] Vaughan Williams eventually wrote his *Thanksgiving for Victory*, a work on a large scale involving speaker, soloists, children's choir, chorus and orchestra. Walton wrote nothing.

chorus parts, it would mean that whatever chorus is going to sing it wouldn't have it before the beginning of July.

So the idea was dropped.

In fact on 4 January Walton had written to Roy Douglas: 'I'm now in the thick of Henry V, about 55 mins' music! and very difficult and I wish I'd never taken it on in spite of the filthy lucre!' On 17 March he told Herbage: 'The vagaries of *Henry V* still continue, though at last the date for the final session has been fixed—23 April. There is still a great deal of music to write.' Combining the music, dialogue, and sound effects with edited film footage took from 11 May to 12 July. The film was first shown in November 1944 and was immensely successful. Olivier said he thought Walton's score was 'the most wonderful I've ever heard for a film. In fact, for me the music actually made the film. The charge scene [Agincourt] is really made by William's music.' It seems that Walton himself scored all the music without assistance, and his model for such an ambitious film score was Prokofiev's *Alexander Nevsky* for Eisenstein in 1938. In the earlier scenes, where Olivier showed the Globe Playhouse of Shakespeare's time, the music is a successful example of pastiche Byrd and the scoring includes tabor and harpsichord (played originally by Roy Douglas). Later, creating moods, he abandoned any 'period' style and wrote two beautiful episodes for strings, worthy to be called Elgarian, the death of Falstaff and 'Touch her soft lips and part', when Pistol says farewell to Mrs Quickly. This is for muted instruments and is in Walton's favourite siciliano rhythm, or nearly. Elgar would have appreciated the subtle humour which based the passacaglia 'Death of Falstaff' on the old drinking-song 'Watkin's Ale'. Another memorable passage accompanied the Duke of Burgundy's plea for peace in Act V Scene ii, when he speaks about the devastation of the French countryside. Walton magically evokes a vision of a past idyll by adapting 'Baïlèro' from Canteloube's *Songs of the Auvergne*. These pieces are now well known in the concert-hall: they are great music independent of the film and the conclusion must be that they succeed so well artistically because they were written in advance of Walton's actually seeing the film. They were a response to a text (Shakespeare's) and an idea, just as Vaughan Williams's music for *Scott of the Antarctic* succeeded because he wrote it as a passionate emotional response to the Scott expedition's fate *before* he saw any of the film. The charge of the French knights at Agincourt is where the *Nevsky* 'Battle on the Ice' sequence is suggested by the throbbing ostinato below the brass fanfares. The climax of this sequence is, unforgettably, not musical—it is the sound of the English bowmen's arrows. Walton did not arrange the suites himself. Sargent devised a four-movement suite in 1945, Muir Mathieson (who conducted for the film) a longer one in 1963. Sargent's, although it omits the Charge and the 'Baïlèro', is the better. It begins with the splendid music that opens the film, when a flute solo accompanies the shot of a playbill fluttering in the wind before the camera shows a panorama of sixteenth century London, closing in on the Globe itself. Wordless voices are used with magnificent effect. Like Mathieson's suite, Sargent's ends with the stirring arrangement of the traditional Agincourt Song but in its more effective choral version.

Olivier told a delightful story about *Henry V*. The film was at some point shown to Walton, innocent of all sound. When he later saw the completed film, Walton said to Olivier: 'Well, my boy, I'm very glad you showed it to me, because I must tell you I did think it was terribly dull without the music!'[4]

---

[4] BBC *Kaleidoscope*, 8 Mar. 1983.

14

Post-war Quartet, 1944–7

It was perhaps just as well that Walton did not proceed with his Henry Wood Te Deum. The jubilee season of 1944 was beset by disasters. This was the year of the flying-bombs (the V1s), popularly known as 'doodle-bugs', which caused considerable devastation to lives and property in London and the south-east and also shook the Londoners' morale. (As Walton wrote to Julian Herbage from Ashby on 7 July 1944: 'I won't pretend that the hazards of the flying-bomb and more so the difficulty of getting on the train haven't anything to do with my not coming to London just now.') The prospect of one of these weapons diving into a Royal Albert Hall Prom was too much for the Government, who ended the season after less than three weeks. Those parts of the remaining concerts which were due to be broadcast were performed in Bedford. After conducting Beethoven's Seventh Symphony unforgettably on 28 July, Wood was taken ill and died on 19 August. Walton wrote a fanfare which was performed at the Wood Memorial Concert on 4 March 1945. This was a revised and amplified version of a fanfare he had written for a Red Army Day celebration in February 1943. Walton's first post-war task was to compose a setting of a text by John Masefield, the Poet Laureate, to be sung at the unveiling on 26 April 1946 of a stained-glass window in memory of Wood in St Sepulchre, Holborn Viaduct, where Sir Henry had played the organ as a boy and where his ashes were buried. Negotiations began in September 1945. Masefield originally selected six verses of his poetry (it is not known which) and sent them to Walton. In January 1946 Walton told Wood's widow that he would prefer to write a piece for strings instead. He had set two Masefield verses but 'it is really not possible, for me at any rate, to make a worthy work out of it. I am loth to make this decision as I have so much respect for Mr Masefield and the last thing I want to do is to offend or hurt his feelings.' The plan then was for the verses to be read at the ceremony, followed by Walton's new work. However, between then and March, Masefield came up with a new poem, 'Sir Henry Wood', which Walton set for un-accompanied mixed chorus, using its first line as the title, 'Where does the uttered music go?' It is a complex setting, and was sung by the BBC Chorus and Theatre Revue Chorus conducted by Leslie Woodgate. Walton assigned all royalties to the Wood memorial fund. The critic of *The Times* called it 'a noble piece of music worthy of the occasion and sure to survive its occasion', but a week later Walton wrote to Jessie Wood: 'I received a batch of press cuttings this morning, & one must admit that it hardly received an overwhelming reception. But I have given up all hopes of critics a long time ago.'

A letter from Walton to Roy Douglas written at Ashby St Ledgers on 30 January 1945 discloses the first sign of his returning to a composition he had pondered in 1939. It also gives an inkling of the difficulties he experienced after writing so much music for war films: 'I'm in a suicidal struggle with four strings and am making no headway whatever. Brick walls, slit trenches, Siegfried Lines bristle as never before. I'm afraid I've done film music for too long! But if I ever break through you must come down here and see it . . .' This is the first mention of work on the String Quartet in A minor. He followed up the military metaphors when writing to Douglas on 5 February: 'I've captured a trench and overcome some barbed wire entanglements—but every bar is a pill-box.' The next letter was on 25 July: 'No very encouraging news of the 4tet. Did you see or hear "Grimy Peter"?' The last reference is, of course, to Britten's opera *Peter Grimes*, which had had its first performance at Sadler's Wells on 7 June. (Walton had a fondness for the English phenomenon of nicknames. 'Belli's Binge' was another he coined, and he was rather proud of 'Arse-over-Tippett', though in his generous moments he as-cribed authorship of this to Lambert. In later life he christened the impresario Peter Diamand, artistic director of the Edin-burgh Festival from 1966 to 1978, 'Double Diamond'.)

This is perhaps the juncture at which to glance at what some of Walton's contemporary fellow-countrymen had been writ-ing during the war years. Since 1939 Walton had composed the music for six full-length films, a major radio-drama, two ballets (one of them a series of transcriptions), a Shakespeare play, and various small commissions such as fanfares. Vaughan Williams, thirty years his senior, had written for three feature films and two short documentaries, but had also completed several short choral works, his Fifth Symphony (performed in 1943), and a string quartet, and, in 1944, started his Sixth Symphony. Bliss had written a ballet, a string quartet, some incidental music, and some songs and had spent two years as BBC Director of Music. Bax, appointed Master of the King's Music in 1942 to everybody's surprise including his own, had remained virtually silent, as had John Ireland. Little came from Constant Lambert; a violin concerto and the *Sinfonietta* from Moeran. More prolific, relatively, was Michael Tippett, whose wartime works included his First Symphony, the *Fantasia on a Theme of Handel* for piano and orchestra, some works for unaccompanied chorus, a second string quartet, the song-cycle *Boyhood's End* for Pears and Britten, and, above all, the compassionate and topical oratorio, *A Child of our Time*, first performed not long after his release from a short jail term in Wormwood Scrubs for failing to comply with the terms of his registration as a conscientious objector.

After Benjamin Britten's departure for Canada and the United States in 1939, first London performances followed of the major works he completed there, *Les Illuminations* in 1940 and the Violin Concerto in 1941. When he returned in the spring of 1942 a spate of first performances followed: in 1942 the *Sinfonia da Requiem* in July, *Seven Sonnets of Michelangelo* in September, *Hymn to St Cecilia* in November, and *A Ceremony of Carols* in December; in 1943 the First String Quartet in April, the *Scottish Ballad* in July, *Rejoice in the Lamb* in September, the *Serenade* for tenor, horn, and strings in October, and the BBC radio-drama *The Rescue* in November; in 1944 the *Introduction and Rondo alla burlesca* in March; and in 1945 the opera *Peter Grimes* in June and the Second String Quartet and *Holy Sonnets of John Donne* on successive days in November. Not a film score among them and scarcely a work, except the *Donne Sonnets*, which could be related directly to the war. It is a formidable list of achievements, crowned by the success of *Peter Grimes*, which, despite the hostility of the company towards its production, was hailed simultaneously as a masterpiece and as the dawn of a new period for English opera—for English music, too, because, although there was a group in the musical establishment which would not accept Britten for extra-musical reasons—he was a pacifist and known to be homosexual—and regarded him as musically facile, 'too clever by half', audiences and a narrow majority of critics appreciated his genius. Thenceforward Britten, thirty-two in 1945, was the 'white hope', and everything he did was eagerly awaited by friend and foe alike, if for different reasons.

Walton, as we know, had seen this, or something like it, coming in 1939. He had not produced a major concert work since the Violin Concerto, and that had been launched in wartime Britain without the glamour of its dedicatee and godfather, Jascha Heifetz. There is no doubt, as far as being the 'white-headed boy' of English music was concerned, Walton was displaced by Britten in 1943 and remained displaced for the rest of his life. Whether they liked it or not, they were rivals and their relationship is worth examining.

After Britten's death, Walton wrote to me that, whatever some people might say, 'we were always on good terms and I much admired many of his works, even if he detested most (not all) of mine'.[1] His encouragement of the young composer in 1937 has already been related, with Britten at first prickly and alert to be snubbed, suspecting condescension where there was probably only a natural reserve. Walton did not condemn him for his pacifist views, as many did, or he would not have spoken for him before a tribunal, and he was certainly not disturbed by homosexuality, having lived a resolutely heterosexual life while remaining on terms of close friendship with Siegfried Sassoon, Lord Berners, Osbert Sitwell, Ronald Firbank, Cecil Beaton, Frederick Ashton, Roy Campbell, and others. To intimates, Walton would sometimes refer to Britten, jocularly, as his 'junior partner', and he was critical of the retinue of hangers-on with whom Britten and Pears were surrounded at Aldeburgh. But he admired what Britten set out to achieve at Aldeburgh and in 1961 sent two manuscripts of *Façade* items—'Old Sir Faulk' and the Polka—to a Christie's

---

[1] Letter to the author, 2 May 1978.

sale on behalf of the Aldeburgh Festival. They raised 65 guineas.

As in the case of Lambert, Walton was not jealous of Britten the man, although Sir Michael Tippett said that he had a 'large-scale chip' about Britten's appointment as a Companion of Honour in 1953 and remarked—not too tactfully to Tippett—that obviously the only right course in English music was to be homosexual. He was envious of Britten's ability to compose, as he believed, at high speed compared with his own laborious method.

I hear the sounds in my head, yes, [he said] and so far I haven't had any nasty shocks when I've heard them played. It's hard work, for me. Sometimes I get stuck over a couple of bars—can't see what to do. Eventually, of course, one works it out and then it seems so simple. The trouble is I wasn't properly trained. I do envy Ben Britten his—not facility, but being able to do it all in his head, like Mozart or Rossini.[2]

He immediately recognized *Peter Grimes* for what it was and told Britten so (although Dallas Bower recalled attending the first performance with Walton and Alice Wimborne. As they left the theatre in Alice's Rolls, Walton said: 'Christ, all those sec-onds!', referring to the repeated use of the dissonant interval of the second). Writing to Britten on 21 June 1945, Walton said:

It is always embarrassing to say things, I find, but I should like to tell you how much I appreciate your quite extraordinary achievement which makes me look forward to your next opera. It is just what English opera wants and it will, I hope, put the whole thing on its feet and give people at large quite another outlook about it. Not, I am afraid, that you will find many other composers, if any, emulating your success. But it may be something if it encourages them to try. Anyway, you are quite capable of creating English opera all on your own. I meant to have written you before but have these last days been overwhelmed by proofs of my Violin Concerto and I fondly but vainly hope to get the score out in time for the 'Prom' performance which I hope you may be able to hear. I say fondly and vainly as I have been four months getting the 2nd proofs. I shall have to move to B. & H.! . . . P.S. This is a 'fan' letter.

But *Peter Grimes* was the cause of some friction between the two composers, when Britten heard that Walton had scotched a

[2] Interview in *Sunday Times*, 25 Mar. 1962, p. 39.

British Council plan to record it in 1946. Walton's opinion was that, as funds were so scarce, something less commercially successful should be chosen. It was a disinterested view, since he had not then written an opera, but Britten could have been forgiven for believing that actions spoke louder than words. The success of *Peter Grimes*, however, had the old catalyst effect on Walton—it stimulated him into wanting to rival it, hence the start on *Troilus and Cressida* in 1947. 'I thought it was not a good thing for British opera to have only one opera by one composer,' Walton said. 'I thought it my duty to try to write an opera'.[3] As a member of the Board at Covent Garden, Walton is said by Lord Harewood in his memoirs to have been one of those opposed to the proposal by Harewood and his then father-in-law Erwin Stein that Britten should become music director of Covent Garden. At a meeting Walton asked whether Britten was 'really a conductor' in the sense that he would conduct Walton's own *Troilus and Cressida* when it was ready. Harewood assured him he would. The proposal was dropped. Leslie Boosey, Britten's publisher, opposed the idea also as a misuse of Britten's talents.[4] Although Harewood implies that David Webster, the general administrator, favoured the plan, Webster's official biographer states that Webster knew Britten was not the right man for the job.[5] Tippett remembers 'Willie going on about "Keeping the buggers out of Covent Garden". To me! People are funny.'

On 10 July 1950, after a visit to the Aldeburgh Festival, Walton told Britten how much he and Susana had enjoyed it. 'I found the general atmosphere most pleasant and stimulating and I am happy that you should think my appearance there helped things a little. I should like to have a shot at a chamber opera or a piece of some kind for you. It seems to me that there is a real future for that medium from all points of view.' It was to be another seventeen years before *The Bear* was produced at Aldeburgh, a chamber opera which begins by poking sly fun at Britten. Another more serious musical tribute from Walton to Britten was the *Improvisations on an Impromptu of Benjamin Britten*, for orchestra, completed in 1969. Walton was often an honoured guest at Aldeburgh, where his seventieth birthday was

[3] S. Walton, *Behind the Façade*, p. 133.
[4] Earl of Harewood, *The Tongs and the Bones* (London, 1981), pp. 133–4.
[5] M. Haltrecht, *The Quiet Showman* (London, 1975), pp. 185–6

marked not only by special performances of *The Bear* but by a revival of some of *Façade* with Pears reciting a selection of numbers. Walton admired Pears's recording of the work made with Edith Sitwell—'the best since Constant'. A Britten opera, to conclude, was the subject of one of Walton's wittiest and most pleasing *bons mots*: 'I hear Ben has written a new opera for television. Now, what's it called? Oh yes, *Godfrey Winngrave*.'[6]

Walton could never resist a joke at someone's expense, but he was generous to his colleagues. Elisabeth Lutyens, for example, in her autobiography recorded her indebtedness to him, describing him as 'the single most beneficent influence on music and his fellow-composers whilst still living in England. There was not a "good musical cause" he would not stretch out an encouraging hand to. Music as a whole and so many composers have reason to be grateful for his generosity and tolerance.' He was 'the best of company, devoid of that "safety first" cautious respectability of the middle classes, backbone, no doubt, of England, but whose very virtues rob musical bread of yeast'. Lutyens, whose husband Edward Clark had helped Walton to obtain Hindemith as soloist for the Viola Concerto in 1929, asked Walton to introduce her to Muir Mathieson so that she could try to obtain a commission for film music. 'Of course I will', Walton replied, 'but I'll do more than that. Write any work you like and dedicate it to me, and I'll give you £100.' The result was her dramatic scena *The Pit* in 1947. Walton had remembered how he had been helped by Lord Berners's 'purchase' of the dedication of *Belshazzar's Feast*.[7] Cecil Gray also wrote of Walton's generosity.

I know of many cases in which he has gone out of his way—and often a long way out—in order to render assistance, sometimes financial, sometimes by recommending publication or performance in influential quarters of works by even formidable rivals, and in other ways as well. (I know of at least one such case in which his help was gratefully accepted and repaid with more than usually shameless ingratitude.)[8]

---

[6] I suppose a new generation may need to be told that Godfrey Winn was a prolific, popular, and extremely able journalist, of somewhat effeminate voice and appearance, who could write on almost any topic in a sometimes flowery style such as women's magazines were then supposed to like.

[7] E. Lutyens, *A Goldfish Bowl* (London, 1972), pp. 147–8.

[8] *Musical Chairs*, p. 287.

For us today to see in perspective the drop in barometric pressure which Walton experienced after 1945, we need to remind ourselves of the high-pitched praise heaped on him in the 1930s (what would now be called 'hype'). First Cardus, already quoted in an earlier chapter, in 1931: 'Nowhere on the Continent at the present time would you be likely to hear a composition of more convincing genius than William Walton's *Belshazzar's Feast*. Hours after the *shock* of the performance, and following a *sleepless night's* study of the score . . .' (my italics). Basil Maine in 1936: 'If in itself Walton's Symphony epitomises the crisis through which English music is passing, it has brought us (I do believe) perceptibly nearer to the time of deliverance. . .' When the First Symphony was broadcast in 1935, the BBC announcer, at a time when any kind of emphasis in tone was severely frowned upon, exclaimed enthusiastically: 'Young William Walton, England's White Hope!' Well might Walton in 1939 have deplored being dubbed a 'white hope'! Their reigns are short, their courtiers fickle, their whiteness all too prone to turn to grey. What a burden was laid on Walton so well-meaningly by Arthur Hutchings (in what is still the best appreciation of the First Symphony) in the *Musical Times* of March 1937 when he wrote: 'Today English music holds a place of dignity and distinction, with promise of a rising school of composers under an exemplary leader . . . one can say for certain that we shall from now wait for every new work of Walton, as we once did of Sibelius, in the certainty of getting something of permanent value.' Certainty and permanent are provocative words.

We left the new string quartet making little progress in July 1945. The year 1946 was comparatively uneventful. Walton joined the Board of the new Covent Garden Opera Trust, of which David Webster was general administrator. [9] A short lease of the theatre was taken by Boosey and Hawkes until 1949 and Ralph Hawkes asked Dallas Bower to persuade Walton to leave the Oxford University Press and be published in future by Boosey's, but the plea was rejected. Work on the

[9] The members at the beginning were Sir Kenneth Clark, Edward Dent, Leslie Boosey, Ralph Hawkes, Samuel Courtauld, Sir Stanley Marchant, Steuart Wilson, and Walton.

quartet continued throughout this year and the first performance was advertised for a Wigmore Hall concert on 4 February
1947. But this was postponed and the première was on 4 May
in a Broadcasting House studio in a broadcast concert on the
Third Programme, which had been inaugurated in September
1946. The performers were the Blech Quartet, to whom the
work had been promised years before, and they gave the first
public performance the following day in the Concert Hall of
Broadcasting House. Desmond Shawe-Taylor wrote in the *New
Statesman*: 'In spite of the lapse of time since his last major work
... it is also in all the essentials the mixture as before; we
recognize the familiar blend of harmonic astringency, rhythmic
and contrapuntal ingenuity, and nostalgic meditation.' The
ominous words were 'the mixture as before', though in this
instance they were meant as a compliment and a reassurance.
The Quartet in A minor revealed a leaner, more muscular,
slightly more classical and just as vital Walton, with a
mysterious solemnity in the slow third movement that goes
beyond the 'bitter–sweetness' of earlier works. The first
movement is in a more-or-less regular sonata-form, the Finale
is a short, swift rondo, and the second movement a fast-moving
Scherzo with rather fewer of the rhythmical snaps and quirks
of its predecessors. Not quite the mixture as before, really, for
there is a new mood. The haunting self-communing that is in
most of Walton's music is here of a deeper cast, less *Angst*-
ridden but no less deeply felt. The poetic end to the first
movement is as spontaneous as anything in the Viola Concerto,
but the gem of the quartet is its dark third movement, a Lento
in F major in which Walton achieves an emotional poise that
testifies to his maturity as artist and man. It is one of his finest
achievements and it was a sure sign that he had thrown off the
trammels of his cinema style and rediscovered his true voice in
an intensive self-exploration.

15
Birth Of An Opera, 1947–8

IN the spring of 1947 Walton went with Alice Wimborne, Alan Bush and Gerald Abraham to represent British music at the Prague Festival. In November he was awarded the Gold Medal of the Royal Philharmonic Society, which was presented to him by Vaughan Williams. But the most significant and far-reaching event of the year was the letter written to Walton on 8 February 1947 by Victor Hely-Hutchinson, BBC Director of Music: 'The BBC has decided to commission an opera and would like you to compose it. I do not at all know how you are placed as regards time, but I very much hope that you will be interested in the idea.'

This was followed up a month later by John Lowe of the BBC music staff, who wrote of the opera being composed 'during the coming eighteen months' and of the 'first studio performance' being given on the Third Programme. Walton accepted the commission. Since the Gesualdo scheme had faded away, he had been on the look-out for an appealing subject and a librettist. At some time in the winter of 1947 he wrote to Cecil Gray to tell him of the commission and that the BBC (in the person of Stanford Robinson, head of BBC Opera) had recommended Christopher Hassall 'to go into the question of the libretto. And I believe that we have at last hit on one in *Troilus and Cressida*, not founded on Shakespeare, but on Chaucer, Henryson and Boccaccio.' When Hassall's name cropped up, Alice Wimborne was immediately able to bring them together. She had known and admired Hassall since 1936 through her friend 'Eddie' Marsh.[1] A letter from her to Marsh dated 13 October 1936 includes the sentence: 'That dear Mr Hassall sent me a copy of his poems with "mine" included. I am

---

[1] Sir Edward Marsh (1872–1953), former secretary to Winston Churchill, editor of the anthology *Georgian Poetry*, friend and patron of Rupert Brooke and many other poets. Hassall wrote his biography (London, 1959).

most proud as I think him the only one of the younger poets, don't you?' It would appear that Walton and Hassall met during February or March 1947. Just before leaving for Prague, Walton thanked him for a letter and enclosures: ' "Troilus" has definite possibilities.' On 31 March Walton wrote: 'I will look forward to receiving the librettos when they are ready. I have had another idea which I will let you have when we next meet.' On 7 April a postcard from Walton at Ashby St Ledgers said: 'Thanks for the synopsis. Just finishing off my 4tet and then will look into them seriously and let you know what I think.' Alice Wimborne wrote to Hassall from Ashby on 26 April:

Dear Mr Hassall. Your kind letter asks for no reply except a line to tell you that I hope you will tell me if at any time we can work together over the opera question in the sense of talking things over quietly or any other way that may occur—I think a great deal about it. I am certain you will find something, even if it takes a little time. I hope being down here was helpful, but once is not enough . . . Dr Walton is so particularly individual I feel sure you need to study him as much as work on finding subjects for him and his music. It will have to be just the right one and what that right one *is* is a most difficult matter, to my thinking. The picture is troubled too by the necessity to keep more than one eye on the public, who must I suppose be able to follow and enjoy an opera written for Covent Garden. It wd. not be hard, I think, to find one that wd. be delightful for a sophisticated audience or shall we say a chosen audience—I sometimes think such a one might get over to any audience, who have a way of liking Walton that is really very striking. His symphony, for instance, has riotous appreciation from the most ordinary Sunday afternoon crowds. But—I wd. not like where opera is concerned to be too certain—it is such a special thing. It was really delightful seeing you here. Yours sincerely, Alice Wimborne.

On 7 June she invited Hassall to lunch at Claridge's—'I have not been there for ages but am told, if you book a table, it is the least awful of these haunts'—because 'we are going to have a grand review of all the librettos this weekend'. The lunch was on 14 June. Next day Alice wrote to him—'Dear Christopher' now:

It is remarkable how much you have observed and understood. There is only this, that I hope you will not feel you are making no headway because things move slowly, you don't hear, and so forth. That means nothing, although very hard for *you*. It means that the Quartet, or rather getting it going, blocked every other activity and thought. He has to write the music for Laurence Olivier's *Hamlet* film, but that is not immediate . . . It is important, I believe, that you should not be perplexed or worried at the unbouncing ball and equally that he should think you are quite unmoved by procrastination! so that he doesn't fear being, as it were, pushed into anything because a decision is expected of him . . . One day the ball will bounce against an electric wire as was the case with *Belshazzar* and *Scapino* . . . The difficulty is to find the perfect subject. Many that would suit his music have other drawbacks, as a case in point, Byron. For apart from suitability for *him* (W.) the story or plot must be so easy and clear and flowing and scenic. Troilus and C. has got that. And it's got, as we agree, the Manon lady which he seems obviously to prefer to the Juliets! Less opportunity for the macabre and sombre than Death's jest book. The picture will clear . . .

Hassall, whose previous theatrical experience had been as author of the 'books' for the popular romantic musicals by Ivor Novello (1893–1951), produced a detailed draft outline of the plot. Walton, who at this date had a Hampstead cottage, 10 Holly Berry Lane, wrote to him on 15 July: 'T & C is excellent. I've been reading it on and off all day and find very little to suggest which might be an improvement and I think now the time has come when the actual text might be got under way.' He then made some suggestions and added: 'It should I think work out at about 3 hours including 40 mins for intervals, or as near as possible to that. Remember that only about ⅓ of the words is necessary for a libretto to what there would be if it was a play.'

There, for a time, the matter rested. Before starting on the *Hamlet* music Walton continued with his self-imposed discipline of 'learning composition' and began a violin sonata. Its genesis was scarcely auspicious. Alice had not been well and had consulted her London doctor, who said nothing was wrong. She and Walton left for a holiday in Capri in September 1947, but on the way, in Lucerne, she again became ill. Walton arranged for her to see a Swiss specialist, who diagnosed cancer of the bronchus, and she entered a nursing home in Lausanne. Because of foreign currency restrictions in that era of austerity and the impossibility of Alice having money transferred from London, Walton was desperate for money to pay medical expenses. By good luck he met his friend Yehudi

Menuhin's wife Diana in a train in Switzerland. She commissioned a sonata for 2,000 francs for Menuhin and her sister
Griselda's husband Louis Kentner.[2]

On their return to England Walton composed the music for
Hamlet. 'Not uninteresting', he wrote to Cecil Gray on 27
December. 'I've had to do nearly an hour of appropriate but
otherwise useless music.' Muir Mathieson, who conducted the
Philharmonia Orchestra in the recording of the score for the
sound-track, considered that Walton excelled himself in *Hamlet*,
'perfect in detail and sure in its dramatic conception'. Walton
used *leitmotif* in this score. An admired episode was the play
within the play, when he exploited the contrasts and correspondence of 'realistic' music, forming part of the action, and
'functional' music, commenting upon it. Thus the Players'
dumb-show is realistically accompanied by a sarabande scored
for a small band of lower strings, woodwind, and harpsichord.
A sinister theme occurs when the poisoner enters and, as the
camera shows the spectators' reactions, this is given a twentieth-
century harmony and the full orchestra is used. When the
actor–king has been murdered and the guilty spectators
become uneasy, the music increases in volume, speed, and
intensity until Claudius's cry of 'Give me some light' is
accompanied by a shattering chord. Walton himself made no
further use of the music, but the opening titles were adapted as
the *Funeral March* by Mathieson in 1963. Four years later
Mathieson arranged a 'poem for orchestra', *Hamlet and Ophelia*,
from the film music ('frightfully dull', Walton thought it).
Some fanfares from the film were grouped into one piece,
Fanfare for a Great Occasion, by Malcolm Sargent in 1962.

The music was recorded for the film in November and
December 1947 and January 1948. Throughout this period
Alice was still ill. 'The doctor's final verdict is due in the next
day or two', Walton wrote to Gray just before Christmas, 'and I
fear it may well be the worst.' Her principal concern in the few
months left to her was Walton's opera, news of which was
given to the Press in January 1948. From Lowndes Cottage on
23 February she wrote to Hassall:

For all I know you may have been expecting a line from me on *Troilus*
for a little time past, but I have been so knocked out by a savage

[2] Letter from Walton to Angus Morrison, 24 Mar. 1969.

treatment, now happily over for the moment, that it has been impossible to write. I am so looking forward to hear you reading it on Wednesday. We must remember, I think, that we are still of very enquiring mind as to the bare subject of Troilus and C., don't you? Owing to *Hamlet* on the one hand and my long illness on the other, we have really not been able to get together any of us which wd. have been so useful as concentrating on the ideas produced thoughts or comments. I think the poetic beauty of your book great. Whether the story is too manufactured is an aspect I expect that will come up. Did you read Newman on *Manon* in the *Sunday Times*? There are so many things to think of if the luckless lovers are to live and move the hearts of an operatic—or any—audience.

After the reading of the libretto, Walton and Hassall decided to send it to Ernest Newman for his opinion and comments. Writing again from Ashby St Ledgers on 13 March, Alice referred first to a bereavement Hassall had suffered:

We thought of you today. Those are always bitter moments however you decide to take them. I hope the peace and serenity of the countryside here and the sense of all the immutable things that this ancient home stands for proved some comfort to you. I hope 'our child' [*Troilus*] will prosper at the hands of Newman. You know that I am not entirely happy about it. The period is not easy. Troy *is* classical. In the broad sense. So is Ancient Greece. Have you looked up *Les Troyens* in *Opera Nights*?[3] I wait breathlessly for criticism, this time constructive. It is lamentable from us who can't construct.

Alice Wimborne died on 19 April 1948 after suffering appalling agonies which haunted Walton for the rest of his life. Through lack of oxygen she turned black. The memory was too much for him thirty-three years later when he spoke about it in Tony Palmer's documentary profile *At the Haunted End of the Day*. With pain visible in his eyes, he described her last days and after a silence said: 'You know, one forgets about it if one can.' For the thirty-five years after her death he had nightmares from which he would waken saying to his wife: 'I have been dreadful to Alice.' She left £100,468. In her will, dated 13 December 1947, she left Walton £10,000 and her London home Lowndes Cottage, excluding the furniture, which belonged to her son. She added a codicil on 12 April 1948 which altered these provisions by additionally bequeathing to Walton a carpet, some furniture, a statuette and her car.

[3] E. Newman, *Opera Nights* (London, 1943).

Newman's 'report' on the libretto arrived just after Alice's death. Dated 18 April, some of it makes pertinent as well as amusing reading:

As a piece of dramatic construction it's admirable, though I have my doubts about the effect of some things in the second act . . . I fancy there would be a touch of the comic, for the ordinary spectator, in the entry of Troilus and Pandarus, and in the episode the morning after, with Troilus 'lying half-dressed on top of the bed-clothes' while Cressida leans out of the window and sings. The correlation with any scene in any hotel bedroom the morning after will be quite involuntary on our part, but it will be none the less there, I fancy. The actual business of bringing T. and C. together is, of course, the most difficult business of the whole story . . . I don't say there shouldn't be a stage setting of the episode; on the contrary, there must be one. But I feel that the way it is handled in the libretto isn't the right one for the stage. Apart from this scene, I have nothing but admiration for the dramatic structure of the piece.

I am very doubtful, however, about much of the diction. In the first place, there are far too many long words *for music* . . . In the second place, I don't think C.H. has always, or even often, realized the difference between verbal speech and musical speech . . . Here is a minor example. In the first act Antenor makes a casual reference to 'the widow Cressida', whereupon Troilus, left alone, ejaculates 'Is her name Cressid? Is that her name?' That is excellent in poetry but, I venture to think, wrong in music, where the order should surely be 'Is that her name? Cressida?' For the cry of 'Cressida' is the highlight, and it should be held in reserve to the last . . .

Throughout, it seems to me, C.H. has worked as a speech-poet working purely in his own medium instead of as a poet writing for music. What is perfectly right in the one case is all wrong in the other . . . what would any listener in the theatre be able to make of 'There stand the satyr-faced plump urns that spill superfluous oleander for the sparrow rootling for grubs among festoons of flowers'? . . . Much of the libretto is too 'literary', too poetical, for music. This is particularly the case in the big choral portions . . . There's one line, by the way, that certainly calls for re-casting. It's a small point, but in its little way it illustrates the difficulty in opera of ensuring that the listener will hear exactly what you intend him to hear. 'Much gloomy good may it do you' will be heard, from the mouth of any singer who can't define consonants—and few of them can—as 'much blooming good may it do you' . . .

Hassall took back the libretto to revise it, working closely with Newman, with whom he became very friendly.

16

Making An Opera, 1948–52

RETICENT as always, Walton wrote no letter in which he expressed his feelings about the death of Alice Wimborne. Work was his therapy. He resumed composition of the sonata— Roy Douglas's diary for 21 June 1948 records: 'William is writing a new Violin Sonata.' He also undertook a considerable revision of the score of *Belshazzar's Feast*. In February 1948 Douglas had told Oxford University Press that there were over eighty errors in the full score. As a result, Walton went through it and in the process took out much of the percussion and drastically rescored the music between rehearsal cue numbers 62 and 65 and between 74 and 77 and also adjusted the scoring elsewhere. The first performance of the revised version was given on 8 March 1950, conducted by Sargent. Later, before the study score was published in 1957, Walton entirely rescored the last fourteen bars, extending them to eighteen, using full orchestra and adding the upper octaves instead of only the lower instruments. He also adopted the suggestion of an organist at a performance overseas that a full organ chord should be added half-way through the last chord of the work.

A severe attack of jaundice put Walton in hospital, after which he went to recuperate on Capri, with Michael Ayrton, the artist, as a companion. Ayrton's painting of Walton, now in the National Portrait Gallery, was done at this time. Although he is depicted casually, with pipe in mouth and glass on table, the face is gaunt and anguished. But Walton's mind was taken off his own grief by his concern for Cecil Gray, who lived on Capri and whose wife Margery had died a few days before he and Ayrton arrived. From there he went for part of August to Blonay in Switzerland, whence he wrote to Hassall on the 19th: 'How are you and how is T & C? I am here till about Sept. 5th and sail from Genoa for Buenos Aires on Sept. 13th and I am

wondering if it would be possible for you to let me have a copy
of the revised version to study on the voyage. It would be a
good opportunity for me perhaps to get some ideas down on
paper . . . I hope to be more or less in at Lowndes Cottage by
the time I return in mid-Nov . . .' Hassall sent him the revised
libretto and his correspondence with Newman.

   The trip to Buenos Aires was undertaken as a delegate to an
international conference of the Performing Right Society at
which it was hoped to persuade Argentina to sign the Berne
convention on copyright. Among the other British delegates
were the composer Eric Coates and the writer A. P. Herbert. It
was to be a turning-point in Walton's life, most pithily and
amusingly described by him in a 1968 interview with John
Warrack:

The first item was a conference to meet Mr Peron [President of
Argentina]. He got up and made a speech of which no one understood
a word, then all his backers got up and screamed 'Peron, Peron,
Peron'. Then Mr Leslie Boosey got up and made a speech in reply, and
when he finished I said: 'Boosey, Boosey, Boosey'. And that became
the cry of the opposition in Buenos Aires for some time, I was told.
After that there was a press conference which I had to give and I saw a
rather attractive girl in the corner. Of course one of the journalists
asked what I thought about Argentinian girls, so I nearly said, do you
see that girl over there, I'm going to marry her, but I controlled myself
just in time. Anyhow, we met, went off to lunch and I proposed to her
the next day. As far as I can remember she said: 'Don't be ridiculous,
Dr Walton.' But we got engaged three weeks after that . . . They played
my Coronation March at the wedding, terrible it was. It was the most
appalling sort of wedding—at nine o'clock at night, in devastating
heat.

   His wife was Señorita Susana Gil Passo, daughter of a
lawyer. She was born on 30 August 1926, so was twenty-two
when she was married, twenty-four years younger than her
bridegroom. She was working as social secretary of the British
Council in Argentina and had arranged the press conference at
which Walton was to be interviewed by local journalists. She
recalled that Leslie Boosey, who had persuaded Walton to
attend to take his mind off Alice's death, answered most of the
questions. He announced (with Walton standing next to him)
that Britten was the most important living British composer.
Señorita Gil's father tried to stop the wedding, but a civil

ceremony was held on 13 December, followed by a church ceremony on 20 January 1949 at which the British Ambassador was present. A telegram of congratulations arrived from Walton's mother in Oldham. The Waltons sailed for Britain on 29 January.

What made Walton take such an impulsive and apparently irresponsible step? It cannot have been any unfulfilled desire to start a dynasty, since he informed his wife (a Roman Catholic) on their wedding-day that he did not intend to have any children and that he would divorce her if she did not agree. Later, when she became pregnant, he compelled her to undergo a dangerous abortion. While the phenomenon of love-at-first-sight is not unknown (Walton claimed it had happened with Alice Wimborne) and the marriage lasted for thirty-five years until his death, one may speculate on Walton's motives—apart from the unpredictable chemistry of sexual attraction—in marrying a girl so much younger than himself and from a background and country which were foreign to him. His life up to that date had been a series of planned withdrawals or escapes, angled to his own advantage, for he retained his Lancastrian shrewdness and steely determination which enabled him to detach himself from his surroundings and view them with a cold, cynical, and sometimes ruthlessly opportunist eye. He had escaped from Oldham to Oxford, from his Lancashire accent, from Oxford to the Sitwells, from the Sitwells to Alice Wimborne. Now he had to escape as far as he could from his memories of Alice and how much further could he go than to an Argentinian girl? Perhaps, too, he wanted to escape from an English musical scene in which he no longer felt he had a major speaking-part, just as he no longer had an entrée to the social world of Ashby St Ledgers and all it brought in train. His decision that they would spend at least six months of every year in Italy is not surprising, since he had loved Italy since the day he first went there with the Sitwells. No houses were available near Amalfi so he agreed to rent one in Forio on the island of Ischia in the Bay of Naples. The Convento San Francesco was not particularly comfortable, but it served them for two years. Walton's primary objective in this self-imposed exile was that he would be able to work on his music without the distractions of London life, which now involved him in boards and committees. It should not be forgotten that in spite of the nonchalant, casual air he adopted, Walton was first and foremost a composer in deadly earnest— one recalls Roy Campbell's remark, applied to Walton, about 'a man living for his art'.

But first they returned to London, where Walton completed his violin sonata, and introduced his wife to his friends. One of the first to invite them to lunch was Osbert Sitwell, offering an olive-branch after the long coolness which had recently again dropped in temperature because each of the three Sitwells resented not having been asked to be the librettist of the opera. Walton also took Susana to Oldham to meet his mother and others of his family. On 28 March 1949 he wrote to Stanford Robinson that the opera libretto was ready, 'so it only now needs the music'. He hoped to have it ready for the 1951 Festival of Britain at Covent Garden, the BBC having agreed that it should be produced in the theatre, not in a studio.

The Sonata for Violin and Piano was played for the first time on 30 September by Menuhin and Kentner in the Tonhalle, Zurich. This seems to have been a trial performance, for Walton then withdrew the work for further revisions. It may be that at Zurich it still included the Scherzetto which was extracted to become the second of the *Two Pieces* for violin and piano published in 1951. When Walton was twenty-four he said to a newspaper reporter: 'When I sit down to write music, I never trouble about modernism or anything else. I certainly never try to write for today or even for tomorrow, but to compose something which will have the same merit whatever time it is performed.'[1] These words apply to the Sonata, one of its creator's greatest works by reason of its sustained inventiveness and mastery of the violin's expressive capacity. It is in two movements, the first in regular sonata-form in B flat, the second a theme and variations in B flat minor. For all its lyrical nature, the music has a strong underlying dramatic tension. As in all Walton's important works, there is an inherent conflict between a melancholy romanticism and rhythmical asperity; and one may surely detect, here, a reflection of the tensions in Walton during Lady Wimborne's illness and death. The piano

---

[1] *Oldham Chronicle*, 1 May 1926.

begins the first movement with a tonic chord in outline; in the
second bar the violin enters with the long and characteristic
first theme immediately recognizable as first cousin to the great

Ex. 10

Walton themes which open the Viola and Violin Concertos.
The first six notes, which form a *gruppetto*, are germinal—
almost a motto—to the whole work. This theme spreads and
grows and is repeated by the piano, which then plays the
second subject, if it can be called that, for it is closely related to
the first. It is the violin's turn now to elaborate this theme, and
the music becomes more dramatic and restless, being dominated by the the often-used Waltonian device of a rising seventh.
The development opens with tonic chords from the pianist. The
themes are worked over very thoroughly, sometimes passionately and sometimes in flowing lyrical style, the violin tender
and expressive over gentle semiquavers on the piano. An
orthodox recapitulation is succeeded by the very beautiful and
profound coda, in which the muted violin muses on the motto-
theme and the piano brings back the second subject in high
octaves.

    The theme of the variations also derives from the first
movement's motto. It is in two parts. Its first half is repeated a
minor third higher, in F sharp minor; the second part is more
rhythmical, the piano adding a codetta consisting of an expressive

descending phrase followed by equal quavers in which some have detected a variety of twelve-note series, though this is not serially used: fragments (cells) of it 'flavour' the seven variations. The first is in 6/8, mainly two-part counterpoint. No. 2 is marked 'quasi improvisando'. This is lyrical and rapturous, with special attention paid to the theme's high notes. No. 3 is a march, with something of a scherzo about it. No. 4 is spiky, living up to its marking of 'allegro molto strepitoso'. No. 5 contrasts piano arabesques (related to the theme) with pizzicato. There is a short violin cadenza just before the end. No. 6, 'scherzando', is brief and lively. No 7, 'andante tranquillo', is the longest and the loveliest, with the theme broadly and expressively treated. There is a short coda, 'molto vivace', which starts like a fugue, decides not to be one, and ends in blazing brilliance.

The *Two Pieces* are a Canzonetta and Scherzetto, the first based on a troubadour melody, 'Amours me fait commencier une chancen novele', from the Chansonnier Cangé manuscript— perhaps Walton found it in his *Henry V* researches. The tune of the Scherzetto is a light-hearted affair, with open fifths in the piano part.

Writing on 24 March 1969 to Angus Morrison, who was about to play the Sonata with Maria Lidka, Walton said:

Odd that you should feel it to be a spontaneous work as it was written at various times and came about by my meeting Diana in the train in Switzerland . . . Between the beginning and the end of working on it, a great deal happened and work was very sporadic. Alice died and I went to Buenos Aires and married Su and completed it in London on our return, so it's surprising that the piece has any continuity at all.

Recommending it to his agent for a hoped-for new recording in 1975, he said the Sonata had been a 'dead flop' in 1950. 'It's very good but in an out of date at that time idiom. It's now back again.'

In Ischia Walton resumed work on *Troilus and Cressida*. A copious correspondence between him and Hassall exists as testimony to the care Walton took over every aspect of his opera and to how much it meant to him. Some of it will be quoted in subsequent pages, although much of it would be tedious and incomprehensible to the reader not equipped with the final libretto and score.[2] But the impression sometimes given by Walton after Hassall's death in 1963 that their collaboration was difficult and touchy is not borne out by the tone of their letters. Both displayed exemplary patience and humour, and Walton was prepared to work again with him. Like every opera composer, Walton altered lines and situations to suit the music and generally 'called the tune' so that the music-drama would be paramount. He knew exactly what he wanted and was determined to attain it. Working on his composition-sketch in short score, he wrote on 20 November 1949:

For one reason and another I have got going on Act II and I've now got as far as I can without consulting you about some changes. The actual point I've reached is where Cres. is about to get to bed . . . As the script stands there is no chance for any of the principal parts to get going or the music either . . . I think the whole Sc. should be planned thus:

Sec. III Scena Concertante. From Pand.[arus], 'Does talking put you off?' to Evadne 'Good night, dear lady'
Sec. II Monologue and aria. Cressida.
Sec. III Pand.'s 'jealousy' aria with interjections—possibly from both Tro. and Cress.
Sec. IV Troilus entrance developing into a Trio for Tro. Cress. and Pand.
Sec. V Love duet and 'avowals' for T. & C.
Orchestral Interlude.

. . .If I could have Cressida's recit. and aria fairly soon so I can get on I should be very grateful . . .'

On 12 December he acknowledged the emendations.

I suggest so that we know, more or less, where we are as regards lengths, timing etc. that you should go to Boosey and Hawkes and buy the libretto (in English) of *Aida* and *Otello* and send me one of each. The 1st Act for instance of *Aida* is exactly the length this 2nd act of ours should be—and I think Cress. aria 'At the haunted end' should correspond to the tenor recit. and aria 'Celeste Aida' in length, about 2'45" so this will give you a model to work on . . . Cress. should be angry with herself for falling for T. and perhaps full of presentiments but she can't help herself. This should come out in the recit. (about 45") and might run something like this 'How can I sleep? I couldn't keep my mind on that silly game of chess for thinking of him, Troilus, who now ever fills my thoughts, blast him, why must I fall for him. I

[2] A synopsis of the opera will be found in Appendix II. This, it is hoped, may be helpful to readers of subsequent pages.

thought never to love again' etc. Then the aria 'Celestial Troilus'! (2'45") . . . I think that 'Oh strange new love' duet should be the highlight of the scene . . . and the climax to the Act should be in the pornographic interlude![3] . . . Thank you for being so sympathetic and helpful. Meanwhile I can score what I've done.

He wrote again on Christmas Day with more changes in the action, but said, 'The new aria for C. is a great improvement and I am forthwith proceeding with it.' In composing the role of Cressida he had in mind the voice of the German soprano Elisabeth Schwarzkopf, wife of his friend Walter Legge.

On 16 January 1950 Walton suggested changes in Troilus's Act II entrance which he found 'a bit weak'. Then:

I've some suggestions for Act I . . . The meeting with T. & C. must somehow manage to get T. something to sing—an aria like C.'s in Sc. 1 Act II. Could it be brought in in this way? After C.'s 'Your name is Tro.' I think is the place. A poem about the length of 'At the haunted'. (The better the poem, the better the music!) . . . You might turn over in your mind this idea for the very end of the opera. Instead of C. killing herself, as she is about to plunge the dagger in her midriff, Diomede (who I should like to think of being on horseback) seizes her wrist and carries her off with the cry of 'To the whore-house with you, my girl!' General confusion. Antenor puts his sword through Calkas— cry of 'To battle' and quick curtain.

It was not to be like that at all, more's the pity.

On 22 February he reported:

I . . . have at last reached the love-duet. I am just 'funking' that for the moment, finding love-music extremely difficult to get any originality into—Wagner, Verdi, Puccini, Strauss always popping their heads round the corner, and I hope to pick up something from Act I. . . I am now starting on Act I. We are in the same predicament over this act as we were over Act II sc. 1 that the hero and heroine especially the latter have next to nothing to sing. T.'s better off with the arrival of the new song, but C.'s position is bad still, consisting as it does of some 15 or 16 short fragmentary sections in the whole act with which little can be done musically speaking . . . We saw *Tosca* in Naples and I timed it Act I 42 mins, Act II 46 mins, Act III 27 mins. Admirable lengths which I keep in view—though I doubt getting the last Act to 27, but Act II won't be more than 40 . . .

---

[3] The 'pornographic interlude' is one of the most colourful parts of the score, occurring midway through Act II, after Troilus and Cressida have gone to bed together. It depicts the storm outside and, at the same time, the storms of passion inside.

In London on 5 February 1950 at the Theatre Royal, Drury Lane, Menuhin and Kentner had given the first performance of the Sonata in its final, revised form. Hassall was there and Walton asked: 'How did it really go? The press seems to be a series of damns of faint praise.' He wrote a few days later to Roy Douglas asking,

Will you let me know how the changes in B.F. [*Belshazzar's Feast*] come off (or not) in M. S. [Malcolm Sargent] perf. next week. He's using 6 trpts a side instead of the bands! I don't much take to opera writing and I fear it may turn out to be a colossal waste of time, especially as I'm sure Cov. Gar. will have been "axed" in the economy drive by the time I'm ready.

To Steuart Wilson, BBC Director of Music, he said (in a letter written on 1 March) that 'I shall be pleased if I get it [the opera] finished in sketch by this time next year. I aim for a Cov. Gar. performance about June '52.'

The next letter from Walton to Hassall, on 17 March, tells of trouble with the opening of the opera. 'I fear I must ask you to re-write this.' He encouraged Hassall to plan a visit to Ischia, adding, 'Can you bring me as much "No Name" tobacco as you can conveniently carry and if you don't need all your lira I will be delighted to buy them from you! The only thing of urgency now is the new version of "Pallas awake" . . .' His confidence in the libretto had been severely shaken by Walter Legge, who visited him in Ischia and, on being given it to read, criticized it savagely.

Walton returned to London in June to conduct a recording of the Violin Concerto for Heifetz, their first collaboration on the work, eleven years after it was written. Heifetz had recorded it in 1941 with the Cincinnati Symphony Orchestra conducted by Eugene Goossens, a recording of some historical importance since it is of the original score and contains some of the percussion that Walton later cut out. In the New Year Honours List in 1951 Walton was knighted.

He returned to the operatic fray in a letter to Hassall from Ischia on 10 February:

I had been hoping that . . . I should be able to tell you that I had completed Act I. This, alas, is not the case, though it is now well on the way and I hope to complete it by the middle of next month or so. There is only one major disaster in the libretto and that is that I'm quite unable to cope with 'Child of the grey sea wave' [in the final version 'Child of the wine-dark sea', Act I]. It evokes the worst type of music from me, real neo-Novelloismo, which I fear cannot be tolerated on the operatic stage. Though the substance is right, it is the regularity, the *tum*-tum-tum, *tum*-tum-tum, which gets me down. To proceed, I by-passed it but I've kept returning to it with, alas, the same results . . .

These were not the most tactful criticisms to make to Ivor Novello's librettist and there was a hurt protest. On 19 February Walton wrote: 'Forgive the neo-Novelloismo but I meant it to refer to the music not the verse'. By cruel fate, the next letter Walton wrote to Hassall on 12 March began: 'I was deeply shocked to read of the death of Ivor N. and I'm sure you are most frightfully upset about it . . . We are in the middle of moving to new quarters [Casa Cirillo] so life is a bit hectic and T. & C. is slightly in abeyance owing to the noise of workmen, etc. . . . It is just as well that you postpone Sicily till next year as we are much on the rocks owing to our move.'

Walton had 'shied off' composing a work for the Festival of Britain. But he helped to prepare the *Façade* entertainment for publication on 26 July 1951, its first appearance in print nearly thirty years after its composition. Walton added a dedication, to his old friend Constant Lambert. This was only just in time, for Lambert died on 21 August. A week later it was announced in *The Times* that *Troilus and Cressida* would be performed at Covent Garden 'next year. It will be produced by Sir Laurence Olivier with scenery painted by Henry Moore.' Two days later, in an interview with Walton, the newspaper reported that 'the first two acts were already in the last stages of completion, he hoped to go to the island of Ischia in November to work on the third and last act. He expected that the opera would be finished by the end of April next year [1952].'

Two major British operas were produced at Covent Garden during the Festival, Vaughan Williams's *The Pilgrim's Progress* in April and Britten's *Billy Budd* on 1 December. Hassall sent Walton a detailed account of the latter, to which Walton replied on 21 December:

I heard with a great deal of difficulty Act 3 and 4 and what I could hear I thought good (I had the score to help) especially the part you don't like. But I think the opening of Act 4[4] is a near thing, but being the genius he is, he managed by the orchestration and a little phrase on the flute to avoid the maudlin bathos which it really ought to have been. Newman,[5] which is the only criticism besides [Eric] Blom that I've seen, must have come as a bit of a cold douche to the hysteria of the earlier press. However, to get down to business. I need hardly say I am doing worse even than usual and am at the moment at a complete standstill and see no signs of a move. To get going I started revising Act 2 from the start and that took me about a month. It is much better than it was and may be said to have reached its final form. 'At the haunted' much better with a new middle section, though I still fear it is the all-time low . . .

Eight days later Walton sent Hassall his thoughts on Act III. He jibbed at the start in Hassall's revised libretto: 'Three and a half pages of dialogue set in recitative or dramatic recitative are going to be stiff both for the audience and for the unfortunate composer. Even C.'s "No sign" is a dramatic scena to be compared with "Slowly it all comes back" in Act I and not aria. I suggest we might open as in the 1st version with the Watchmen singing a homesick song for Greece etc. . . .' He was still trying to find the right end for the opera:

Do you think the following idea is worth going into? After D. [Diomede] says 'Shame on Argos', Cress. realising what is in store for her hysterically appeals to each in turn, Pandarus to help her, Diomede to forgive her, to the chorus to protect her, and to Priam to pardon her, all in vain. In fact an ensemble with chorus and everyone and finally Calkas giving her the dagger. Then she can bellow 'Open the gates' etc., collapse and immediate curtain . . .

This, with slight modification, was adopted, but Walton was still to have other ideas about it.

By the time he wrote again, on 9 January 1952, Walton had received the new version of the love duet. 'Alas! I fear it will not quite do.' He then explained how he wanted to set it, making suggestions and adding: 'Continue in this strain for some 12 or so lines, same scansion more or less. These should be so impassioned as to set fire to the typewriter as they are

---

[4] *Billy Budd* was originally in four acts, later revised to two. The first scene of the old Act IV was Billy's aria as he awaits execution, now generally regarded as one of the highpoints of Britten's operas.

[5] Ernest Newman, in the *Sunday Times*, described *Billy Budd* as 'a painful disappointment . . . the least notable of Mr Britten's four operas'.

being typed!' He was still 'in two minds', he said, about a
quartet in Act II.

I believe it may be too static and spoil the impetus for the curtain. Also
the words as they stand won't do. The verses I suspect should be of
varying length, otherwise it could become like a verse of a hymn with
each voice saying different words, if you see what I mean. Get the
libretto of *Rigoletto* and look at the words for that—there couldn't be a
better model. Also it might not be a bad idea to look at the libretto of
Un Ballo in Maschera though I doubt if it's in English (both Ricordi).
There's a good love scene between Amelia and Riccardo in Act 2 . . .
    One thing which we mustn't lose sight of is the question of form,
particularly the musical form of Act 2, and should be borne in mind.
We've more or less proceeded on the plan of recit. and aria, the latter
being more or less self-contained such as 'How can I sleep' followed
by 'At the haunted end of the day' (incidentally, still more improved—
it will end up by stealing the act, but we must make the love scene the
best and highwater mark not only of the act but the whole op.) and 'Is
anyone there' culminating in 'jealousy'. I feel the same pattern should
be more or less kept up throughout.

On 2 February he reported:

The dark night of the soul is darker than ever. I made no progress with
'No sign'. I've got a theme for it but can't force the words to fit.
Unfortunately I think it is the right tune and atmosphere for the piece.
Perhaps I'd better copy it out and send it to you and you can strum it
out on one finger and see if you can fit some words. About the quintet,
there is no bounden duty for us to have any at all in the whole op., but
if it seems to fit . . .

Twelve days later:

I've received the love-scene and like it much better. The iron-curtain
has descended on my inspiration . . . Once I'm on the other side of the
Interlude the rest of the act should be plain sailing as I've sketched
pretty well to the end. But the wait [for libretto alterations] wasn't
time wasted as I did some chunk of scoring, the only thing about that
though is that it is something I can do anywhere.

In the previous letter Walton had asked for a 'shantyish' song at
the start of Act III—'Shenandoahish'. He returned to this now:
'Not as you've noted a "Rolling down to Rio" but a good old
sentimental nostalgic yearning one with a lot of alls-welling
from the watchmen echoing backstage from post to post and an
odd remark or two from the other soldiers . . .' The letter then

diverted to other matters. King George VI had died on 6 February, hence the mention of a forthcoming Coronation. It also emerges that further attempts were being made to woo Walton away from Oxford University Press:

I'm glad you appreciated *Wozzeck* and *Mathis*.[6] There's no doubt they are two masterpieces of our time—one really can't mention *Budd* I fear in the same sentence—not that mind you it hasn't its moments, but it hasn't the consistency or substance of greatness that the other two have. *Wozzeck* I've not seen on the stage for twenty years but have heard 3 or 4 concert perfs. since and I've got the score with me just in case! *Mathis* we heard in Rome last year.

About Alan Frank [at this date music editor of the OUP and later to succeed Hubert Foss as head of the music department] and the O.U.P. This is a tricky business. The first point being that I've been with the O.U.P. right from the beginning of my career and they have always behaved well and I don't exactly have the face to leave them now for no reason at all. I was approached about 6 months ago by big-boy Boosey himself and I wrote him what I've just said. Now this new fumbling by [Anthony] Gishford ... means that I didn't fob off Boosey as well as I imagined. Now, without doubt there is a lot, in fact everything, to be said for B & H for this particular work. The O.U.P. possess no operas—that's not quite true, they have the original and uncommercial version of *Boris* which is hardly ever done in that version, nearly always the Rimsky—and that great commercial proposition VW's Pils. Prog. Now B & H possess a large number of successful operas, Strauss, Bartók as well as more popular ones.

I believe the basis for launching a new opera is this kind of bargaining—we'll perform *Budd* for instance 4 times at £25 a time if you'll let us have 6 perfs. of *Rosenkavalier* for £10 a time instead of £60 a time. Not only has B & H that card, but they are highly organised, operatically speaking, the world over and doubtless could guarantee T. & C. (that is if it isn't a most awful flop) 200 in the 1st season—like they have with the Stravinsky [*The Rake's Progress*] and doubtless with *Budd*. The O.U.P. have neither of these cards, and Frank is inclined to exaggerate how well known my name is amongst the operatic world—but the fact is that it isn't at all—in fact my reputation abroad is uncomfortably small as far as I can gather. Whether that is so or not, my name is not enough to launch an opera on and we need an operatic 'Helen', though God forbid we write 1,000 operas.

I pointed this out to A. F. when I told him about B's approach and it just dawned on him that the operatic field is quite a different one to the symphonic one. He nevertheless felt that some special effort would

[6] *Wozzeck* was performed at Covent Garden, conducted by Erich Kleiber, on 22 Jan. 1952.

have to be made and on the whole I'm inclined to stay with the O.U.P. even if we don't get quite the same out of them as from B & H. Also being suspicious-minded, wouldn't B & H do the dirt on one and sit on the bloody thing as being possibly the only rival to BB? Maybe I'm too suspicious! . . . I think we ought to keep an eye open on the Coronation. Another march! & this time a really singable tune in the middle, with words! I ought to have done it with the old one and then I was young and innocent and high-minded in 1937. Now I approach my 50th birthday March 29th things have changed! . .

They had indeed.

PART III

Sir William

17
Coronation Interlude, 1952–3

THE *Troilus* correspondence shows that it was the composer—as it was Verdi, Strauss, Britten—who was constantly concerned about the timings of scenes and arias and the practical business of staging the opera. Walton would challenge Hassall about a character's entrance, for instance: why bring them on at that point and give them nothing to sing for ten minutes? It was Walton who kept altering the stage action and demanding lines to cover the new situations. For a man writing his first opera, he showed an astonishing appreciation and knowledge of what could or should be feasible on the stage, all of which suggests he was potentially the dramatist Tovey had suspected as a result of hearing the Viola Concerto and *Belshazzar's Feast*. On 6 March 1952 he was still worrying about the love duet and asking for two extra verses.

If you could build it so that there is a climax in the 2nd of the new verses, it would be all to the good—then the last would be quiet. Now a word on the Interlude. Perhaps you'd reconsider it, in fact I wish you'd write the music as well! because I'm fair flummoxed by it. I think there should be some musical reference to what is going on, or should be, in the bed, some kind of orgasm or not? And to fit it in the outline as it stands is the difficulty. And the length—it can't be too short or it will be referred to as 'premature ejaculation'. And the timing for the 'distant drum'? . . .[1]

Hassall sent the new verses and Walton's next letter is dated 29 March, his fiftieth birthday.

I'm still apprehensive about the Interlude but will see what turns up when I get to it. What I meant about the timing of the drum was how long when it should start and stop. You suggest when Tro. says 'Why

---

[1] In the morning after the love duet, a drum announces the approach of soldiers—the arrival of Diomede to effect an exchange: Cressida for the return of a captured warrior, Antenor.

P[andarus]' but I've done a little Waltonese there and at T.'s satire and P.'s reply. P. comes in and says 'Who waits' etc., goes to the window (by the way there was no glass in those days!) looks out and continues 'Great heavens! They're Greek—Greek soldiers. They've halted in the yard. But why here? Why at my house? Maybe 'tis news of C.' O.K.? Thank you for your letters and congrats. They are mutual—it's the next 50 years that count! I hope to heaven I finish Act II by the time you arrive. In fact if by the middle of April I see no end in sight I shall start on Act III. I should like to get it under way . . . I'm still not very happy about the new opening sc. of Act III. I feel the 'shanty' should echo or rather pre-echo what Cress. is feeling in 'No sign'. The soldier is longing for his girl in Greece, Cress. is longing for her T. in Troy if you see what I mean. And it needn't be too shantyish or we shall be accused of committing 'Buddery'! As it is, I just noticed that the 'Good night, sweet uncle' etc. sc. Act II is rather too closely like the 'good night' sc. at the end of the 1st act of *Lucretia*.[2] It's the association of the words more than anything else. But I don't see what can be done about it—in fact nothing at all. Let it stay and risk it being noticed . . .

During May Hassall went to Ischia with his wife, who was recovering from an operation, and he and Walton worked hard at the opera. The correspondence resumed on 10 June:

I've not done anything further on Act II but will bear in mind a canon quartet, probably a fugal one might be better. Anyhow it's an idea and it's about time I brushed up my counterpoint! Act III hasn't gone too badly. I've cut the soldier's song in the distance completely, substituting 3 All's wells with fanfares getting further and further off . . . I had a talk with D. Webster[3] from Milan. Owing to Harewood's royal connections he has arranged that the Queen has commanded an opera for the Coronation season. It is, I need hardly say, not T. & C. but a new one on 'Elizabeth & Essex' by Billy Britten. How he is going to get it done in 9 months (it must go into rehearsal at the end of March) I don't know. But genius will out. Plomer, Auden tells me, is doing the libretto which is not ready yet. Auden's just returned from Paris where he'd seen both B.B. and said he thought the opera absolutely the end and hadn't gone down too well.[4]

So D.W. as far as I could gather tactfully hoped that T. & C. wouldn't be ready for the Coronation season, but if it were could it open the season? That of course means having everything ready by the end of next Jan. instead of end of March, a vital two months' difference and I

[2] Benjamin Britten's opera, *The Rape of Lucretia* (1946).
[3] Sir David Webster, general administrator, Royal Opera House, Covent Garden, 1945–70.
[4] *Billy Budd* was performed at the Théâtre des Champs Elysées, Paris, in May 1952 as part of the Festival of Twentieth Century Art.

doubt if it can be done even working 16 hours a day. So I think it is better not to hurry it and wait till the opening of the autumn season. It is all slightly irritating. B.B. has to give up all his engagements and Cov. Gar., which is already broke, has to compensate him. But there it is. We've no friends at Court, so we must put on a smiling face and pretend we like it. But we shall not have such a glamorous opening as we might have had in next June what with all the visitors, distinguished and otherwise, for the Coronation. Anyhow, budder them! We dined with Auden last night. Very amiable and wants to go through T. & C. with me. I think I will, as it will be interesting to get an outside opinion, though I hate doing it, playing the thing as badly as I do . . .

He kept Roy Douglas, who was copying the score of the opera, in close touch with progress. 'I sent more yesterday', he wrote in the summer of 1952, 'and hope to be through with Act II by Xmas. Am still having libretto trouble with Act III but Auden, who is here till the end of Nov., is having a go at it so I'm hoping it may turn out not to be quite so awful in the end.'

Auden and Chester Kallman first rented a house in Forio, Ischia, in 1948 and went there almost annually thereafter, becoming deeply interested in their vegetable garden, among other things. At some time during June the playthrough of *Troilus* took place. 'He seemed genuinely enthusiastic about the music', Walton wrote to Hassall, 'and liked best all the parts you like, especially "Child of the wine-dark wave", "Slowly it all comes back", "At the haunted", so perhaps I'm not a very good critic of my own music!' Auden made several quite drastic suggestions for changes of action and motivation which Walton passed on to Hassall.

From my point of view [he wrote], it is to say the least a bit of a bore to have to start the Act all over again but I think it may be worth it . . . Unfortunately as W. had to go we were unable to discuss fully the last scene. But he said these things stuck out to his mind: that he felt Pand. who after all is a very important character should not be lost sight of too easily, that Cress. should not fall on Tro.'s body. At all costs must we avoid any resemblance to Tris. & Is. [Tristan and Isolde]. That Tro. should roundly curse her with his dying breath. That Cress. could as part of her dress have a dagger on her belt so as to avoid risk of it dropping if handed to her by Calk., anyhow a far less clumsy way. That her last words are not right. Having been shunned and cursed all round by all, she would be almost mad with terror, horror and hysteria—in fact a short sort of Lucia . . . I do hope, in fact I know, you won't take it all the wrong way as he was really fearfully amiable about the libretto but thought it useless to just make vague uneasy compliments. We needn't necessarily agree on all his criticisms and suggestions, but on the whole I'm of the opinion that they are good . . .

'Wystan is here till the end of Nov.,' Walton wrote to Hassall when he returned to the island after a visit to London in July for performances of *Façade* and *Pierrot Lunaire* at the Festival Hall.

He has a lot to say about Act III. But more of this anon when he's put down his ideas on paper. His main idea is that there should be a grand quintet so that everyone has a look in. I've an idea that the curtain should be much quicker after that and it should come thus: Diomede says at the moment 'as for C. she has her uses'. I propose he should say to the soldiers 'Take her—she's all yours' or something like that and the curtain comes down as they are about to pounce on her. Perhaps a bit brutal but makes it more dramatic and less Isolde-ish . . . I've written D.W. about H.M.'s refusal so all is clear for you to have fun with I.L. and get her to make some sketches quickly.

The last sentence refers to the decision by Henry Moore, a friend of Walton, not to undertake the designs for the opera. He had also been invited to design a Covent Garden *Ring* cycle but eventually decided against theatre work. There was then a plan that Constant Lambert's widow Isabel would design *Troilus and Cressida*.

Towards the end of the summer of 1952 preparations for the Coronation of Queen Elizabeth II on 2 June 1953 began to take shape. The Westminster Abbey organist, Sir William McKie, was director of music for the event. He met Walton in London on 16 September and extracted a promise of a new Te Deum, assuming that both royal and archiepiscopal approval were forthcoming. Walton also agreed to the use of the new march he intended to write. McKie wrote later that he was 'most impressed at the thoroughness of the briefing which he asked for. He came down to the Abbey, asked for most precise information about size and composition of the choir and the orchestra and how they were to be placed. He also wanted to know what would be happening in the ceremonial just before the *Te Deum* was to be sung and what music would precede the *Te Deum*.'

In October Walton declined a request to contribute a madrigal to *A Garland for the Queen* because he didn't 'want to be accused of making a corner in the Coronation music.' He began to compose the Te Deum in November. On the 28th he wrote to Hassall:

I've got cracking on the *Te Deum*. You will like it, I think, and I hope he will too. Lots of counter-tenors and little boys Holy-holying, not to mention all the Queen's Trumpeters and sidedrum. I hope to finish it by the end of next week and then will get going on the March, so hope to be clear of all that by the New Year which I hope will see the last of T. & C. Have you any poetic suggestions for the big tune in the March?

During December he wrote to McKie: 'Though I hesitate to hazard an opinion when I am so near to a work, I think it is going to be rather splendid. I have made use of the extra brass, but arranged it so that it can be dispensed with, if unpractical for any reason. There is quite an important and indispensable organ part!'

On 29 December Walton told Hassall: 'After a spot of bother with the "Virgin's womb" (the kind of trouble I always seem to get into—don't tell the Archbishop!) the *Te Deum* is complete and both full and piano scores dispatched. Quite a lot of work. It is not too bad for an occasional piece and should be right for the ceremony. The March is under way but not too bright as yet.' This letter then reverts to detail about the action of Act III of the opera and indicates that the revised libretto had been sent to Ernest Newman. The march, *Orb and Sceptre*, was completed in mid-January 1953 in spite of his confession on 5 January to Alan Frank:

I've not been feeling in a regal mood and the March has been making rather heavy going, but I've arrived at last at the Trio. It at the moment doesn't seem too good—not as good or popular as 'Crown Imp.' and perhaps it is a warning not to tempt providence twice over. However I daresay it will improve with a little gingering up and will certainly be good enough if not outstandingly good enough for the Ceremony.

Permission was obtained to dedicate the March to the Queen.

In the second half of January Walton wrote the variation on 'Sellinger's Round' which he contributed to the Aldeburgh

Festival at Britten's request.[5] Giving Hassall this news on 4 February—'I'm itching to get back to T. & C.'—Walton advised him to 'get tickets now' for Heifetz's first public performance in Britain of the Violin Concerto on 9 June.[6] He then sent two pages of suggested changes to the libretto of the opera. He was still fashioning Cressida as he wished:

Having made her decision, she should positively throw herself at Diomede. In fact she should positively let rip with all her pent-up sex which has been accumulating in the past weeks, so there is now no doubt at all in the minds of the audience that she intends to be unfaithful but at the same time she intends to get married as well . . . Read the life of Puccini by George Marek. Most interesting and instructive. I'm an angel in comparison to the time he gave his librettists.

Hassall's friend Sir Edward Marsh, on whom Hassall modelled elements of Pandarus, had died in January. 'It was sad about Eddie', Walton wrote, 'but he had a fairly good innings, though everyone these days seems to live to be 90 and never stop writing chilly symphonies![7] It may be just as well he didn't live to see Pandarus'. On 22 February he reported:

I finished orchestrating the whole love duet Act II but have not taken the plunge yet into the Interlude. I've also orchestrated all to date of Act III and am now back on the grind, so to speak. Eve [Hassall's wife] will enjoy this riddle rejected by the *New Statesman* for 'This England', sent in by someone from the Children's Corner (!) of a provincial newspaper:

> I am long and thin
> I am covered in skin
> My head
> Is sometimes red
> I am in three parts
> I go into tarts
> What am I?

Rhubarb, believe it or not, is the answer.

[5] *Variations on an Elizabethan Theme 'Sellinger's Round'*, first performed at Aldeburgh by the Festival Orchestra conducted by Britten on 20 June. The other variations were by Arthur Oldham, Michael Tippett, Lennox Berkeley, Britten, and Humphrey Searle.

[6] The performance was given in the Royal Festival Hall, with the Philharmonia Orchestra, conducted by Walton.

[7] Vaughan Williams's *Sinfonia Antartica* had had its first performance on 14 Jan. 1953, three months after the composer's eightieth birthday.

Walton flew to London in March to record *Orb and Sceptre* with the Philharmonia Orchestra on the 18th, its actual first performance. From Forio on 5 May he wrote his last letter about *Troilus* for some months:

I've been scoring Act I so have had no reason to pester you. It's pretty tough going and rather slow going and needless to say I've not got as much done as I'd hoped ... I should like to finish the scoring of Act I and II before we come back here at the end of Sept. And if everything is fixed Xmas should see it more or less in the bag. However I've been asked to do the music for the 'Romeo' film[8] and to fill my depleted purse I'll do it and they will pay most of it here. But as yet I don't know the dates ... but I fear it may cut into T. & C.

In the interim, Hassall took soundings from Laurence Olivier, to whom on 23 June he sent a copy of the libretto at Walton's request together with a report of 'a long session I had recently with Ernest Newman'. At this session the ending of the opera had been discussed:

I feel the solution reached by Newman and me is the answer, but if you and William can improve on it then so much the better. The problem arises from three things. First we must build up sympathy for Cressida to the end, and so far from cheapening her, she must grow into a tragic, or at least an intensely poignant, figure. Two, she is universally known as 'False Cressida', and it's as important not to whitewash her legend as it is vital to give her tragic dignity. Three, the existing end of her legendary story is essentially unoperatic, so a change is necessary anyway; but any departure from custom must be of a kind in keeping with the first two of these points. (You may remember that the legend is that Diomede, having grown tired of her, rejects her, and she goes to the next man, then the next, sinking by degrees to a common drab. We are never told what happens to her at the last.)

For the Opera it's enough to show that she abandoned all hope of Troilus and yielded to Diomede. In an earlier version this was easy, for there were two scenes in Act III and at the end of the first we saw Diomede, in triumph after his seduction dialogue, going off into the tent with Cressida ... But William wanted the 3rd Act in one scene, for very good reasons that Newman wholeheartedly supports ... William has suggested that Cres. should suddenly surrender with passion and abandon. This would be exciting in itself, but would it really be in keeping with the woman we want to present at the end? And is it in character? Newman thinks it would be a grave mistake. He

8 An Olivier–Shakespeare project which did not materialize.

says she should yield 'with dumb resignation'. The circumstances are too much for her. He thinks the existing words for her here are right . . . He said Tro. should die almost at once. At all costs he should not survive . . .

And so on. After six years the libretto still had no satisfactory conclusion!

Walton was back in London in May 1953. He conducted a rehearsal of the Te Deum at the Abbey Song School, with William McKie at the piano. He also attended the final rehearsals in the Abbey itself. Walton and his wife attended the Coronation. The Te Deum was the last music in the service, apart from the National Anthem, being sung after the Blessing before the Queen's Recess. Walton's setting made a thrilling climax to what had in any case been a musically distinguished service, with contributions by Vaughan Williams, Parry, Howells, Dyson, and Stanford, not to mention Handel, Byrd, and Gibbons. Yet Adrian Boult, who conducted some of the music at the service, was shocked by the 'pagan' sound of the Te Deum.[9] No doubt this was a Freudian reaction to resemblances to *Belshazzar's Feast*. The Te Deum shows Walton at his ceremonial best, secular rather than sacred. The 'allegro vigoroso' orchestral opening, a motif denoting the idea of praise, is immediately echoed by the organ before the choir enters with its shout of praise in D minor. For 'To Thee all angels cry aloud' the key changes to E flat and the choir is divided into two. For 'Holy, holy, holy', in A flat minor, a semichorus is used (counter-tenors rather than altos where possible), with the other semi-chorus answering and the tonality moving to C for 'Lord God of Sabaoth', triumphantly for full choir. The praises of the Apostles and Prophets are sung in A major to a new, flowing melody. This gentler section is succeeded by Waltonian fanfares in G major for 'Thou art the King of Glory' and ends ('We therefore pray Thee help Thy servants') with music which embodies the Anglican tradition. First and second subjects are recapitulated before the inspired final cadences.

Walton himself described the Te Deum as 'rather splendid' and in the same letter said the march, *Orb and Sceptre*, was 'neither better nor worse than the last',[10] but there his objective

---

[9] Letter to author, autumn 1977.
[10] Letter to Angus Morrison, 24 Jan. 1953.

judgement faltered, for it is certainly inferior to *Crown Imperial*, lacking as fine a central tune and the vigorous panache of its outer sections. Boult conducted it in the Abbey on 2 June and Barbirolli the first concert performance five days later in the Royal Festival Hall. Walton's commercial idea of a 'Land of Hope and Glory'-type song with words from the tune of the trio evaporated. On 8 June he was at the gala first performance at Covent Garden of Britten's *Gloriana*. His arrangement of the National Anthem was played before and after the opera. The performance on the Queen's entry into the royal box was disastrous, half the orchestra waiting (correctly) until all the royal party had taken their places before beginning to play. Such was the atmosphere of professional partisanship and hostility in regard to *Gloriana* that some people believed Walton had planned some sort of sabotage and these false accusations caused him intense distress. He also went to Aldeburgh on 20 June for the first performance of the 'Sellinger's Round' variations. The names of the six composers were printed in the programme in alphabetical order and Walton suggested that the audience should try to guess the identity of the composer of each variation, with free seats at a 1954 festival event for the winners. No one guessed correctly. (The correct order was Oldham, Tippett, Berkeley, Britten, Searle, and Walton.) It was Walton, too, who had suggested that each composer might introduce a short quotation from one of his own works into his variation. He quoted *Portsmouth Point* in the closing bars of his contribution.

In August the Waltons crossed the Atlantic in the *Ile de France*. Walton conducted *Belshazzar's Feast*, the *Sinfonia Concertante*, a *Façade* suite, and *Orb and Sceptre* at the Hollywood Bowl. While on the West Coast he was entertained by Heifetz, Rubinstein, and Stravinsky; and his brother Alec, who worked in Vancouver, joined him in Los Angeles. But the busy summer had entailed a four-month hiatus in the composition of *Troilus and Cressida*.

18

Finishing An Opera, 1953–4

WALTON'S first letter to Hassall since May 1953 was written on 8 December. 'At long last I'm getting down to write to you. I've been scoring away like mad and have completed Act I and am now immersed in Act II which I fear will take me to the end of June (a month behind schedule!).' He was still altering small details of the plot, cutting and omitting lines and verses of the libretto. Auden had again addressed himself to Act III.

The thing with Wystan was not too satisfactory as I didn't like to pester him too much as he was working. But the enclosed [a draft of a sextet] turned up just before he left . . . Whether we decide to have a sextet or not, I'm not at all sure, but if we do, this is a model of how it should be worked out, the parallel rhymes in each part etc. . . . The end, I feel, as it stands is anti-climactic and the opera is all over when Tro. is killed. Anyhow I will leave it to you to cudgel your brain over this . . .

In January 1954 Walton acknowledged receipt of a new version of certain arias. But he also had to pour soothing-syrup over Hassall's affronted anger about Auden's contribution:

I see now from my letter of Dec. 8th that I was singularly inexplicit. That he did anything is entirely due to my asking him to look at Act III as I wasn't entirely happy about it. He took it and when next I saw him he said that both he and Chester had separately reached the conclusion that a quintet or sextet or some concerted piece was needed. After some discussion I said granted that a sextet is the thing, how and what is the proper way to lay it out because I don't know how it should be and am not at all sure that Chris H. (that's you) knows either (perhaps presumptuously assuming that you didn't though you remember we failed miserably when we tried to lay out a quartet towards the end of Act II?). I continued—can't you do it now—just sketch out how the rhymes etc. should balance etc. and C.H. will catch on in a trice. No, he said, he wouldn't do that and that in any case he would not dare to presume on to somebody else's territory and that he had a high regard for you and wouldn't risk hurting his relations with you.

Anyhow in the end I persuaded him that it would be a great assistance to us both if he would just do an outline of how he thought this sextet could come about and he reluctantly in the end said he'd try but that it was to be considered nothing else but what is known in the film world as a guide-track and that is how and what I hoped you would take it for. He never meant that I should set it and I never for an instant have ever thought of doing so and even less so now ... In fact it is all my fault, but I thought it would be a help and I hope this somewhat tardy inexplicit explanation will clear things in your mind ... But now to come down to it, if there is to be a sextet you will have to write it. On the other hand, as I said in my last letter but one, if you are not convinced about it (and being in uncertain mind about it myself you will now have to convince me) don't do it. It's sure not to be right if you are unconvinced ...

Walton wrote in this letter of 'being fairly optimistic about finishing by the time we are due to leave at the end of July', and he ended with a postscript: 'Though while trusting dear Ernest N.'s judgment to a large extent, one mustn't forget that he can't help seeing things through rose-Wagnered spectacles. I can't help but remember that he passed the script as being perfect God knows how many years ago, when there were only sixteen insignificant lines for Cress. throughout the whole of Act I!!'

One of the reasons Walton hoped to finish the opera in July was, as he wrote to Alan Frank on 11 December, that he had been 'offered the music for Antony & Cleo—the film is being made and paid for here. As the amount mentioned is something like 8 million lire, it would put me out of the clutches of the Treasury for some years to come. That would be roughly Dec. '54 to March '55'. But nothing came of the plan. In the same letter Walton reported that Herbert von Karajan had conducted his First Symphony on 5 December in Rome.

Of course no one let me know, so we didn't go to Rome nor were we able to listen in as it was on short-wave nor have I heard anything about it. All very irritating ... The Violin Concerto wasn't a wild success at the San Carlo on Nov. 15th ... [Aldo] Ferraresi is really excellent—far better than Campoli or anyone in England—he gave a really splendid perf. But the orchestra. God help me—it was the worst perf. I've ever heard—the tuttis were quite unbelievable. Rodzinski was very apologetic—he had put down what he thought was a familiar programme for the rest of the concert, *Oberon* and Tchaik. 4. To his horror, when he thought he would be able to devote the major part of two rehearsals to the concerto, he found that Tchaik. 4 had never been played before in Naples. So that dished everything.

This letter ended with a reference to Bliss's appointment as Master of the Queen's Music in succession to Bax, who had died in October 1953. 'You see I was right about A.B. I must write him a line or he will think I disapprove! I can't say how glad I am that it was not offered to me.'

The amount of trouble and delay caused by amendments to the *Troilus* libretto which Walton realized were needed while he was scoring the opera, is evident from his next letter to Hassall, written some time in January 1954, in which he drafted the text for the 'We were alone' duet that ends Act II and took issue with Hassall, who had replied to a query about Pandarus's second entrance in Act I with a reading of the situation totally at variance with the composer's.

All these years [Walton expostulated] I've been under the disillusion that P. comes out just to comfort his niece and to carry on the good work he has undertaken (i.e. that of forwarding Tro.'s suit) . . . To me it would seem that if he had overheard anything that has been going on in his absence . . . he has heard all or nothing and I should have thought for him to have heard nothing was the simplest solution. However, if you don't approve of him . . . finding her in tears, there is ample room for him to make any desperate gesture you may feel he should make . . . All I can add is that if I've been stupid about it all this time, how much more so would be the audience, and I may say that it was from no comment of Wystan's, as I only showed him Act III, but purely from a musical one of having two unconnected bits of recit . . . (Evadne's and Pan.'s) to deal with and finding that they seem so to speak to cancel each other out . . .

Chivvying Hassall for 'a bit of dialogue' on 14 January, Walton wrote: 'It has been well worth re-doing Act II as far as the Interlude, though it has taken some six precious weeks. I've tightened it up considerably . . . The whole love scene is now much better proportioned and is fairly straightforward . . .' In particular the end of the opera was still a problem. Sending some Act I corrections to Roy Douglas on the same day, he wrote:

I fear it will be possible to save few of the pages already done from Act II. When I began looking at it again, I found more and more things which could be bettered and in the end came to the conclusion that it was quicker to do a fresh score than to try and start patching the old. You will, I think, admit that it has improved immensely from its tightening and pruning . . . The libretto of Act III is still in a fearful muddle, but I'm making him come here to clear it all up.

He wrote to Hassall:

If you don't like the idea of Cress. being seized by the soldiers for garrison hack, would this idea be better? After Dio. says 'Soldiers, take her' or something, Cress. should scream 'Father' and Calk. should break away from his guard and stab her, followed by the curtain. Evadne is a bloody bore and I don't see the absolute necessity of her treachery being brought to Cress.'s notice, but knowing nothing of the rules of dramatic construction, I'm doubtless wrong . . .

A few weeks later he wrote again to Roy Douglas:

Delighted to get . . . your approval of the changes in Act II. Bore that it was to do, and how much more so for you, I feel it has made all the difference to the chances of success or failure of this Act, though I wish I had the time it has taken in doing it . . . John Warrack seems to be bothered about Fl. 3 doubling Picc. and Fl. in G. Surely Fl. 3 generally does so and likes it as he gets an extra 10s. for each instrument or am I wrong about this?

Prospects for the production at Covent Garden now began to be discussed. With Henry Moore's defection as designer, his replacement became urgent.

Larry—without absolutely saying so—says in so many words that . . . he would prefer [Roger] Furse for the obvious reasons. [Furse had worked with Olivier in stage productions and on the films of *Henry V* and *Hamlet*.] I think we had better agree, though I'd prefer Isabel, maddening though she may be, and after all Furse is very good and knows his job especially in conjunction with Larry. Of course it will be too irritating if after all Larry finds himself unable to produce it for one reason or another and we find ourselves saddled with Furse without Larry. But we shall just have to bear it. He's not heard from Webster nor indeed have I had a reply to my letter of the end of Oct. regarding casting.

On 14 February he reported:

Alas, things have not been going too well. The Interlude, which as you

know I always had a 'phobia' about, hasn't turned out too well, I
fear—but it will, I hope, 'get by'. It is finished and Franz R.[1] will
doubtless be able to give you some idea of it . . . Its form is like this: 48
sec. storm to where the gauze comes down, 40 sec. fumbling and then
back to the 'storm' for 40 sec. and the 'effing' and storm work up to a
climax for another 40 sec. (in fact, if without offence I may say so, they
come together) followed by 45 sec. dawn . . .

He was still rewriting the end of Act II and asking for new
lines, and still doubtful about the need for a sextet in Act III.

In his next letter to Hassall, on 23 February, casting was at
last discussed.

The more I think about the Schwarzkopf situation, the less I like it.
And I'm seriously thinking of postponing the 1st perf. till she is
available, which alas doesn't appear to be before May '55 which does
seem a long way off. On the other hand, being here I've no way of
judging whether Wilma Lipp or any other singer who is put forward is
right for the part or not. And I don't by any means trust Cov. Gar.
judgment even on a work they should know, let alone on one which
they have not an inkling. Ring up Alan F. and see what he thinks.

Alan Frank wrote to Walton on 1 March:

I had long talks over the weekend about casting with Christopher,
Legge (who was extremely helpful), Enid Blech in the absence of
Harry, and Joan Ingpen—hence my two telegrams to you . . . Without
exception all these opinions said that Lipp was an excellent light
coloratura, but absolutely unsuitable for a part written for Schwarzkopf. Everyone in fact seemed astonished that she should ever have
been thought of. And the whole episode leaves me rather horrified at
the casual way in which apparently Webster and/or his Covent
Garden associates treat casting. Anyway it's a lesson for us all. Apart
from Enid's staggering but doubtless impracticable suggestion of
Callas, so far Eleanor Steber seems the best bet and would I think be
very good. What a pity that Lisa Della Casa doesn't speak English. It is
indeed maddening about Schwarzkopf, but you evidently see the
disadvantages of postponement and already I think she is only free
from 1st—12th May. Obviously if we cannot get a first-class cast for
late 1954, we may have to postpone the opening, but I very much hope
this won't happen.
   Everyone backs Gedda as first choice for Troilus, and I can't quite
gather what is going on. Webster says he turned it down because he

---

[1] Franz Reizenstein (1911–68), composer and pianist. He played the opera,
as parts of the score reached England, to Hassall. He also made the piano
reduction of part of Act III; Roy Douglas did the rest of the opera.

1. As a schoolboy, Christ Church
Cathedral School, Oxford

2. With Dora Foss at
Symonds Yat, Herefordshire,
in the 1930s

3. Musical evening in Wimborne House, Arlington Street, London, in the late
1930s, painted by Sir John Lavery. In the foreground are Ivor and Alice
Wimborne, Walton, and Garrett Moore (later Earl of Drogheda)

4. Walton in 1949

5. Congratulating Sir Malcolm Sargent at a Royal Albert Hall Promenade
Concert, 1 September 1965

6. A drawing made by
Gino Coppa in 1975
and inscribed to Alan
Frank

7. Susana and William Walton, 1971

8. At the Ritz on his seventieth birthday, 29 March 1972

9 and 10. The only surviving sketch of the Third Symphony, inscribed to
André Previn, with (opposite) a transcription

*Written in C.

11. In his study in Ischia, 1981

doesn't want to learn anything new, which may be so, but also slightly hinted that a contributory reason was that Gedda had quarrelled with Legge (he didn't put it as directly as that). But this doesn't fit at all with Legge's attitude, who immediately promised to talk to Gedda if we wish ... The best young English tenor is Alexander Young, whom you once mentioned, and he is singing extraordinarily well. But his voice is said to be definitely not big enough for Covent Garden ... Uhde[2] for Diomede is, as you say, the best suggestion ... It is obviously going to be a tricky business assembling this sort of international cast unless CG works with a bit more efficiency than it has shown so far.

By the time Walton wrote again to Hassall on 9 March he had seen David Webster—'not very satisfactory, except that I pointed out that Lipp was a coloratura and he tried rather unsuccessfully to think up some other names. The only things that seem to be fixed are the date Dec. 3rd and Malcolm S. to conduct ... Meanwhile the San Carlo are very keen on getting it and I've had to be fobbing them off, not too hard in case La Scala falls through.'

Covent Garden's first approaches to Elisabeth Schwarzkopf on Walton's behalf met with the answer that she was booked elsewhere far ahead. Walton commented to Hassall on 17 March: 'About Walter and Elis. I don't attach any blame to them whatsoever because it would be asking too much not to accept, as there is still an element of doubt as to whether I shall be ready in time. This, I think, is also at the back of D.W.'s mind and accounts up to a point for his dilatoriness.' Reizenstein was to play through the score as far as it was ready in London on the 25th.

I'm very pleased about F.'s opinion about Acts I and II [Walton wrote]. I feel rather like John Ireland (you don't know him but he invariably replies if one says how much one liked, say, the last movement of one of his works 'Oh, I suppose that means you don't like the other movements'). F. doesn't mention my most sensitive point almost, 'Child of the wine-dark wave', you might stealthily get his opinion on this! ... I like about 'the acute responsiveness to the text', he is quite right and that is where your appalling responsibility lies, when you're good I'm better, when you're not so good I'm much worse—so, rather late in the day to say so, bear that in mind! No one has had such a rotten time as you since Puccini's maltreatment of his

[2] Hermann Uhde (1914–65).

librettists' lives! However look at the result! And don't think that you are going to have a moment's peace till the last bar is finished! What a life!

There followed several pages of suggestions and queries about Act III. Hassall was due in Ischia in May and Walton ended his letter with some unorthodox practical advice:

And you can bring me some tobacco and no nonsense about showing it to the customs official and having to pay thro' the nose, we can't afford it! What you do is this. Buy 2 lbs of 'No Name' and 4 plastic pouches from Woolworths—empty the tins (a bit of a bore this, but Eve will do it), pack 8oz into each pouch by shoving hard, this is quite possible as I know from experience. Put a couple of elastic bands round the pouches. To travel you put one pouch in each of your overcoat pockets, the other two in your bag in the middle, surrounded by shoes, etc. When you are asked if you have anything to declare, don't go white in the face and say 'Si' but smile in a superior manner and say 'Niente' and you'll be left in peace. If you open your bag, they only poke around the sides and if Eve is up to form with her deceptive genius nothing will be found. But in any case they wouldn't bother you as it is open and in pouches and for your personal use . . .

On 25 March ('the day of judgment') he told Hassall that Act II had now only to be orchestrated 'and I hope that Franz will get it in time for April 5th so the two Acts can be heard complete by Malcolm S. etc.' He was still rewriting the section of Act II after the love duet and the later duet at the end of the act.

Incidentally I believe that one of the difficulties of 'We were alone' is the length of the lines. Short lines are much easier to cope with, one can always expand if one needs to, but one can't contract a long line without as a rule making nonsense of it, and I think it is a thing to discover rather late in the day . . . You will have heard, I expect, from Alan Frank that Schwarzkopf is out as she goes to America early Oct. This is a bitter blow. Also it seems Larry will be filming so we shall at least be able to have Isabel. I wrote to D.W. suggesting postponing it till Schwarzkopf is back, but I gather from A.F. that that will lead to all sorts of complications. On the other hand it is more than possible I shan't be ready anyhow—a terrifying thought. In fact at the moment I'm full of gloomy prognostications.

Olivier having declined to produce the opera, George Devine was appointed, with Sir Hugh Casson as designer. According to Walter Legge in his appreciation of Schwarzkopf, she 'regretfully declined' the role of Cressida 'because the text was Ivor Novello-ish and in English'.[3] Lady Walton, in her biography of her husband, states that Schwarzkopf

was not allowed to sing Cressida . . . The excuse was that her English was not good enough and her voice not suitable for the wide range that William required. In retrospect Walter may have been right, because when William recorded *Troilus* later—and in this recording [of extracts] she was allowed to sing the part of Cressida—the recording sessions had constantly to be interrupted to allow her voice to rest, as well as to check her English diction.[4]

In his biography of Walton Neil Tierney quotes Legge as writing that Schwarzkopf 'always disliked singing in English because of the impure vowels'. She also 'feared that her slight German accent [*sic*] might invoke hostile criticism from diehard purists. Additionally, having carefully read the libretto, she found little to admire in either the story-line or the characters and . . . she could not accustom herself to Walton's melodic line, however hard she tried.'[5] The fact remains, though, that in 1951 in Venice (and later in Milan) she had sung, in English, the role of Anne Trulove in the first performances of Stravinsky's *The Rake's Progress*. She did not, however, sing in any British stagings.

At the beginning of April Walton's principal concern was the scene in Act III between Cressida and Diomede. By the 8th he had received Hassall's account of 'Judgment day II' (5 April, when Reizenstein had again played the opera to Sargent).

It doesn't seem to get us much forrader as regards casting. P.P. [Peter Pears] is an excellent idea for Pand. and I think there is just a chance he would accept. Perhaps B.B. should be attacked first! For instance you might slip a word to him at a P.R.S. meeting saying how 'ripping' it would be etc.[6] but is [Richard] Lewis booked up already so far ahead? Tenors indeed must be in demand. I think he'd be good but of course rather a similar voice to P.P. but that perhaps would not be of vital importance. I've set inquiries on foot to find out if Sena Jurinac would be willing and if her English is up to it. Della Casa I'm all for too, but what about her English? I've no idea. I've a feeling that

---

[3] E. Schwarzkopf, *On and Off the Record: A Memoir of Walter Legge* (London, 1982), p. 143.

[4] *Behind the Façade*, p. 146.

[5] N. Tierney, *William Walton: His Life and Music* (London, 1984), p. 133.

[6] Britten had a fondness for schoolboy words such as 'ripping'.

someone told me she didn't speak it at all. But with either I should be delighted, if they do. No sound from the shadowy depths of Cov. Gar.—not even a watchman's cry! I'm rather for M.S. [Malcolm Sargent] coming here for a few days. It would be a good thing to get the tempi right, right from the start.

I suspect F. is playing it on the slow side, it is very difficult not to on the pfte, but it is well metronomed and the duration marked ... I don't want to appear hidebound about this, but the tape will be very useful if accurate in that way, the notes don't matter so much, for répétiteurs etc. Act I is 43', Act II 45'. Thirty sec. either way don't much matter, but there shouldn't be more of a discrepancy than a minute at most. No one except me seems to realise how long a minute is and of what vital importance it is not to have too many of them!

Shortly after this, Covent Garden suggested that the Hungarian soprano Magda Laszlò should sing the role of Cressida. Walton made jocular reference to this in his next letter to Hassall on 18 June, when he badgered him about what he called the 'pomposo' scene in Act III when trumpets sound off-stage and Greek soldiers and women enter, followed by Diomede who has come to claim Cressida.

I imagine it thus. Trpts and chorus more or less together 'Hail—Argos' backstage. Then Tro.: What is this (sudden?) uproar? These voices? What are they calling? The orch. in front then takes up whatever tum-tum-ta-tum God gives me during which Cress. sings It's too late—Fate followed by Pan. and Tro ... These latter will occur in the quintet bits of 'pomposo' which presumably is not for orch. alone but mixed up with a few Heil Hitlers etc. If this is the right idea just O.K. it and I'll proceed ... I wanted and have written an extra exquisite line after 'Will he like me thus, hair loose and flowing'.[7] I doubt if you can better mine in the ancient Minoan style: 'Will he like me thus, tits boldly showing?' I'm sure Laszlò has beauties and it would liven things up no end. And how this brightens up our correspondence when published! So send me your alternative or I'll leave it as it is ...

Six days later Walton wrote again. This time he was more concerned. Hassall had heard Laszlò in Gluck's *Alceste* at Glyndebourne and had not been too favourably impressed.

I find your excellently vivid account of Laszlò slightly disturbing. Her lack of English is the most disturbing feature. Do you really think that she can learn it parrot-fashion and sing without a foreign accent in the

---

[7] In Act III before Troilus returns, Cressida sings, 'Will I please him thus, hair loose and flowing? Do the gold leaves match the gold of my girdle?'

4 or 5 months remaining to us? Have you ever come across a similar case? If so, well and good. I think from looks she's 100 per cent Cress. Of course you couldn't have heard her in anything less like T. & C. than *Alceste* and it can hardly be a fair criterion and I can't help feeling that someone who looks like that can't pull it out, or have it pulled out for her, so to speak, at the essential moments, but naturally a good deal of that will have to be the duties of Malcolm and George Devine. I can't believe a Hungarian (Jewess?) lacks temperament—it can only be dormant and perhaps the music will help her to it. My music on occasion does have that effect on the coldest fishes—like Adrian Boult for instance!

But the point is, if we don't have her who in heaven's name are we to have? You might speak to David W. about possible alternatives. You can say my only doubt is about her English, the rest I'm sure we can bring out to our satisfaction.

He then switched to the casting of Troilus.

[Richard] Lewis I've not heard for some time except in the *Rake* on the air but I thought excellent, but when I saw him on the stage he seemed an awful clod, but he's probably improved since then. I'm quite prepared to accept him if Nicolai Gedda falls thro. How much of the music has Laszlò seen? Enough to show what she's in for, I hope. And she must know whether she can honestly sing it or not. If she is fixed, can you see that she gets a copy of the German translation as then she can at least learn what she is singing about. I once heard Joan Cross sing in P. Grimes in German and she had not the remotest knowledge of the language and she'd learnt it parrot-fashion and one couldn't have been more convinced and the Germans, or Swiss rather, were too. So I suppose it can be done.

Throughout July Walton was still working on the last scene of Act III. He told Hassall he expected to finish the opera at the end of August. On 5 August he put forward an interesting candidate as designer of the costumes:

I enclose some designs which have been done by a young but experienced French designer Jean-Pierre Ponnelle. These are some of the costume designs for *Nathan der Weise* by Lessing. He has also worked at the opera in Milan, Rome and Naples, Venice, Berlin, Paris, Munich etc. so I'm not recommending someone who doesn't know his job as anyhow you can see for yourself . . . He is very adaptable and would of course fit in his designs with the stage sets. He wouldn't, I think, be expensive . . . I have seen other designs than these and consider him pretty good. There is also the advantage that he has already done dresses for Laszlò. Anyhow, find out how George and Casson feel . . .

On 22 August he was still tinkering with Act III—and the very end of the opera was still unsettled.

The end will, I think, have to be vaguely similar to the end of Act II. 'And by this I'm still your Cress.' being identical almost with Tro.'s last words in Act II and therefore being almost obliged to use the old T. & C. theme. I think highly dangerous to resurrect the 2nd most awful piece in the work at the last moment, when I hope everyone will have temporarily forgotten it!

The 6tet is more or less satisfactory and will doubtless improve when I score it. It looks ridiculously chaotic on paper from a literary point of view . . . This afterbirth of the 6tet has been difficult and is not really right yet. . . I trust the next 10 days will see me thro'. I'd like to be able to leave with all complete by Sept. 11th which is when ma-in-law goes . . .

On 7 September he wrote:

I expect to be sending off the last sheets at the end of the week. The last bit is proving a bit tiresome—it is not that it is really difficult, but that I'm just ever so slightly exhausted . . . George [Devine] may or may not have had time to show you my solution to Act I. He's probably too engrossed in *Nelson*.[8] I hope it will satisfy you both . . . As I have said before I'm feeling a slight holiday would not be out of place, so we have decided to stay on another week, not here, but with some new friends of ours, the Duca Camerini who has that nice house you will doubtless remember at Porto which looks like a Moorish mosque . . . The weather here is glorious after having been not so good thro August, in fact it's the first good this year so it seems stupid not to take advantage and have a slight break seeing what one is in for during the next months. So I don't suppose we shall be in London before the 27th which may well be too late for me to be of help on the galleys [proofs of the libretto for publication] but I don't think I'm needed if you check carefully from the full score, where incidentally you will see (crescendo) that I have not buggered up (crescendo) your fucking verses in the way they (crescendo) bloody well deserve!! As ever, William.

The manuscript score is dated 13 September 1954 and the opera was dedicated 'To my wife'. Walton wrote to his librettist: 'Dear C. Thank you for everything. I'm finished. Next please. See you on the 27th. You will find all changes such as there are in the scripts enclosed herewith. As ever. W.' Seven and a half years had passed since Alice Wimborne, Walton, and Hassall had first discussed the libretto.

[8] Lennox Berkeley's opera, produced at Sadler's Wells Theatre on 22 Sept. 1954.

19

Troilus and Cressida, 1954–76

WALTON arrived in London at the end of September 1954 to prepare for the first performance of *Troilus and Cressida*, but first he had an operation at the London Clinic. He soon realized that his expectation of a trying time at Covent Garden was to be fulfilled. Magda Laszlò's English was so poor that she had to be coached by Susana Walton, an Argentinian. The final choice for designer of the costumes was Malcolm Pride. The scenery was the responsibility of Sir Hugh Casson, assisted by two ex-students of the Royal College of Art, David Gentleman, who provided a painting on the temple wall in Act I, and Sheila Mitter, who designed the curtain in Pandarus's house in Act II. Casson was also helped by Jocelyn Herbert and William Glock's wife Clement. Rehearsals began in November. The first orchestral rehearsal broke up in chaos, most of the time being spent in correcting errors in the band parts, of which Roy Douglas found 238. Walton was furious and rang Douglas at midnight to say he was determined to break with Oxford University Press and to sign a contract next day with Boosey and Hawkes. Douglas persuaded him to wait before he took such a drastic step. Next day, Douglas met the Publisher of OUP and Alan Frank to tell them of Walton's justified anger; after this, OUP effected a *rapprochement* with Walton.

But the major cause of frustration was Sir Malcolm Sargent, who had not conducted opera at Covent Garden since he conducted Charpentier's *Louise* in 1936, nor anywhere else either, except for D'Oyly Carte's Gilbert and Sullivan productions. As rehearsals progressed, Walton became increasingly disenchanted by his conductor. Chivalrous efforts to defend Sargent's name in this affair are well meaning, but, alas, he cannot be defended, even by his sensitive biographer Charles Reid, who wrote:

There were rumours of difficulties, contentious words and hard feelings at rehearsal . . . What unnerved some of the singers was that Sargent declined to beat certain of their unaccompanied passages. 'Without the beat it's terribly difficult', they would say. To which Sargent would reply that a beat wasn't necessary. These bars were silent for the orchestra; he would join the singers at the end. 'But', came the objection, 'the rhythm goes on for the orchestra as well as for us whether they're silent or not, and sometimes the metre changes on the way. A beat would help us all.' Deadlocks of this kind were not easily resolved . . . The conducting score, not yet engraved, was in parts hardly legible. To make matters worse, Sargent refused to wear spectacles or, at any rate, had not reconciled himself to doing so . . . This above all other factors is said to have been the one that made the atmosphere at rehearsals acrimonious . . .[1]

Richard Temple Savage, bass clarinettist in the Covent Garden orchestra from 1946 to 1963 and Librarian of the Royal Opera House from 1946 to 1982, has described these rehearsals even more vividly, from first-hand experience:

Sargent had not changed since the pre-war days when he had tried his hand at *Louise*. He still did not seem well acquainted with the score, he was still an inveterate fiddler and tamperer with other people's works, and he still did not relate well to the singers. He always addressed them by the name of their role in the opera, 'Troilus' for Richard Lewis, 'Pandarus' for Peter Pears and so on. It must have made them feel like mere ciphers, and certainly gave offence . . . Like the pre-war Italian conductors, he refused to beat unless the orchestra was playing but, whereas in Italian opera it is often a matter of pure recitative, here Sir Malcolm was leaving whole ensembles to fend for themselves without setting a tempo; he would let them start and then join in with the orchestra. Peter Pears came forward at rehearsal and begged him to give them a beat and he still refused, saying he would feel such a fool, conducting without the orchestra. The acid murmured comments from the pit can be left to the imagination. As for the alterations, they were legion. Every afternoon after the rehearsal all the material . . . had to be carried up four flights to the library, where Walton would join me with the score to look through all Sargent's recommendations and then make alterations which we had to transfer to all the parts. Sargent had a particular aversion to anything written for two harps and was constantly cutting out bits of the second harp part. One day Walton came upstairs looking particularly disgruntled and suddenly burst out: 'I'm not making any more alterations. It's my fucking opera and I'm going to write some *more* for the second harp!' Which he proceeded to do, and this seemed to have the desired effect.[2]

---

[1] *Malcolm Sargent*, pp. 381–4.
[2] R. Temple Savage, 'A Voice from the Pit, 5. From 1953 to 1958', *Opera*, 38/

The first night on 3 December was a brilliant occasion. As Charles Reid wrote: 'There were more smart people in the Opera House who had never been seen there before than veterans could recall. Hundreds stayed on for what is said to have been the biggest first-night party ever held there.' Frank Howes in *The Times* next day wrote that *Troilus and Cressida* had 'formed itself into a great tragic opera.' Philip Hope-Wallace in the *Manchester Guardian* referred to 'this fine opera, a grand opera in the true sense', but generally there was a feeling of disappointment that Walton had written an old-fashioned opera in a tradition then held to be outmoded. Covent Garden, as Lord Harewood admits in his autobiography, was divided into factions at this time, those who admired Walton's opera and those who regarded it as totally irrelevant compared with Britten and with Tippett's *The Midsummer Marriage*, which was also in preparation for its première before a mystified audience on 27 January 1955.

It will be convenient here to abandon chronology and follow the history of *Troilus and Cressida* in Walton's lifetime. There were further performances at Covent Garden in April 1955 with the original cast. Walton had by now made cuts in each act. On the Covent Garden provincial tour from February to April 1955 the opera was performed in Glasgow, Edinburgh, Leeds, Manchester, and Coventry, conducted by Reginald Goodall. Only Lewis as Troilus represented the original cast. Cressida was sung by Una Hale, Pandarus by Raymond Nilsson, and Diomede by Jess Walters. This cast and conductor performed it at Covent Garden in July 1955. During 1955 Schwarzkopf and Lewis, with others, recorded extracts from each act, with Walton conducting. This certainly demonstrates that it would have been wiser for Walton to have conducted the whole opera himself at Covent Garden, for he imparts more passion and drama to the music than anyone else.

The first American performance was in San Francisco on 7 October 1955 with Dorothy Kirsten as Cressida and Lewis as Troilus. Erich Leinsdorf conducted and Walton flew over to hear the production, which he admired. It was also performed by New York City Center Opera on 21 October 1955 (and again

1. (January 1987), pp. 23–5. Reprinted in *A Voice from the Pit* (Newton Abbot, 1988), pp. 151–2.

in March 1956) with Phyllis Curtin as Cressida and Jon Crain as Troilus. The conductor was Joseph Rosenstock. As Walton had told Hassall, the Italian opera houses were anxious to produce the work and it was duly staged at La Scala on 12 January 1956 in a translation by Eugenio Montale. The American soprano Dorothy Dow sang Cressida and David Poleri (later to be killed in a helicopter crash) was Troilus. The conductor was Nino Sanzogno and the producer Günther Rennert. On 24 December 1955 Walton wrote to Hassall from Milan:

Rennert is splendid, full of imaginative ideas which makes everything and everybody become intensely alive and vivid, in fact I've never realised what a producer can really do. The singers are, to the most minor parts, first-class, especially Dow who acts almost as well as she sings . . . So far there have been only pfte rehearsals and the orchestral, of which I think there are at least 11, start in earnest at the end of next week. There are two general rehearsals on the 9th and 10th so you had better be here on the 8th. We've seen *Il flauto magico* and *Norma* with Callas who is fantastic but all the same I prefer Dow for Cress . . . Xmas greetings and what not, as ever, William. P.S. Your genius for double meanings has even penetrated into Italian. 'Impregnable Troy' when translated has an alternative meaning, to whit 'you unfuckable sow'!

The optimism in this letter soon faded, according to Susana Walton.[3] The scenery was over-lavish, Rennert 'kept looking for hidden meanings in the action', and Sanzogno was rehearsing elsewhere, so Walton had to conduct the sessions with the singers. He heard the first-night performance seated between Callas and Schwarzkopf. In the final scene, someone inadvertently removed Troilus's sword with which Cressida stabs herself and the unfortunate soprano panicked and all too obviously tried to find it. The opera and its composer were booed and hissed by a small section of the audience at the end and there were derisive whistles during the performance. Eugenio Montale, writing in the *Corriere d'Informazione* on 13 January, recorded an 'ambiguous reception . . . There were three calls after Act I, three after Act II and five after Act III.' Montale described the main characters as 'figures realised and composed by a musician who is a man of the theatre and really

[3] *Behind the Façade*, pp. 154–5.

knows his craft. It is not always easy, however, to follow them in the tortuous line of the vocal writing which demands almost superhuman intonation.' He suggested that Walton had 'glanced at the thematic content of our recent "opera veristica" ... He has done it, however, in an entirely individual way, always breaking off with a cut, an abrupt leap, in places where some development would conventionally occur'. The veteran Guido Pannain, in *Il Tempo*, in a very long and thoughtful notice, found Cressida's music

not sensual nor romantic but tender and sorrowful ... Opera presents itself to Walton, before any natural intuition, as the 'idea' of opera, that is to say as a preconceived form. And it is the form of Italian opera of the 19th century, in other words Verdian melodrama. To Walton's essentially retiring, dry, musical temperament there is added a superimposed rhetorical style, an accent that is truly operatic and which lies somewhere between Verdi and so-called *verismo*—realism. This was unexpected from a composer like Walton ...

Giulio Confalonieri of *La Patria* was more hostile.

In Walton's opera we find no melodic or rhythmic phrase or other characteristic which we could call original ... The external structure of *Troilus and Cressida* resembles Strauss at his worst; but a crowd of other models appears in the score ... The public evidently gave the work a rousing reception—but your critic hastens to say that mingled with the loud applause were hisses and catcalls and he is not sure that the opera really made a good impression.

Nothing more was heard of *Troilus and Cressida*, apart from a radio performance in Canada, with Jon Vickers as Troilus, until 23 April 1963, when it was revived at Covent Garden with Marie Collier as Cressida, André Turp as Troilus, Otakar Kraus as Diomede, John Lanigan as Pandarus, and Josephine Veasey as Evadne. The conductor was again Sargent, although Walton had pleaded with David Webster to find someone else. Walton cut about eight minutes of music. Collier was probably the best Cressida there has so far been. She and Walton recorded a brief extract from Act II, with the original Pandarus, Peter Pears. Sadly, Christopher Hassall died from a heart attack while running to catch a train to attend one of these 1963 perform-ances. Desmond Shawe-Taylor, in the *New Statesman* of 3 May 1963, remarked that, while in the original production 'every-thing was at least under control, on this occasion many of the exits and entrances were simply a means of getting on and off the stage. The lighting and the operation of various stage machineries were often haphazard and some of the singing sounded perilously like sight-reading.' Marie Collier sang Cressida, with Richard Lewis as Troilus, in seven performances at the Adelaide Festival in the spring of 1964.

The failure, for such it undoubtedly was, of *Troilus and Cressida* was a source of distress and discouragement to Walton. He felt it had not had a 'fair crack of the whip' from Covent Garden and he hoped for a new production that might cause judgements to be revised. In his highly self-critical way he also decided on revisions and began these in earnest in 1972, the year of his seventieth birthday. Covent Garden wanted to stage it in this year, but failed to find a suitable Cressida. With Hassall dead, Walton now had no compunction in altering parts of the libretto. He said: 'If Hassall had been alive, I would not have been allowed to re-write *Troilus*. Every time I took out a word or a note, there was a row. I think his style had been ruined by Ivor Novello.'[4] As previous chapters have shown, this was unfair. By no means was there 'a row' over every alteration to the libretto.

According to Walton's friend, the critic Gillian Widdicombe, she suggested the major feature of the revision to him over Christmas 1971. She had fallen and hurt her head while staying in Ischia. Together she and Walton played a tape-recording of the Covent Garden *Troilus* and some records Miss Widdicombe was reviewing, among them one of English songs sung by the mezzo-soprano Janet Baker. After a Quilter song she remarked: 'There's your Cressida'. She recounted:

This seemed such a peculiar suggestion—in the light of daily letters from Covent Garden in which dramatic sopranos who had just sung Lulu in Wales were hopefully mentioned—that for half a second Walton thought that brain damage had indeed been sustained ... Within five minutes he had decided that it would be very sensible to make the tessitura more accessible, and at the same time tidy up a few odd corners about which he already had reservations. The next time he came to London, he saw Baker on the stage for the first time—and was captivated by her emotional projection as much as her superlative vocal qualities. She accepted the idea gladly.[5]

[4] Interview in *The Times*, 12 Nov. 1976, p. 9.
[5] *Music and Musicians*, Nov. 1976, p. 35.

No doubt this is true, but the facts are that Walton had heard Janet Baker in July 1970 when she sang his *A Song for the Lord Mayor's Table*. Alan Frank wrote to Peter Hemmings, then general administrator of Scottish Opera, on 30 July 1970 to say that Walton had been 'bowled over by Janet Baker' and thought she would be the ideal Cressida. Would Scottish Opera consider a production? On 17 August of the same year, Walton wrote to Frank: 'I talked to Janet about T. & C. She was all for it, but I fear it is too high for her.' On 25 September he wrote to Diana Rix, one of the artists' managers at his agent's, Harold Holt Ltd: 'What about trying your charms on double-crossing [Peter] Diamand and get him to put on Tro. with Janet and the Scottish Opera for '72 [Edinburgh Festival]? The part won't need a great deal of readjustment for Janet. I did ask her to look at it and she may for all I know have rejected the idea altogether as not being suitable for her voice.' By December 1970 his favourite for the role was Anna Moffo. On 1 January 1972 he wrote to Frank: 'I've toyed with the idea of Janet Baker but it would entail an enormous amount of drudgery and expense to transpose many parts down a tone.'

Act II of the opera was the first to be revised and was performed in its new version at a Henry Wood Promenade Concert in September 1972, conducted by André Previn who, as conductor of the London Symphony Orchestra since 1968, had made something of a speciality of Walton's music and had recorded the First Symphony with outstanding success and the composer's warm approval. Casting for this performance, linked to the possible Covent Garden revival, had had its problems. On 10 December 1971 Walton wrote to Previn:

It was very good of you to try so valiantly to convince Leontyne Price about T & C, but I am not really surprised at her refusal. However, I see in the *Times* of the 9th that Placido Domingo is singing in *Tosca* with Gwyneth Jones at Cov. Gar.—an ideal couple for T & C. I've accordingly cabled Tooley[6] and written him and have also written Glock[7] to induce him to engage them for the Prom on Sept. 14. It might be a help in inducing them to take it on at C.G . . . I've been doing penance in the good cause by listening to the records of *Midsummer Marriage*. It has its moments, but not often. The musical comedy bits are the best, but not up to V. Youmans! The sentiments

[6] Sir John Tooley, who had succeeded Sir David Webster as general administrator of Covent Garden in 1970.

[7] Sir William Glock, BBC Controller of Music 1959–72.

are identical, but the music alas not. However it was all in the good
cause to hear the singers! I thought neither Remedios and Carlyle, or
Burrows and Harwood were really good enough, but of the two, if
driven to it (which we may be) Remedios and Carlyle are the best. I
think it is a marvellous idea about the Prom . . .

Ideas changed quickly over the next few weeks. On 9 February
1972 Walton wrote to Diana Rix:

About Troilus. [Heather] Harper has refused a bit doubtfully and Cov.
Gar. Tooley wants [Joan] Carlyle. I don't, nor does Previn. What about
getting in touch with Glock and putting forward Kiri te Kanawa? If
she's as good as she looks, she'll be marvellous. Glock might (or your
pal at the BBC) put her in for the Prom on Sept. 14 . . . I've a feeling
that they are trying to fob me off with anything at C.G. I'm probably
wrong.

Eventually Jill Gomez was chosen to sing Cressida at the Prom
performance, with Richard Lewis as Troilus. 'We have found
our Cressida for Covent Garden, Janet Baker,' Walton wrote to
Previn on 8 May, 'but I'm behind the time as to the Proms.
Gomez seems to be alright, but I'm wondering if she will want
the transpositions as much as Janet does. We are looking
forward to your coming. I am having the piano tuned so that
you can settle down to do eight hours a day!' When Walton
heard a tape-recording of the Prom performance of Act II, he
wrote to Previn: 'As usual, you've got it all right.'

Acts I and III were revised and cut during 1973. Changes
involved later in adjusting the role of Cressida to Baker's voice
were not numerous. Her Act I aria 'Slowly it all comes back' was
lowered by a whole tone as was her final aria 'Open the gates!' In
Act II 'At the haunted end of the day' went down a minor third to
G minor, to enable a greater application of tone-colour.

Otherwise the major changes were cuts, amounting to about
half an hour of music overall. In Act I the speaking part
(through a megaphone) of the Voice of the Oracle was deleted
and Calkas's long aria 'Launching at dead of night his frail
craft' was excised. Consequent changes to the libretto involved
a small amount of new music for Antenor on his first entry.
There were alterations to Cressida's music and words in the
second stanza of 'Slowly it all comes back', and one of Evadne's
lines was reallocated to her. Further small cuts were made in
Cressida's and the chorus's part at the news of Antenor's
capture.

In Act II Scene i the principal change was the elimination of the quartet of ladies in attendance on Cressida. This meant that the break-up of the supper party was considerably shortened and some of the ladies' music was reallocated to Cressida and Pandarus. The ladies' three-part serenade, 'Put off the serpent girdle', as they prepared Cressida for bed, was deleted (surviving as a separate publication). The vocal layout of Cressida's aria 'How can I sleep?' was slightly changed. Pandarus's 'Jealousy' aria was deleted, necessitating changes in Cressida's part when Troilus enters (now saying 'Enough of this scandalous nonsense' instead of 'Enough of this damnable lying'). Walton changed Pandarus's excuse for Troilus's arrival so late at night from 'On jealousy's hot grid he roasts alive' to 'On *desire's* hot grid': his explanation that 'there's nothing like a sharp misunderstanding' for bringing lovers together was changed to 'nothing like a little friendly cunning' which is more believable and more consistent with Pandarus's character. Troilus's aria 'If one last doubt' was shortened. In Scene ii, apart from changes to Cressida's tessitura, lowering it by two tones, the only revisions were a cut in Pandarus's solo when the Greeks arrive, cuts in the Troilus–Cressida duet, and a three-bar cut in the final orchestral postlude.

The off-stage trumpet part in Act III was simplified. Evadne's 'Night after night' was shortened and Cressida's recitative before 'O tranquil goddess' much abbreviated. Calkas's aria 'Cressid, daughter' was nearly halved, and there was a small cut in Cressida's 'You deathless gods'. The exchanges between Diomede and Cressida were substantially cut and Diomede's aria 'O Cressida let it be forever' lost its second stanza. Cressida's dialogue with Evadne, 'Will I please him', was shortened. Henceforward the revisions were minimal, but the stage business before the final curtain was simplified and the orchestral postlude shortened by a few bars.

Walton would have preferred the revised opera to be produced by English National Opera at the Coliseum, but its managing director Lord Harewood, not then an enthusiast for the work anyway,[8] persuaded him that Covent Garden had a

---

[8] In the synopsis for the opera which he wrote for the tenth edition of *Kobbé's Complete Opera Book* (London, 1987), 1131–4, Lord Harewood stated: 'Revival has suggested that the music's strong lyrical impulse, from the same source as that of the composer's Violin Concerto, puts it into a longer-lasting category than that of "well-made" play.'

priority. Once again, though, it must have seemed to Walton that the opera had a jinx on it. He himself was in poor health. The conductor, André Previn, developed bursitis a week before rehearsals were due to begin and withdrew. He was replaced by another American, Lawrence Foster, who had to learn the score. (One suggestion for a replacement Walton described as 'worse than a resurrected Malcolm Sargent'.) Alberto Remedios, who had been selected to sing Troilus, also backed out and EMI, who had agreed to record the opera 'live' during the six performances, refused to accept the Swedish tenor Gösta Winbergh as a replacement because he had no 'recording reputation'. Richard Cassilly eventually sang the role. Colin Graham, the producer, was informed by Covent Garden that it would authorize expenditure of only £15,000—the equivalent of the cost of renovating the original production, which was by now so derelict that it could not be used. With most of this paltry sum swallowed up by the cost of costumes, the scenery had to be devised from borrowings from other productions and what furniture could be salvaged from the old *Troilus*. (In the same season Covent Garden spent £150,000 on a new production of Puccini's *La Fanciulla del West*.)

The first performance was on 12 November 1976, with an all English-speaking cast headed by Baker and Cassilly, with Richard Van Allan as Calkas, Gerald English as Pandarus, Benjamin Luxon as Diomede, and Elizabeth Bainbridge as Evadne. The critics were still lukewarm rather than enthusiastic, while some, such as Bayan Northcott in the *Sunday Telegraph*, pulled no punches: 'Have I got to be frank?' he wrote on 21 November. 'Very well . . . there were moments during the evening of 12 November when I caught myself wondering why on earth Covent Garden was bothering itself at all with such a hopeless old dodo as *Troilus and Cressida*.' But there were full houses for the series of performances and much enthusiasm. This made Walton happy.

Walton's attitude to the 1976 version of his opera was unequivocally stated in his conversation with Arthur Jacobs in November 1977. Asked if he thought there were now two versions of the work, he replied: 'Oh no, that's a final alteration, a final thing. I hope they aren't going to start mucking about later on and do two versions.' Jacobs asked why he didn't write it for a mezzo originally. 'Because there weren't any I knew at the time. I felt Janet Baker was the right kind of temperament for it, more than other people I heard sing it.' 'And you don't want people to start making a choice between Version 1 and Version 2?' 'They won't be able to. They'll have to have the orchestral part rescored by somebody else. It won't be me.'

The recording was issued in the latter part of 1977 and Walton sent a set to Roy Douglas for his seventieth birthday— 'It is I may tell you much better to be 70 than 75 which doesn't suit me at all in all sorts of ways, mostly unpleasant . . . I would like to thank you for the help you have been when I was harried by film producers etc. As ever affectionately.' In February 1978 Walton replied to a letter from Douglas:

I quite agree with all you say of the recording of T. & C. Personally I still would like to cut even more, but it was difficult enough to do, not having Chris Hassall to help cutting the libretto. He disapproved madly about cutting any of it & I suppose we should have had much fighting over it. In fact both the first and the last acts need re-modelling and re-writing, but it was beyond me to do both on my own, so I did the best I could with it. It is, to me at any rate, much better than it was originally, and I don't miss at all 'Put off' & 'Who will pick up', in fact they had both been in my mind for years as being the bits I'd cut first. Actually the best perf. of it was of the second Act at the Proms a couple of seasons ago [actually six], with Richard Lewis (a bit old but alright nevertheless) & André Previn conducting. It was a great blow that A.P. was too ill to conduct at Cov. Gar. but Foster did well under the circs. But the whole business was for me a great trial— in fact it made me ill.

'If my aim here', Walton wrote in a note accompanying the 1955 recording of extracts from *Troilus and Cressida*, 'was a close union of poetic and music drama, it was also my concern to recreate the characters in my own idiom as an example of English *bel canto*, the parts carefully designed to bring out the potentialities of each voice according to its range—in the hope of adding another "singers' opera" to the repertory.'

Did he succeed? If at all, only partially. Much as one wants to like and admire *Troilus and Cressida*, repeated hearings afforded by the complete recording of the revised version, combined with experience of seeing both the original and later Covent Garden productions, confirm a reluctant opinion that the opera is a failure. Not, I hasten to add, because it is written in an unfashionably romantic idiom. 'Unfashionably' is a relative term, since in 1988 neo-romanticism is itself the height of fashion. No, the fault lies first and foremost with the choice of subject. Troilus and Cressida never began to be great tragic lovers like Tristan and Isolde, nor like Radames and Aida. Alice Wimborne was right, at the very beginning, to be 'not entirely happy' about the choice of subject and to be concerned that 'there are so many things to think of if the luckless lovers are to live and move the hearts of an operatic audience'. The Walton–Hassall Troilus and Cressida do not move our hearts, in fact it is difficult to care much at all about their fate. Neither is well drawn. Troilus is no heroic figure—only a 'wimp' would, at the end of Act II, so feebly consent to Cressida's exchange: 'We cannot, dare not disobey.' Cressida is never the Manon figure to whom Walton was initially attracted. Even such a sympathetic portrayal as Janet Baker's could not arouse much sympathy for her plight. In any case Walton made a cardinal error in transposing the part for a mezzo. If Cressida is to have any credibility, it must be as the dramatic soprano of his original conception.

Musically the opera is disturbingly uneven. This can safely be attributed to Walton's dissatisfaction with the libretto. For all his work on it, it remained a flowery, self-consciously poeticizing effort. Newman's final verdict that it was 'the best poetic opera text since Hugo von Hofmannsthal' is a preposterous statement, as can very soon be gathered if one considers lines such as these, uttered by Troilus:

> I bring you life that withers like a rose,
> but while it blooms the glory overflows.

Neither composer nor librettist had previous operatic experience and both had painfully to feel their way towards their goal. Walton had a much clearer idea of how to attain it, but even he, when he started work, had only two choral works to his name and scarcely any songs. Compare this with Britten whose *Peter Grimes*, though it too had libretto troubles, was preceded by *Paul Bunyan* and several song-cycles. What about Puccini, then? He did not need to apprentice himself to opera by writing song-cycles. But he grew up in an operatic environment and he wrote tentative operas before he created a

real masterpiece. Walton was expected to produce a masterpiece at the first attempt, as he had done with symphony and
cantata. Nor was he helped by those of his admirers like Frank
Howes who compared *Troilus and Cressida* with Wagner's *Tristan
und Isolde*. Its milieu is that of Italian opera, Puccini and Verdi,
but, though much is made of Walton's romanticism, the trouble
with *Troilus and Cressida* is that it is not romantic enough. It is
not full-blooded. Walton did not go the whole hog: reticence
and restraint lead to some beautiful music but not to com-
pelling theatre. But that it is the opera he wanted to write
cannot be doubted. What may be speculated is what sort of
opera he might have written if he had collaborated with W. H.
Auden, who would perhaps have drawn a less conservative
musical result from him.

The loveliest episodes of *Troilus and Cressida* are certainly
memorable, if few and far between. The recurrent theme for
Cressida, used as a *leitmotif* for the love-token of her crimson
scarf, was described by William Mann in *The Times* in 1976 as
'the most beautiful and haunting love-theme Walton has ever
invented'

Ex. 11

and few would disagree with that judgement. Cressida's arias
'Slowly it all comes back' and 'At the haunted end of the day'

Ex. 12

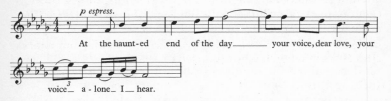

At the haunt-ed end of the day____ your voice, dear love, your
voice____ a - lone_ I __ hear.

have a pervasive melodic melancholy that grips the heart of the
listener—although, ultimately, not as unforgettable as the
comparable music of the Violin Concerto—and, while Troilus's
'Child of the wine-dark wave' in Act I is feeble, his 'Sooner
would leafless boughs' in Act III has the ring of genuine
operatic love-music. The Act II love duet, on which Walton
placed so many hopes, is convincing enough from 'If one last
doubt' onwards, but it is the orchestral interlude that remains
in the listener's mind and this is excelled by the touching 'O
tranquil goddess' section of Cressida's 'No answering sign' at
the start of Act III as she awaits a word from Troilus.

These all represent Walton in his strongest lyrical vein and
one wonders if perhaps Cressida is Walton's operatic portrait of
Imma. Some of the lighter passages, too, are remarkably
effective, such as the delightful 'Then it's time we returned to
rest' in Act II. Pandarus's music has been highly praised, but
with repetition wears very thin. Pandarus is a mincing,
probably homosexual, epicurean such as the late Max Adrian
would have effectively played in the theatre, but Walton's
music for him, with its tedious repetitions and melismata, is an
obvious parody of Britten that is not sufficiently acute. In the
'pomposo' episode in Act III, the Greeks enter to a Waltonian
film-music march, a sad let-down at a crucial point, and at two
other important junctures his inspiration fails him: the sextet
never really catches fire (to compare it with either the
Meistersinger quintet or the *Rosenkavalier* trio is ridiculous) and
Cressida's final aria, where a great melody is required, is
commonplace. Apart from Pandarus, none of the other characters is well drawn. Diomede is cardboard, Calkas much the
same, and Evadne just a device for the plot.

In the colour and delicacy of its orchestral writing, *Troilus and Cressida* showed that Walton's hand had lost none of its cunning, even gained in the skills of poetic restraint. But as an attempt to inject new life into the nineteenth-century conception of opera as a conflict between the public and private lives of heroic protagonists, it was less successful than Samuel Barber's *Antony and Cleopatra*, which met with comparable critical scorn when it was produced in 1966.

20

Cello Concerto, 1955–7

T‌HE reason Laurence Olivier withdrew from his intention of producing *Troilus and Cressida* became evident in May 1954 when he announced that he was to make a film of Shakespeare's *Richard III*, a part he had played on stage with conspicuous success. Walton had agreed to compose the music, which he wrote in some haste during February and March 1955, having been ordered not to produce a score until the recording of the music was being undertaken at Shepperton Studios. He enjoyed working with Olivier.

I did learn some facility with film music [he said in 1962], but hated writing it until I came to work with Larry Olivier. The thing about film music is that it needn't necessarily be good or bad, only it must fit. They all say the great characteristic is that it shouldn't be heard . . . There was one patch in *Richard III* that I just couldn't face—only some two minutes. Well, I filled it in somehow. Larry was furious. 'You really have played it down this time', he said. But the engineers said it was just what was wanted.[1]

The *Richard III* score is the least impressive of the three Shakespeare films, although the Princes in the Tower music is another of his touching and delicate miniatures. Walton felt there was a limit to the number of ceremonial fanfares and battle charges he could manufacture for these historical chronicles and he wrote on the score the instruction 'Con prosciutto, agnello e confitura di fragole' (with ham, lamb and strawberry jam). This applied to the music for the film's titles, a Coronation march for Edward IV, later published (in an arrangement by Muir Mathieson) as a separate piece, *Prelude, Richard III*. Mathieson also arranged six items of the music as *A Shakespeare Suite* for orchestra. Nos. 6, 5 and 2 of the Suite—

---

[1] Interview in the *Sunday Times*, 25 Mar. 1962, p. 39.

'Trumpets Sound', 'I would I knew thy heart' and 'Music Plays'—were also published as *Three Pieces for Organ* (March, Elegy and Scherzetto). These were originally written for organ and were played thus in the film, having been recorded in Denham Church. There was to have been another Shakespeare-Olivier–Walton film, *Macbeth*, in 1958, for which Walton was prepared to adapt and expand the music he had written for Gielgud in 1942, but Olivier could not raise the money for it. Walton set aside time for composing the score during 1958–9.

Richard III apart, Walton had no major composition on his desk in 1955. He received honorary doctorates of music from the Universities of Cambridge and London and made arrangements of the British and United States national anthems for the Philharmonia Orchestra's American tour under Herbert von Karajan in October and November 1955. The flamboyant arrangement of 'God Save the Queen' had its first performance at an 'eve of departure' concert in London on 18 October. According to Stephen J. Pettitt, historian of the orchestra, 'it featured the horn section very prominently, and at a signal from Dennis Brain they raised the bells of their instruments in the air for maximum effect. It was a popular arrangement and was still pasted inside the music folders ten years later.'[2] The arrangement of 'The Star-Spangled Banner' is said never to have been played.

While in Milan for the première of *Troilus and Cressida* in January 1956, Walton received a commission for an orchestral work from the director of a festival which was to mark the seventieth anniversary of the city of Johannesburg in the autumn of that year. He wrote the *Johannesburg Festival Overture* between February and 31 May 1956, having sent to the African Music Society for a selection of recordings of African music. The six-minute overture is in an elaborate rondo form and is fast throughout—'presto capriccio' becomes 'vivacissimo' and finally 'prestissimo'. It has been described, with some justification, as a modern Rossini overture and the device of a sudden change of key, as in the march in *Belshazzar's Feast*, is a brilliant *coup* that always takes the listener by delighted surprise. Its large orchestra includes maracas and rumba sticks, and the

---

[2] S. J. Pettitt, *Philharmonia Orchestra, A Record of Achievement, 1945–1985* (London, 1985), p. 78.

snatches of African melodies, one of them identified as by the Congo guitarist Mwenda Jean Bosco, occur in an episode where the horns are accompanied by the exotic percussion instruments. Sargent conducted the South African Broadcasting Corporation Symphony Orchestra in the first performance in Johannesburg City Hall on 25 September 1956. Some of the South African press were quick to make the point that 'a local composer would have written something more indigenous . . . at half Walton's fee'.[3] It was introduced to Britain in a Liverpool Philharmonic concert on 13 November, conducted by Efrem Kurtz. But a more significant Liverpool connection with Walton occurred in this month: the Liverpool Philharmonic Society commissioned a second symphony to celebrate the 750th anniversary of the granting of the charter of incorporation to the borough. A first performance was planned for some time in the anniversary year, 1957.

In a curtain-speech at the end of the Sadler's Wells Ballet's 1954–5 season at Covent Garden, the artistic director, Ninette de Valois, announced there would be a new three-act ballet by Ashton and Walton in the following year. She did not mention a subject, but it was to have been *Macbeth*, for which Ashton produced a detailed scenario.[4] However, the project foundered, not least because Margot Fonteyn did not warm to the role of Lady Macbeth. Nor was she attracted by the idea of Miranda in *The Tempest*. Ashton then settled for the story of the water sprite Ondine. During 1956 he invited Walton to write the music, but, having another commission under way, Walton suggested Hans Werner Henze, whose *Undine* was produced as a result in 1957. Walton had first met Henze at Auden's house on Ischia in 1953 and a friendship developed. Henze went to live in Ischia shortly afterwards.

The commission—$3,000—with which Walton was engaged throughout 1956 was a cello concerto for Gregor Piatigorsky, who admired the Viola and Violin Concertos and deputed the pianist Ivor Newton to approach Walton on his behalf. Walton's reply to Newton was, 'I'm a composer. I'll write

---

[3] *Forward*, Johannesburg, 5 Oct. 1956.
[4] Ashton's full synopsis and his instructions to Walton for the score may be read in David Vaughan's *Frederick Ashton and his Ballets* (London, 1977), pp. 434–8.

anything for anybody if he pays me . . . Naturally, I write much better if I'm paid in dollars'.[5] When Piatigorsky met Walton he said he had 'never met such a rare combination of greatness and simplicity'. The Cello Concerto, Walton's first large-scale concert-hall work for eighteen years, was chiefly composed between February and October 1956, relatively fast for Walton. But a letter to Roy Douglas dated 10 August 1955 about some revisions for *Troilus and Cressida* includes the sentence: 'Have got the Vlc. Conc. under way but that is only just. Rather a good opening'. There was a continual exchange of letters between Ischia and the cellist's home in California or wherever Piatigorsky was playing, such as Japan. As always, there were problems, delays, and doubts. When Piatigorsky cabled his acceptance of the completed work, Walton wrote on 4 November 1956 to 'dear Grisha' that he was

so happy that you should think the whole work wonderful. It is to my mind the best of my, now three, concertos. But don't say so to Jascha . . . I must thank you in the first place for having commissioned the work, but more so for the patience with me in my darker moments, and some were very dark indeed . . . I only hope it will come up to your expectations when you come to play it with the orchestra. If anything in the orchestration—that vibraphone, for instance—should irk you, just cut it out, because it's not absolutely essential (though I might miss it).

Then, as Walton wrote to Alan Frank on Christmas Day, 'P. wrote that he was not happy about the end, or rather Heifetz to whom he had been playing the concerto wasn't. Accordingly, as I always do what I'm told, I've written two new endings and leave them to choose. Both are I think better than the original.' But Piatigorsky reverted to the original ending.

The first performance was scheduled for Boston, with Charles Munch conducting, on 7 December 1956, but Piatigorsky was ill and it was postponed until 25 January 1957. The first London performance was less than three weeks later, at a Royal Philharmonic Society concert on 13 February, with Sargent conducting the BBC Symphony Orchestra. Queen Elizabeth the Queen Mother attended and the concert was

[5] I. Newton, *At the Piano: The World of an Accompanist* (London, 1966), pp. 128–9.

televised by the the BBC. Walton planned to attend the London première, but on his way he and his wife were injured in a car crash in Rome. Walton broke a hip and was in hospital for three months. On 11 February Roy Douglas spent two hours with Piatigorsky at the Savoy Hotel discussing alterations to the solo cello part; Walton's approval was required and was received later that night by telephone from the hospital. Tapes of the Boston and London performances were sent to Walton and he wrote to Piatigorsky with some suggestions: 'I do so hate asking you to do this, and I know you won't think it is because I don't appreciate your playing of the work as a whole; but it is just in these parts that the performance could be tightened up. While I, of course, don't expect you to adhere rigidly to my tempi, I do feel the discrepancy of your timings and mine is a little too much.' He wanted the interpretation to be 'altogether more tough and rhythmical'. When Piatigorsky's recording was issued in 1958, Walton thanked him for 'an absolutely superb interpretation and performance. Everything about it is just as it should be, and your playing magnificent!' He wrote to Alan Frank on 10 July 1958: 'I must say the recording is pretty superb both as a performance and as a work—and I say it who shouldn't—and to hell with Messrs. Heyworth and Mitchell [the critics Peter Heyworth and Donald Mitchell]. They should, by the way, be both blacklisted so don't send them *any* scores of my works!'

There is an intriguing postscript. At Christmas 1974, to encourage Piatigorsky to return to the concert platform after his long illness, Walton wrote another new (twenty-three-bar) ending for the concerto (starting five bars before rehearsal cue 19 on p. 112 of the printed score). There had been a long exchange of letters, Piatigorsky wanting 'a less melancholy ending'. This in itself was ironic, since in his original instructions to Walton in 1955, quoted by Ivor Newton, he had asked that 'the ending should be brilliant and the coda fairly long; all the existing concertos end abruptly, even the Dvořák'. Piatigorsky approved the revision in 1975, whereupon Walton wrote: 'I don't know why it bothered me so much. I'm losing confidence in my powers and find composing increasingly difficult. I never found it easy but now it is the very devil.' Piatigorsky never played again, for he died of cancer in August

1976. The revised ending has never been performed. In his 1977 conversation with Arthur Jacobs, Walton recalled Piatigorsky's original doubts about the end, 'but I didn't have any till many years later. Then I thought "Perhaps he's right".'

Critical reaction to the concerto in Britain was mixed, but generally favourable. Desmond Shawe-Taylor wrote of 'a major Walton work. . . new both in form and spirit . . . The work is curiously one and indivisible from beginning to end.' Colin Mason, who had his doubts about the Violin Concerto, declared the new work to be 'entrancing, a work of a man refreshed in spirit.' To others, notably Peter Heyworth in *The Observer*, it was final proof that the *enfant terrible* of the 1920s— was he ever that?—had become a pillar of the musical Establishment. Peter J. Pirie, writing in 1972, described the concerto as 'insipid, and few people coming to it in ignorance could identify its composer at all'—a statement that cannot be taken seriously whatever one may think of the music. In a thoughtful article in *The Listener* on the eve of the concerto's first London performance, Donald Mitchell cited Adorno's theory that *Angst*, 'the element of anxiety', was the distinguishing feature of the 'heroic epoch' of modern music—Stravinsky, Bartók, Schoenberg, Berg, and Webern—and that this element had, by 1957, been repressed. 'Tension has relaxed and imagination grown feebler . . . Afraid of being afraid, modern music is allowing itself to grow old.' Walton's Cello Concerto, Mitchell thought, was an example of an idiom growing old.

Sargent sprang to Walton's defence. He was incensed by Peter Heyworth's statement in *The Observer* that 'there is something rather alarming about the stagnant quality of Walton's recent music.' 'This statement', Sargent said in a letter to the editor, had 'no justification whatsoever.' Walton was 'discarding the superfluous and growing in sincerity. I wish the same could be said of more of our modern composers.' He challenged Heyworth's notions of composers' development by inviting him to place Beethoven's symphonies in chronological order as if he did not already know it—'No. 4 and No. 8 would surely come before No. 3 and so on.' Heyworth replied that 'Beethoven is hardly a happy example of a composer who did not in each work extend his idiom . . . I never suggested that a composer has to develop systematically in a straight line.

Walton's recent music seems to me to show no vital development at all.'

Walton found himself out of joint with the times in 1956 as
he had done in 1945. It was a time of immense change and
transition in the British musical outlook in influential circles.
There was a reaction against Vaughan Williams, now with only
two years left to live, and his followers, who were regarded as
having made England into a parochial musical country, turning
in on itself and ignoring the developments of the previous fifty
years. The cold-shouldering of Schoenberg and the Second
Viennese School had gone on long enough, it was decided.
Serialism was the preoccupation of the young. Maxwell
Davies, Goehr, and Birtwistle were clamouring to be heard; so
were Boulez, Berio, and Nono. With the appointment of
William Glock as Controller of Music at the BBC in 1959, a
more catholic approach to the repertoire was encouraged.
Britten was unaffected, but many composers, like Rubbra, Bax,
Ireland, and others, found themselves out in the cold. Walton
was among them. Yet 1960 was the centenary of Mahler's birth
and ironically precipitated over the next decade a period of
interest in the post-Wagnerian romantics, including Elgar. The
eventual result was a much more open-minded attitude to
compositions of all schools and temperaments than had ever
before existed. Gone was the day when a critic would say, as
Eric Blom had said: 'We don't want Mahler here.' So no one
now would say: 'We don't want Boulez here.' And in due
course Walton and his like were admitted back into the fold.
But it took twenty painful years for this reversal of trends to be
accomplished.

Out of this general situation arose a personal relationship
which reflects high credit on the two men concerned, Walton
and Peter Heyworth. The latter had been appointed chief music
critic of *The Observer* in 1955. He allied himself with those who
wanted to see a broadening of taste and a widening of the
repertoire. This was the period of the 'angry young man', a term
applied to John Osborne, author of the play *Look Back in Anger*,
which epitomized a generation in revolt against the shibbol-
eths of the past. Heyworth, in his quiet cultured way, was an
angry young critic. It enraged him that the musical Establishment—itself a term which had just come into vogue—still

regarded Schoenberg's *Five Orchestral Pieces* as decadent and music of the 'sick room', that Bruckner was still written off as an imbecile, Mahler barely tolerated, Stravinsky dismissed as a composer who changed his style with every work, and Vaughan Williams and Walton regarded as the apex of advancement where British music was concerned. It was a complacent attitude, but it also had a nasty side, such as the barely disguised hostility to, and only grudging acceptance of, the genius of Britten. In November 1955 Heyworth wrote a disparaging review of an all-Walton concert at the Royal Festival Hall which Walton conducted. In it he said: 'I confess to finding little that is of interest in either the Violin Concerto or the Symphony; that is not because they are not "progressive" (horrible word) but because to my ears they contain little distinction of idiom.' The symphony was 'in some ways an imposing work and the scherzo has real impetus and rhythmic élan', but its Finale was deplorably weak. The concerto Heyworth found 'permeated by a pale lyricism dangerously close to a cosmopolitan version of the pastoral meandering which has sapped the roots of so much English music.' Two birds with one stone! Heyworth felt that Walton was a classic case of a traditional composer who lacked a tradition. Walton had only Elgar as a traditional model, since he was not inclined to the folk-song school. Yet a composer of Walton's generation in Austria had a whole and continuous tradition within which to manœuvre his style.

Walton was aghast at Heyworth's strictures and amazed by them, too, for this was the first sustained assault on his music which singled out some of the pre–1939 works as targets. Heyworth then went to Milan in January 1956 for the Scala production of *Troilus and Cressida*. He did not like the work but thought it was 'magnificently conducted by Nino Sanzogno' and that Sargent's Covent Garden reading appeared in retrospect as 'unpleasantly glib.' Afterwards he met Walton at the lavish party given by one of Puccini's descendants and found that they 'clicked.' Walton did not like his views, naturally, but he found himself rather liking the man. Then came the Cello Concerto. Heyworth's *Observer* notice began by remarking that the concerto contained little that would have startled an audience in the year that the *Titanic* met its iceberg. But he found it 'more personal in accent and less strained in manner than *Troilus and Cressida*, where an attempt to work on an imposing scale lured the composer into borrowing too many *grandes toilettes* for the occasion.' Nevertheless there was 'something rather run-down and enervating about the work as a whole'. Then followed the remark about 'stagnant quality' which has already been quoted and Heyworth concluded by using the works of Walton as a stick with which to beat the critic of *The Times* (Frank Howes) and his followers: 'There is something uncomfortably Canute-like in the manner in which they are used by musical orthodoxy as a refuge from the fierce currents of our time. Certainly this blinkered outlook is in striking contrast to Sir William's own open-minded interest in present-day music.' Walton's friends made sure that his convalescence after his car accident was enlivened by a ferocious anti-Heyworth campaign. On his next visit to London, walking with crutches, he went with Alan Frank to the Wigmore Hall. Seeing the assembled critics, he demanded: 'Where's that bugger Heyworth?' and brandished his crutch threateningly. But he was incapable of sustaining real hostility and a friendship developed to the extent that in December 1958 he wrote to Alan Frank: 'By the way, Peter H. and I had a grand rapprochement during my last visit to London and we are now buddies. Though I expect I even asked him to continue to be rude about me in the press!' Protesting to Heyworth about his views, Walton said to him on one occasion: 'You know, you don't only attack my music, you reduce my income!'

Two events drew them together during 1958. They hatched a scheme for a series of concerts of modern music in which they tried to interest Granada Television as sponsor, and Walton did his best to effect a reconciliation between Heyworth and Hans Werne Henze, who had taken umbrage over an *Observer* review. 'Sorry about Hans', Walton wrote to Heyworth on 22 November 1958, 'but after your experience with me you ought to know how quite impossible composers are! I must say I thought it a mild and reasonable notice—one I should have been delighted to have! I expect by now he's back in Naples & hope you will come to see us when I will pour the requisite amount of soothing oil, I hope with due effect.' Later, Walton invited Heyworth to Ischia to try to mend the rift. When Henze continued to be unco-operative, Walton remarked: 'Some people don't know when they're lucky!'

The Cello Concerto is written to a formula similar to its predecessors, with the first movement slow and reflective. Its long opening theme, comprehensive and chromatic, with ambiguously unstable tonality, is a seductively amorous invention:

Ex. 13

Yet for all the vacillation between major and minor and the pull of G against A flat, there is a relaxation of tension here compared with the opening theme of the Viola Concerto. The melody is drenched in melancholy but it is also drenched in Italian sunshine. It is lazily content to repeat itself rather than to undergo development. But it is also a magnificent structural feature, with its rising intervals, dotted-note figure, and firm rhythmic outline. All the other themes in the concerto are derived from these elements. A second subject makes particular use of a descending scale of semiquavers, and towards the end of the movement the main theme reappears in passionately expressive guise, scored for high strings and richly harmonized.

The Scherzo is a typically Waltonian firework display, especially notable for the lightness and transparency of the orchestral scoring (as indeed is the whole work: the soloist is never obscured and the percussion, which includes celesta and vibraphone, is discreetly and colourfully used). There are resemblances to the music for Pandarus in *Troilus and Cressida*; and the alternation of two quick and lively bars with two of slow song-like melody is a device that has been justly described as operatic.

The Finale is a theme and improvisations. It reverts to the mood of the opening movement and the theme, stated by the cello in its high register over a pizzicato on the orchestral strings, is elaborately constructed. A passage for solo cello links the theme to the first variation in which the cello's semiquavers are accompanied by a shimmer of tremolando strings and the exotic interventions of xylophone, vibraphone, celesta, and harp. The second improvisation is for cello alone, a bravura affair doing duty for a cadenza, the third an energetic orchestral toccata. Cello alone has the fourth variation, rhapsodic, expansive, and ending with high trills which merge into the epilogue, where the richly harmonized episode from the first movement recurs and the Finale's principal subject is heard again in its original form. It is a stony heart that cannot respond to the urgings of this work's coda.

Cellists have been quick to add the concerto to their repertoire. It is beautifully written, grateful to play, and presents the listener with few problems. But there are few challenges too, and one is bound to concede that the concerto, for all its incidental beauties, is often too relaxed for its own good, too episodic (the start–stop of the Finale can grow wearisome), and inclined to droop into mannerisms when Walton retraces old melodic patterns which have lost for him the zest associated with new inventions. It is indulgent music; and if in most performances one is only too willing to listen indulgently, that is the measure of Walton's superb craftsmanship and his ability to spin out a web of sound with all his old skill, if not with all his old flair. For a composer with a restricted idiom, it was perhaps dangerous to write concertos for three stringed instruments. That Walton never composed a woodwind or horn concerto is a matter for everlasting regret.

21

Partita and Second Symphony, 1957–60

TOWARDS the end of 1955 Walton was one of ten[1] composers
invited to write works for the fortieth anniversary season of the
Cleveland Orchestra, of which the conductor was Georg Szell,
who had for many years been a champion of his music. The
result was the *Partita*, completed towards the end of 1957 and
one of the unqualified successes among his post–1945 works, a
jeu d'esprit which is witty, brilliant, and, in its central 'Pastorale
Siciliana', delicate and touching.

Although Walton is often regarded as a 'slow' composer, he
was businesslike and methodical in his attitude to his work
and planned far ahead, as is demonstrated by a letter to Alan
Frank on 9 November 1956, in which he detailed his schedule:

Flute Con. for Mrs K. [Elaine Schaffer, wife of the conductor Efrem
Kurtz]. Very remote at the moment and not to be thought of till
everything else I've on hand is out of the way. Liverpool [Second
Symphony] . . . we talked of 57/58 season. I shall aim for that . . . The
Cleveland work I doubt finishing by Feb. as we leave here mid-Jan.
and shall be in London most of Feb. . . . I can then get started on the
Symphony . . . Other works I plan to do in the next couple of years or
so are a Sinfonietta for Birmingham . . . Variations on a theme by
Hindemith (for H. I've long promised him a piece—the trouble of
course is to find the theme. In the end I shall probably have to think of
one of my own). A Sonata for Strings for no one in particular. Fl. Con.
if sufficiently induced. After that I could (having 7 or 8 comparatively
new works to keep one going) take time off for another opera if Chris
has found a suitable subject by then. Of course all one's plans will be
blown sky high if—if—! Anyhow that is the general overall plan and I
don't propose to take too long about carrying it out.

A week later he asked Frank to send him scores of Roussel's

---

[1] The other nine were Gottfried von Einem, Bohuslav Martinů, Henri
Dutilleux, Boris Blacher, and the Americans Alvin Etler, Paul Creston, Howard
Hanson, Peter Mennin, and Robert Moevs.

Third and Fourth Symphonies and Suite in F, Berlioz's *Roméo et Juliette* and *Nuits d'été*, Prokofiev's Sixth and Seventh Symphonies, and Mahler's Ninth Symphony and *Das Lied von der Erde*. A few months later, in response to a suggestion by the impresario Ian Hunter, he added to his list a double concerto for Heifetz and Piatigorsky, though he told Alan Frank: 'Starting from Jo'burg it is 10th! . . . Of course a double con., however spectacular the kick-off might be with such stars, has not much of a chance in the ordinary way. How often is the Brahms played? Not very often and one could hardly expect to equal that. However if the spirit was moved and the dollars jingled who knows?'

The first delay to his plans was caused by his car accident and it was not until 8 April 1957 that he could write to Frank: 'I've started a few bars of the *Partita* and hope to swiftly gather momentum.' On 21 October he wrote:

This morning I dispatched the last pages of the P. When you have the chance to see it as a whole, I think you will agree with me (I refer to III) that I have been sailing far too near to the wind, in fact one could say perhaps, that one has G.T.F. [gone too far]. However, if it does come off, it should make a rousing and diverting finish to the work. It is meant to divert, and also to annoy, and I shall be intensely disappointed if I get a kind word from either P.H. [Peter Heyworth] and D.M. [Donald Mitchell] or anybody else. 'Vulgar without being funny' (in the words of the late Sir. G. Sitwell in reply to some inane remark of Osbert) is the best I can hope for. Incidentally the lyric that goes with the middle tune (A maj.) is not, as you might at first think, 'The flowers that bloom etc.' but (if you should have to dispute it with anyone) 'There was a young woman of Gloster, whose parents thought they had lost her' etc.

And now I have to look a 2nd Symph. in the face! I shan't do that, at any rate, for the next 10 days as our case comes up on the 29th [a suit for damages following the car accident]. If we win, it should be a pretty hilarious one, if not, full of the popular 'angst' and gloom. Apropos the symph., the other night we had dinner with some people who included an Italian painter called Pagliaccio (I need hardly add he's known as Rusticana) and his American wife (who I need again hardly add is in fact a Russian). After dinner she, through the aid of a tea-cup, got in touch with the spirits and to her guide in particular and invited us to ask questions. My questions were these—Q. Shall I finish my new S. by April 1st 58? A. No. Q. When will it be finished? A. June. Which seems to me a fairly accurate estimate. Q. Why can't I finish by April? A. Because of difficulty with the last movement (typical!) but you could finish by April if you cheat and so throw it away. Q. Will it be better than the 1st? A. Yes indeed. Q. Will my not finishing by April cause a lot of trouble etc.? A. Not at all—don't worry. I then switched to T. & C. Q. When will T. & C. be revived? A. Never (which I have been suspecting for some time to be the truth). However it continued to say not to bother as in a few years' time I should produce a real masterpiece! Then I asked Q. Shall we win our case? A. Yes. So one can have a better idea by the 29th as to how to judge his other answers. [They did win]. One more question—Q. Have I also got a guide? A. Yes and her name's Salome! Next time I get through I shall ask her what she thinks of Richard Strauss! . . .

The Waltons' case was heard in Civitavecchia. On leaving there they went to stay for two weeks with Osbert Sitwell at Montegufoni. 'He was quite keen on doing a libretto for the oratorio for Uddersfield', Walton told Alan Frank on 7 November. This was one of the first mentions of a commission for the Huddersfield Choral Society's 125th anniversary in 1961. He added: 'I feel a "classical" approach to the Symph. is the right one, so would you send me some Haydn or Mozart scores. I leave it to you what—the M. G minor, etc., say 1/2 a doz. of each!'

Walton contributed a programme-note for the first performance of the *Partita* in Cleveland on 30 January 1958 which reflects the spirit of the work:

It is surely easier to write about a piece of creative work if there is something problematical about it. Indeed—so it seems to me—the more problematical, the greater the flow of words. Unfortunately from this point of view, my *Partita* poses no problems, has no ulterior motives or meaning behind it, and makes no attempt to ponder the imponderables. I have written it in the hope that it may be enjoyed straight off, without any preliminary probing into the score. I have also written it with the wonderful players of the Cleveland Orchestra in mind, hoping that they may enjoy playing it . . .

The critic of the *Cleveland Plain Dealer* received the work in the spirit it was offered: 'It is masterfully orchestrated and concise in form . . . it is filled with boisterous blasts of colour and sturdy uninhibited tunes . . .'

The *Partita* was first played in England during another orchestra's anniversary celebrations. The Hallé had had its hundredth birthday on 30 January and its conductor, Sir John Barbirolli, had invited several British composers to conduct their own works. Walton conducted the *Partita* in Manchester on 30 April and Sir John its first London performance two days later. It has few rivals among English compositions as a short *divertissement* for virtuosos only, and it is Walton at his most relaxed. Its opening Toccata is a new etching of *Portsmouth Point*. It is no longer Rowlandson's, more like Le Corbusier's. The high-rise flats have gone up on the waterfront, the quayside tavern has a chromium-plated bar, but Scapino is a regular customer. The Pastorale Siciliana is time for siesta, with wistful romantic dreams and languorous looks at Daisy and Lily, lazy and silly, while the Giga Burlesca has a deliciously vulgar tune,

Ex. 14

as Walton intended, almost as good as that in the Finale of the *Sinfonia Concertante*. The orchestration is as apt and sparkling as the craftsmanship is impeccable, and the flame of inspiration burns high. The *Partita* is not a work about which to be patronizing or condescending: it is music in which Walton's genius for burlesque in a lyrical framework may be heard at its most highly developed and captivating, the whole work, it has been suggested, being a brilliantly creative paraphrase of Roussel's Suite in F.

From 19 May 1956 Walton had officially been designated for tax purposes as a British citizen resident wholly abroad. He sold Lowndes Cottage, which he had rented for two years to the Oliviers, and invested the proceeds in building a hilltop house at Forio on land he had bought. This was to be La

Mortella, his home for the last twenty-one years of his life. Building began in November 1961 and the Waltons took possession in August 1962. Susana created a magnificent garden. Because he lived among them, the Italians were interested in Walton's music even if they did not like it much when they heard it. Thomas Schippers conducted the first performance of *Belshazzar's Feast* in Italy in Rome in April 1959. Walton suggested that the Santa Cecilia Chorus should sing it in English because he had been so impressed by the impeccable diction of the Radio Audizione Italiana Chorus when he had heard a broadcast in 1957 of the first Rome performance of Elgar's *The Dream of Gerontius* conducted by Barbirolli. In that broadcast, too, he heard Jon Vickers as Gerontius—'magnificent, better than Lewis'—and told Alan Frank that 'if he can act, he would make a splendid Troilus.' There was talk at this time of a 1958 Covent Garden revival of the opera and Walton wrote to Maria Callas asking her to sing Cressida. She replied 'very sweetly, saying "no", that she hasn't a date free next year etc. I must say it is only what I expected.' Frank assured Walton that Vickers was excellent on the stage and reminded him that he had sung Troilus when the complete work was broadcast in Canada. Early in January 1958 Walton reported arrival of Osbert Sitwell's text for the Huddersfield oratorio. 'It's called *Moses and Pharaoh*. Not very satisfactory at first glance—very diffuse. It's about the plagues and the exodus through the Red Sea etc. In fact a bit of Cecil B. de Mille! But I think that something can eventually be made of it.' Nothing was, though. Perhaps he decided against a new oratorio after he had heard Fricker's *The Vision of Judgement* at the 1958 Leeds Festival. He had studied the score beforehand and wrote to Alan Frank: 'It looks extra good, in fact I'd go so far as to say it's the best choral work since B.'s F. (not uninfluenced by it, incidentally) if not better than that.'

During 1957 and 1958 Walton sketched ideas for the Liverpool symphony commission. 'Glimmerings of the Symph. are beginning to stir slightly,' he told Alan Frank on 28 November 1957. By January 1959 the first movement was virtually complete. Henze thought well of it, Walton reported on 15 April, and he sent it to London on 13 July. Towards the end of 1958 he accepted a commission for a new opera from the

Serge Koussevitzky Music Foundation in the Library of
Congress. During 1959 he thought seriously about composing a
piano concerto for Louis Kentner. He had toyed on and off with
the idea for some time, he told Alan Frank, but there were 'so
many notes.' Two other commissions arrived in 1959. ABC
Television had planned a series of one-hour programmes based
on Churchill's *History of the English-speaking Peoples* and Walton
was asked to provide a march to accompany the opening and
closing titles and credits. This, an Eric Coates-like affair with
an excellent tune, was recorded on 25 May 1959 but, for
reasons unknown, was never used in the series. The other
commission was from Peter Pears and the guitarist Julian
Bream. During the 1950s the guitar was enjoying a boom in
popularity in both serious and popular (skiffle) music. The
example of Andrés Segovia had inspired younger classical
guitarists such as Julian Bream and John Williams, and
composers were not only writing guitar concertos but using the
guitar (and the lute) as a substitute for piano accompaniment in
songs. Britten in 1957 had written his superb *Songs from the
Chinese* for Pears and Bream, who were then anxious to add
other new works to their repertoire. Walton asked Christopher
Hassall to select some poems for him. Six sixteenth- and
seventeenth-century lyrics were chosen from Gerald Bullett's
The English Galaxy of Shorter Poems and Walton gave the sequence
the brilliant title of *Anon. in Love*. To help Walton know what
was feasible for the guitarist, Julian Bream provided him with
a long diagram of the fingerboard. *Anon. in Love* was first
performed as part of the Aldeburgh Festival on 21 June 1960 in
Shrubland Park Hall, Claydon, Ipswich. Walton rescored it for
tenor and small orchestra in 1971. The first three songs are
'lento', the last three 'allegro'. All are splendid settings, Walton
especially enjoying the bawdiness of the last three. The cycle's
chief interest is as the first of several late Walton works which
are overt tributes to Britten. The repetitions and roulades are
designed to suit Pears's vocal style, but they go deeper than
parody in this case. As the older masters used to compose
'homages' in the style of the composer whom they were
flattering by imitation, so Walton here presents his first homage
to Britten.

Walton completed the second (slow) movement of his Second Symphony in January and February 1960, the Finale on 22 July. He rewrote the first movement and sent it to Alan Frank in April. The Liverpool anniversary for which the work was commissioned had long since passed and the city now agreed to yield its claim to the first performance to the Edinburgh Festival provided that it was played there by the Royal Liverpool Philharmonic Orchestra. This duly occurred on 2 September 1960, with John Pritchard conducting. It was another unhappy experience for the composer. Rehearsals were held in a schoolroom, where Pritchard could not obtain a satisfactory balance. The critics, almost to a man, were lukewarm—'the mixture as before'. As a result EMI cancelled its plans to make an immediate recording.

It is easy to understand how and why the Second Symphony was underestimated on its appearance, as it perhaps still is. Everyone was expecting another emotional blockbuster like the First. Instead they heard a lighter, much shorter, three-movement work that might have been more suitably called a Sinfonietta. Elgar's two symphonies were built to a similar scale and emotional groundplan. Walton's are not and to compare one with the other is a pointless exercise. They are, moreover, separated by twenty-five years. By 1960 Walton was a more accomplished composer, perhaps less spontaneous but no less driven by white-hot inner compulsions for which his music is a release. The Second Symphony is better scored than the First, which can sometimes sound callow and blatant in its more emphatic climaxes, and its expressive content is more subtly conveyed. Almost too subtly, because a general complaint, to which I subscribed for many years, was that it concentrated on manner rather than matter, that Walton concealed a depletion of imaginative invention by wrapping it up in a masterly display of technical expertise, so masterly that it could seem glib, unambitious, and mannered. The truth is that the Second Symphony is curiously reluctant to yield its secrets and inner meaning through a few hearings. Not that it is difficult music, but it does need concentrated and frequent listening before, suddenly, the veils part and one is admitted to the inner circle of its highly distinctive sound-world. Then one discovers how one has hitherto been deceived: this is more than a divertimento.

Where the First Symphony is expansive and rhetorical, full of dramatic and melodramatic gestures, the Second is compact, almost elliptical. It suddenly boils with rage or passion rather than hammers its clenched fist. There are a lot of notes, but none is wasted. Its concision is a remarkable technical feat in itself. Everyone at first was so busy identifying the familiar features of the Walton landscape, the rising sevenths, crackling rhythms, and tense fugato, that they overlooked the new sounds, the crystalline percussion and dark-hued contrapuntal tapestry of the slow movement, which continues the intro-spective lyrical vein of the comparable movement in the Quartet.

The first movement is a novel construction—it is virtually monothematic, since all the themes are closely related or founded on a common stem, and it combines the functions of an opening allegro and a scherzo. The principal subject is the essence of Walton, a distillation, familiar yet new, of the themes which open the Cello and Violin Concertos:

Ex. 15

As Frank Howes's penetrating analysis of the symphony shows,[2] the harmonic ambiguity and the uncertain tonality,

[2] *The Music of William Walton*, pp. 40–53.

hovering and darting round a tonic G, are calculated to a nicety. They impart a tart pungency to the music, which is enhanced by the orchestration, with flute, harps, E flat clarinet, celesta, and piano contributing to the glitter of the sound. Because of the brevity of the themes and the rapid, explosive method of their treatment, the impression is given of a lightweight structure, of pattern-making in notes that has no fundamental driving force. And who can tell, since Walton did not, if there was not within him at this point a distressed and angry sense of frustration at the reluctance of his music to provide him with the melodic flow of old? Ejaculatory outbursts of brass, with trilling woodwind, are followed by a terrifying percussive fury which seems to be railing against the vagaries of creativity. The episode is repeated before the movement subsides on to a G minor chord coloured by a note from the vibraphone.

The Lento assai belongs to the sound-world of *Troilus and Cressida* and might almost be an orchestral portrait of Cressida. The principal theme, for woodwind, is not unlike Cressida's aria as she waits for news of Troilus at the start of Act III. The colours are dark, the design seamless, as tension mounts over a low throb from harps. A big string theme seems always likely to develop, does so briefly, but sinks back into the sinister rumbling of timpani playing intervals of major seconds and minor thirds, followed by a more agitated passage. What I call the Cressida theme returns and becomes impassioned. The movement then calms down in scoring that suggests Ravel and Debussy in its fastidious concern for the most delicate yet precise forms of colouring. It is not an elegy, but a deeply personal reverie, smouldering with repressed emotion.

Much has been written about the Finale's opening theme being a twelve-note series, but there is nothing serial or atonal about the use to which it is put in ten variations, a fugato and a

Ex. 16

coda. If Boyd Neel had shouted 'Scheherazade' on first hearing it, as he had done with *Scapino*, it would have been understandable. The variations are mostly short and mostly fast, supremely well scored and exemplary exercises in the art of composition, for the ingenuity displayed is considerable. As in the First Symphony, the problem of how to reach the end is solved by fugue, leading to a 'scherzando' coda. The movement is the weakest in the work inventively, but in a good performance its effect can be as exhilarating as was intended. The use of variation-form was shortly, as will be seen, to lead to a particularly fine example.

Far from 'marking time', as one critic put it after the symphony's first performance, Walton had in fact progressed towards a more compact method while remaining true to the musical morality and integrity which pervade his best music. Peter Heyworth in *The Observer* likened the slow movement to Elgar and added: 'Walton cannot be accused of suffering from the national disease of castrated good taste. There is no lavender water in his lyricism.' But he found the new work

a more adroit affair than its rather overrated predecessor, but it never gives the impression of any extension or deepening of musical language, such as a composer arrives at, not by arbitrary decision to 'do something different' but simply in the creative struggle to pin down and develop an inner concept of the imagination . . . There leaps from almost every bar an intense sense of character, compounded of that odd assortment of jauntiness, irony and an underlying melancholy. A creative artist often reflects the society that gives him birth, and who are we to object if Sir William, like most of us, prefers to look backward provided that he does it in his own way.

That was the majority view of the work. The symphony is notable for its physical energy and excitement; and these seem to me to convey an air of desperation. We may only guess at the cause, but, whatever it was, it gave the work the enigmatic character that makes it so deceptive a piece on first acquaintance.

Walton went to New York in February 1961 to hear Szell conduct the symphony in Carnegie Hall—and was overjoyed by the performance which, in his view, vindicated a work he himself regarded as better than his First Symphony. When in the following year he received the Cleveland recording, he wrote to Szell:

Words fail me. It is quite a fantastic and stupendous performance from every point of view. Firstly, it is absolutely right musically speaking, and the virtuosity of the performance is quite staggering, especially the *fugato*. But everything is phrased and balanced in an unbelievable way, for which I must congratulate you and your magnificent orchestra.

A drop in Walton's American income during 1959 precipitated this letter from Alan Frank to the manager of Oxford University Press in New York, John Ward:

I warned him about the ASCAP drop, and in fact it has been pretty severe—not much more than a third of what he got the previous year, which was, admittedly, a pretty good one. As a result he writes that he is giving six Italian gardeners notice to quit: how many gardeners that leaves him I don't know, but I am sure that you will agree that this sort of thing cannot be allowed to continue.

It also precipitated this from Walton to Alan Frank on 4 February 1960:

It would seem that I may have to contemplate doing another film. I had either forgotten or not realized that they were such moneymakers after the initial payment. Fenn wrote to me not long ago saying that Wilcox would like me for 'The Reason Why'. Charge of the Light Brigade! I more or less turned it down, but perhaps should see what it is all about. Maybe I could discover another 'Colonel Bogey' of the 1855s. And that would be out of copyright and with no widow to take the cream. It wouldn't be till Nov. But I expect someone else has got it by now.

They had.

During the latter part of 1960 Walton composed his Gloria, commissioned to celebrate the 125th anniversary of the Huddersfield Choral Society and Sargent's thirtieth year as its nominal conductor. The work was in short score by 23 December and the orchestration was finished by the following September. Sargent conducted the first performance in Huddersfield on 24 November 1961 and the first London performance on 18 January 1962 (before which Walton made his customary amendments and alterations). At Huddersfield Walton again concluded that Sargent had not sufficiently studied the score beforehand and the two men were scarcely on speaking terms during rehearsals. The anonymous critic of *The Times* wrote of 'no sensation of old ground retilled, but of new

ground made fruitful by an old, still vigorous hand', sentiments with which it is not difficult to disagree. The piece is
skilfully constructed, but the invention simmers on a low light.
Writing to Malcolm Arnold on 5 November 1970, Walton
commented on the Gloria:

Glad you didn't come to Uddersfield. Though Wyn Morris[3] is by no
means the bad conductor he has been made out to be (in fact he's v.
good) how can anyone hope to obtain a performance of any worth if
the chorus & orch. meet for the 1st time at the actual perf. which is
what happened. So it was not very impressive but v. depressive.
Anyhow I'm not sure that it is at all a good work, in fact not perhaps
worth the hard labour necessary to make it really come off.

The Gloria is written for three soloists (no soprano) and a
large orchestra with a *Belshazzar* percussion section. Dissonance
is ensured by the use of block harmony with clashing
semitones and seconds, and chords in which a minor third,
diminished fourth and major ninth are reiterated emphatically.
It is in eleven sections, beginning with a 'maestoso' orchestral introduction, dominated by a pentatonic motto-phrase on
trumpets and trombones, before the choir's entry with its
jubilant fanfares of 'Gloria'. This first section ends with a hectic
orchestral passage before the calmer C major episode of 'In
terra pax' with its descending figure for flute and oboe and
antiphony between men's and women's voices leading to an
unaccompanied unison. The 6/8 and D major tonality of the
first section return in 'Laudamus te', which verges on the fugal.
The basses' 'Glorificamus te' is almost jazzy, leading to an
elaborate climax at 'Dominus Deus', where the brass motto-
theme from the introduction is the basis of the melodic
material. A quiet slow unaccompanied section follows at the
first mention of Jesus Christ.
The soloists enter individually with 'Miserere', a despairing
cry over an agitated orchestral accompaniment. Tenor, contralto,
and bass are heard praying above a choral muttering. Before the
'Quoniam', trombones recall, though not exactly, the introduction. The choir's eight-part statements lead to the unaccompanied 'Tu solus altissimus Jesu Christe'. The 'Cum sancto spiritu'

---

[3] Wyn Morris, b. 1929, was conductor of the Huddersfield Choral Society
from 1969 to 1974 and trained the chorus for this performance of the Gloria.

is set 'maestoso', with a brilliant orchestral passage involving timpani, bells, harp and organ. Again, the motto-phrase from the beginning of the work is the germ from which the music is derived. For the excited coda, the Amen is expanded to include 'Gloria in excelsis Deo'.

In 1961 Walton returned to Oldham to become its fourteenth honorary Freeman on 22 April. His mother had died in 1954, but the conferment ceremony was attended by his sister, Mrs Nora Donnelly, on holiday from New Zealand, and his elder brother Noel. The frontispiece of the illuminated resolution contained the opening bars of the Marcello anthem 'O Lord Our Governor', which Walton had sung as a test-piece nearly fifty years earlier to qualify for admission to Christ Church Cathedral choir school at Oxford. After the ceremony, Walton felt moved to declare that he would dedicate a composition to the town as soon as possible. The intention remained unfulfilled, although he thought at one time of honouring his promise with a clarinet concerto for the Oldham-born Sidney Fell. The following year, the year of his sixtieth birthday, he was appointed *accademico onorario di Santa Cecilia* in Rome. He visited Canada for the first time and heard the first American performance of the Gloria in Los Angeles. Some of his friends might have been surprised to read one of the replies he sent to sixtieth birthday congratulations. To Peter Heyworth he wrote on 13 April:

My dear Peter (or tormentor-in-chief!) I have only just got round to getting through my b & b [bread and butter] letters, so forgive my tardy but nonetheless most grateful thanks for the delicious pot of caviar which was highly appreciated by us both. I did as a matter of fact twice telephone you but with no success. Once on Sun. morning April 1st pointing out you had split an infinitive and hoping you would have to bore yourself blue by reading through your contribution to find it. However you weren't there. It is about time you came to see us again. Surely a book on the deficiencies of Italian opera houses is within your scope. Blessings, and don't pull your punches! As ever, William.

Walton had by now been Peter Heyworth's guest in his London home. A few months later Heyworth received a postcard on which Walton had pasted a cutting about Stravinsky's comment to the editor of the *New York Herald Tribune* following a review by Paul Henry Lang of his television opera *The Flood*: 'Of hundreds of reviews of my New York work most of them, like every opus since 1909, were gratifyingly unfavourable. I found only yours entirely stupid and suppurating with gratuitous malice. The only blight on my 80th birthday is the realization my age will probably keep me from celebrating the funeral of your senile music columnist.' Walton added: 'Even "big fleas" suffer from their P.H.s! See you soon.'

Later in the year he went to London for an engagement which appealed to his sense of fun.

Malcolm Arnold has asked me to conduct an extract from B.'s F. for the Hoffnung Astronautical concert on Nov. 28th [he wrote to Alan Frank on 25 October]. Top secret! After the chorus and orch. pompously filing in, including myself, I bring them all in on the word 'Slain'. Nothing more. It might be rather funny. Anyhow it will be the easiest £20 I shall have earned ever! No objection I trust![4]

Work on the completion of three commissions awaited him on his return to Ischia. The first City of London Music Festival had paid him £500 (subscribed by the Goldsmiths' Company) for a cycle of six songs, *A Song for the Lord Mayor's Table*. Again, Hassall selected the poems for Walton—Blake, Thomas Jordan, Wordsworth, Charles Morris, and two eighteenth-century Anon. 'Busy on the songs', Walton wrote to Alan Frank on 23 February 1962. 'I must say I find writing for the pfte very irksome and have spent a lot of time on No. 1 which is really for orch.' Three songs were written by the end of March 1962 and the rest by early in June. The first performance was given in the Goldsmiths' Hall on 18 July by Elisabeth Schwarzkopf, accompanied by Gerald Moore, who later wrote: 'The quick songs with so many words would present problems to a foreigner but E. S. is a fine linguist and sang them as if she were an Englishwoman.'[5] Walton must have smiled ruefully as he thought of Cressida. But when he orchestrated them for the 1970 City of London Festival, he set a precedent for the course he was to take with his opera and made them suitable for Janet Baker, who sang them in the Mansion House, with the English Chamber Orchestra conducted by George Malcolm, on 7 July

---

[4] The result can be heard in the recording of the concert, issued by EMI in the album 'Hoffnung's Music Festivals', SLS 5069.

[5] Letter to Stewart Craggs, 23 Apr. 1976.

that year. Walton excelled in ceremonial works of this kind and these songs are a happy example of his lyrical vein at its most fluent and witty. In the first song the Walton of the Coronation marches is not far away in this description of a Lord Mayor's banquet, with corks popping in thirds. The setting of Wordsworth's 'Glide gently' is the gem of the cycle, a beautiful and imaginative evocation of the Thames. 'Wapping Old Stairs' is a jilted girl's complaint, with a time signature of one minim per bar. In 'Holy Thursday' Blake's description of a St Paul's charity children's service involves Walton in using a different figuration for each of the three stanzas. 'The Contrast' of Charles Morris's poem is between the Londoner who is bored in the countryside (staccato accompaniment to patter) and the sustained chords and sheepbells of rural peace. The final song, 'Rhyme', is a merry cadenza of bell-chimes, beginning with St Clement's ('Oranges and Lemons').

Walton then completed his *Granada Prelude, Call Signs and End Music*, commissioned in June 1961 by Granada Television, delivering the score in August 1962. The *Prelude* was a march. Granada never used the *Call Signs and End Music* and gave the march to Gilbert Vinter to reorchestrate for symphonic wind band. In this form it was recorded and was transmitted at the beginning of the day's transmission from 1965 until September 1973. One of the tantalizing might-have-beens of operatic history was mentioned by Walton to Alan Frank in a letter dated 23 August 1962: 'A dead secret for the moment is that Patrick [Spike] Hughes has written suggesting *The Importance of Being Earnest* as a libretto. I think a very possible idea but I don't see getting down to it till after the round-the-world trip.' No more was heard of what would have been ideal Walton territory.

The third commission was on a larger scale, an orchestral work for the 150th anniversary in March 1963 of the Royal Philharmonic Society. Walton decided that this was the occasion for which to compose the *Variations on a Theme by Hindemith* which he had had in mind for some time, taking his cue from Elgar by making it a tribute to a friend pictured, and quoted, within. He had never forgotten how Paul Hindemith had come to the rescue of his Viola Concerto in 1929. Their friendship went back to Salzburg in 1923 and Walton had continued to admire the German composer's work. He probably realized, too, in 1963, that they were both experiencing chill blasts of critical disapproval and wished to express his fellow-feeling. So he took a theme from the second movement of Hindemith's Cello Concerto of 1940 and wrote the *Variations on a Theme by Hindemith*, dedicating it to Hindemith and his wife Gertrud. 'The Vars. are beginning to move a bit,' he wrote to Alan Frank on 22 December 1961. 'If you know the theme you will see that it is not entirely my fault if it turns out to be a late Vic. or early Edwardian work!' Completed on 6 February 1963, it was a relatively speedy effort for Walton. He conducted the first performance in the Royal Festival Hall on 8 March 1963. A private recording was made two days earlier and this was sent to Hindemith, who wrote to Walton from Vevey on 29 July:

Finally our criss-cross journeys came to an end and we could sit down in front of the exhaust [*sic!*] of our gramophone and play your piece, score before us. Well, we had a half-hour of sheer enjoyment. You wrote a beautiful score and we are extremely honoured to find the red carpet rolled out even on the steps to the back door of fame. I am particularly fond of the honest solidity of workmanship in this score— something that seems almost completely lost nowadays. Let us thank you for your kindness and for the wonderfully touching and artistically convincing manifestation of this kindness (even old Mathis is permitted to peep through the fence, which for a spectre like him seems to be some kind of resurrection after artificial respiration!)—I am glad that Georg Szell had a great (and well deserved) success with the piece in the States. I also shall put it on my programs as soon as possible . . . I shall do my best to become a worthy interpreter of WW . . .

Hindemith never conducted the *Variations*. He died, in Frankfurt, on 28 December 1963.

The *Variations on a Theme by Hindemith* is one of the finest of Walton's works, sometimes described as a 'conversation' between the two composers. Walton was concerned to impose Hindemith's personality upon the work, since the theme he used was not merely a melody but a thirty-six-bar extract lasting nearly two minutes, stated in Hindemith's own harmony, with the solo cello line distributed to other instruments. Part of this extract was 'naggingly familiar' to Walton, as he told Frank Howes, and it was some time before he realized it was similar to a theme in Hindemith's opera *Mathis der Maler*.

Ex. 17

The opera theme occurs in the second scene of the sixth tableau, depicting the temptation of St Antony, and Walton quotes it, within quotation marks, in his seventh variation:[6]

Ex. 18

This resemblance also came as a surprise to Hindemith. It derives from the use of triads equidistant from each other by a major third.

The pre-war Walton would have been incapable of writing so elegant and poetic a composition as the *Variations*, its poise giving no suggestion of emotional detachment, its economy displaying consummate mastery. Here is all the flavour of the complete Walton, familiar yet new, enticing in its civilized reconciliation of form and content in perfect symmetry. The form is Introduction (Theme), nine variations, and Finale. Walton used the serial order of Hindemith's first twelve notes for the tonalities of the variations. The first two variations are fast, 'vivace' and 'allegramente', scored with the acuity and brilliance which characterize the whole work. The 'larghetto' third variation is mainly for strings, a siciliano in lighter vein, followed by a 'moto perpetuo' variation with harmonies in thirds. The fifth variation is 'andante con moto', closely following the Hindemith theme. It has a bass in quavers ('legatissimo') which develops, on flute and horn, into the BACH motif. Variation 6 is 'scherzando', in which fragments of the theme are tossed about the orchestra. The percussion section is prominent—side-drum, tambourine, harp, and xylophone. Various parts of the theme are the basis of the 'Lento molto' seventh variation, first on the horns, then on strings. The

[6] P. Hindemith, *Mathis der Maler*, vocal score, p. 221, four bars before [47].

theme is heard fully, culminating in the direct quotation from *Mathis der Maler*. This is the most expressive variation and is followed by a chattering 'vivacissimo' eighth variation. The last variation is a short 'maestoso', which is more of a fanfare-like introduction to the Finale. This is a fugue, typically electric in its unorthodox entries. After its excited climax, a tranquil coda begins with the theme restated. A trumpet, imitated by horn, plays the *Mathis* quotation, stressing its kinship with the concerto theme. Having made this point, the work ends quickly.

The *Variations* was received with politeness but no marked display of enthusiasm. A few noticed its exceptional quality from the first, but it took the best part of twenty-five years for it to be valued as the masterpiece it is, and in that time only one conductor besides Szell has chosen to record it (Vernon Handley in 1988). When Walton heard Szell's recording, he wrote to John Ward, the representative of Oxford University Press in New York: 'I don't know whether you have heard him perform the work. I have, in Amsterdam a few months ago. It was stupendous. After rehearsing it, he turned to me and asked me for any comments. I could make none as there was nothing to say except a rather tame "thank you". How can one comment on a performance which is flawless in every aspect?'

22

The Bear, 1964–7

WALTON laid aside composition for almost a year while he went on his travels. He visited Israel in July 1963, conducting performances of *Belshazzar's Feast* in Tel Aviv, Haifa, and Jerusalem. From there he flew to Chicago, where he conducted three programmes of his music at Ravinia Park, summer home of the Chicago Symphony Orchestra. In February 1964 he went to New Zealand, leaving Lady Walton (who had broken her leg) in Ischia. He visited his sister, Mrs Nora Donnelly, at Hastings and conducted his works in concerts in Auckland, Dunedin, Wellington, and Christchurch. ('My own experience in New Zealand was comparatively pleasant, chiefly because the personnel of the orchestra was so nice,' he wrote some years later, 'otherwise there is precious little to be said for it or its inhabitants as far as I could pick up.')[1] He then spent ten weeks in Australia conducting his orchestral works. At the Adelaide Festival there were seven performances of *Troilus and Cressida* with Marie Collier as Cressida and Richard Lewis as Troilus. Joseph Post conducted. Walton reported to Oxford University Press:

Though the stage was rather cramped, the production and sets were more than adequate. Unfortunately the orchestral pit was also very small and the strings had to be reduced to seven 1st violins etc. However, despite these handicaps, it had a tremendous impact in the smallish theatre. Joseph Post knew the score backwards and his tempi were impeccable, resulting in lively, spirited and dramatic performances.

Walton caused a minor rumpus in Australia by objecting to the custom there of playing only a few bars of the National Anthem at the start of each concert unless it was an important occasion,

---

[1] Letter to Malcolm Arnold, 1 Oct. 1970.

when it was played in full. This, he said, was undignified and
it would be better not to play it at all. These remarks, of course,
were interpreted by the Press as 'Walton trying to stop the
playing of *God Save the Queen.'*

When he returned to Ischia, late in May, Walton wrote on the
27th to Alan Frank: 'I may say I got to know my own works
pretty thoroughly and I must say that I thought that I should
become absolutely sick of them, but both symphs., the Vars.
and the concertos wear very well and go down well also, rather
to my surprise, but I found myself getting irritated by the
smaller works, Jo'burg, Partita, Façade etc.' He settled down to
begin to compose a work for small orchestra on a concertante
plan which Georg Szell had requested. On 14 September he
wrote to Alan Frank:

Been having a life and death (more of the latter) struggle with mov. I
and shall have more or less finished it in sketch by the time we leave
for London on 19 September. It is not, I fear, at all good—when I tell
you that in a passage of some 26 bars there are no less than 14
sequences you can see I've sunk pretty low and what's worse is that I
can see no way round. But not only that, the piece is not really suited
to the medium and might be anything for str, 5 w.w. and a smattering
of brass, harp and perc. now and then. I hardly dare face G.S. with it!

Walton showed his sketch of the work to Szell and Frank in
Amsterdam some weeks later. Frank wrote to Szell on 26
November: 'I was careful not to quote any opinion as coming
directly from you. In fact, however, Walton said at the end of
our conversation that he was entirely in agreement with our
assessment of the sketch material for the new work and of the
general situation . . . I have the feeling he may scrap the
material.' Which he duly did. Szell later asked for a third
symphony.

November 1963 saw large-scale celebrations of Britten's
fiftieth birthday on the 22nd. His fame was at its zenith,
following the success of his *War Requiem*, written for the
consecration of the new Coventry Cathedral in May 1962. The
exchange of letters between Walton and Britten at this time
sheds a touching light on their relationship. Walton wrote from
Ischia on 23 November 1963:

Dear Ben. You must almost by now be suffering from a surfeit of adulation and praise, so I won't add to it. All the same I should like to tell you that I celebrated your birthday in my own way by playing my favourite works—*Spring Symphony*—*Nocturne* and *War Requiem*—each in its different way a masterwork, particularly the latter—a non-stop masterpiece without blemish—in fact on a par with the two great Requiems of the 19th century or for that matter any other century.

In the last years your music has come to mean more and more to me—it shines out as a beacon (how banal I'm becoming!) in, to me at least, a chaotic and barren musical world and I am sure it does for thousands of others as well. I know that I should understand what is going on, but it is a matter of age—old age maybe; but there it is—I don't. But I do understand, appreciate and love, I hope, nearly everything about your music, not only the ingenuity and technique but the emotional depth of feeling and above all the originality and beauty of sound which permeates these works. The *War Requiem* is worth hundreds of Lord Russells and Aldermaston marches and it will surely have the effect which you, possibly sub-consciously, have striven for, for you have made articulate the wishes of the numberless inarticulate masses.

Now I must stop before I descend to complete drivel, but I write only (purely selfishly but from the heart) because I should like you to realise that I am (this being a 'fan' letter) one of your most enthusiastic 'fans' and I look forward to your next works, especially *Lear* [a projected opera on *King Lear* which was never written].

Please don't bother to answer this (unless you should happen to want to) as I know that you must be snowed under with letters to answer, and if by some mischance you should be again afflicted with 'conductor's elbow' or whatever it may be, and think of taking the 'cure' here, we should be more than delighted to entertain you. Su joins me in all good wishes to you and Peter. As ever, William.

Britten replied from The Red House, Aldeburgh, on 16 December:

My dear William, I am most grateful to you for having written so warmly & generously for my birthday. I had already been so pleased to get the telegram from you and Sue & then was overwhelmed to get the letter as well. It was a wonderful tribute from a colleague & you know, I think, how much this kind of thing means—especially in those (very frequent!) moments of depression when one can't find the right notes, & also loathes every thing one has ever done. (I don't think any composer has ever felt less confident than I—especially somehow when the public praise seems to have got rather out of hand!) I do thank you most warmly.

I don't know if I ever told you, but hearing your Viola Concerto & Portsmouth Point (works which I still love dearly) was a great turning point in my musical life. I'd got in a muddle; poor old John Ireland wasn't much help, & I couldn't get on with the 12-tone idea (still can't)—& you showed me the way of being relaxed & fresh, & intensely personal & yet still with the terms of reference which I had to have. It comes, I'm sure, because the ideas were fine & clear, which is all that matters.—And, I've never forgotten your noble & generous support of me in a very low moment in the War.

One day I'd love to come & see you both in Ischia. We are having a sabatical [*sic*] 1965 (if we survive that long!)—no concerts at all, & maybe then we could visit you. It would be so nice to see you again, & that wonderful place . . .

Much gratitude again—& love to you both. Ben.

Walton replied to this on 2 January 1964.

My dear Ben. I was so very, very pleased to receive your letter and to learn that I have been of some help to you both in your work and in other ways in your early days. It seems hard to imagine your ever being depressed about your work—perhaps that is partly why it is what it is, for at its best, it always to me gives an impression of complete spontaneity and freshness of inspiration. As for myself, I must admit to have been suffering from a prolonged bout of depression and for months have hardly put down a bar worth keeping. All my later works always receive such a drubbing from the press, especially my last one, which incidentally I consider one of my best, 'The Variations on a Theme by Hindemith'. I am rather diffidently sending you the score, which is just out. His death is a great shock to me . . . We shall be delighted to have you both to stay during your sabbatical year . . . Ischia is now so near—you leave London at 9–30 and are here for lunch and vice versa. Yours, William.

Writing a year later to another friend and fellow-composer, Malcolm Arnold, on 20 February 1965, Walton said:

Sorry to have been so dilatory in answering your letter, but believe it or not I've been having a bout of work after a very long interval—an Anthem with words by Auden for our Alma Mater Ch. Ch. Oxford. Also a Missa Brevis (very brevis) for Coventry. But don't think I've got religious mania! I was pleased and proud to hear about your conducting the Spitfire music and that it made such an impression. But I was furious to read what the damn fool of a critic said about your new piece. They are really insufferable—all of them.

Walton's friendship with Arnold began after the younger man had congratulated him on the Cello Concerto. They soon discovered they had a similar sense of humour and, although politically poles apart, this made no difference since they avoided the subject. Both liked good wine and enjoyed expeditions to night clubs, and there was a deeper element in their mutual understanding, because Arnold's music was considered by many critics to be out of touch with the preoccupations of the leaders of musical thought. Arnold had also written a good deal of film music, so this too was common ground.

The anthem to which Walton referred was the idea of Dr Cuthbert Simpson, Dean of Christ Church Cathedral 1959–69, who was concerned by the dearth of good suitable anthems for the feasts of the Apostles. He asked Auden for the words probably as early as 1962. Auden sent Walton his text and later saw him in Ischia. Walton's version of the work's genesis was given to Alan Frank in a letter dated 2 January 1965: 'Wystan Auden sometime last year at Oxford let himself and me in for writing an Anthem for Ch. Ch. Choir. He said he must have been in his cups! Anyhow a few days ago what he calls "this bloody anthem" arrived, so I suppose I must do it. It is a somewhat obscure and difficult-to-set text.' Three weeks later he wrote that it was 'difficult to keep from being difficult to sing and I know b— all about the Organ!' The manuscript score arrived at Christ Church in parts, the last bit in time for the first performance of *The Twelve*, as it was called, at evensong on 16 May 1965, when Dr Sydney Watson conducted the cathedral choir in which Walton had sung as a boy and the organist was Robert Bottone. The anthem is dedicated 'To Christ Church, Oxford, and its Dean, Cuthbert Simpson'. It is for mixed chorus, with solo parts for soprano, alto, tenor, and bass. Auden's text is in three parts—a description of the Apostles' work, a prayer for support, and a meditation on their work. The organ part is difficult and Walton orchestrated it (for double woodwind, heavy brass, plenty of percussion, harp, and strings) in time for a concert on 2 January 1966 celebrating the 900th anniversary of the founding of Westminster Abbey. He conducted on that occasion, when Martin Cooper in the *Daily Telegraph* described it as a 'miniature cantata' and remarked on its 'Belshazzar-like final jubilation' after a big fugal chorus.

The *Missa brevis* was written at the request of the Friends of Coventry Cathedral for the new cathedral's choir. It was to have been sung there for the first time on Easter Day 1966 but the performance was cancelled because of illness among the choirboys. 'I'm also on to the Missa Brevissima,' he wrote to Alan Frank on 17 March 1965. 'I doubt if there will be more than 8 to 10 mins. of it. Remembering the boredom I suffered as a dear little choirboy, I've made it or am making it as brevissima as poss. It should be v. popular among Communion takers. But how uninspiring are the words!' It lasts barely seven minutes and is for double choir, unaccompanied except when the organ is used in the Gloria. The most enchanting music is in the Agnus Dei, with antiphonal soprano and tenor soloists and the choir in four parts. The rising motif which begins the Kyrie in bare two-part writing is a diminished fourth which is imitated and inverted in other voices.

In spite of the discouragement of *Troilus and Cressida*, Walton was anxious to compose another opera. In an interview printed in Covent Garden's magazine *About the House* in May 1963, he went so far as to say that he was 'beginning to work on a new opera, again with Christopher Hassall as librettist. Sadler's Wells and Covent Garden both asked me to write one during more or less the same week. Sadler's Wells got in first, but perhaps I shall write two.' This may have been wishful thinking, and in any case Hassall's death rendered any such projects impossible. Prodding came from another, perhaps unexpected, quarter—Aldeburgh. Writing to Alan Frank on 6 January 1965, Walton said: 'By the way, will you look at Chekhov's "Bear" or "The Proposal", suggested by P.P. as a one-acter for Ald. The Bear I think is the more feasible and the action could be put into the "Hunting Shires". But I don't quite know who to ask to pull it together.' He had an opera commission 'in storage', as it were—in November 1958 he had accepted one from the Koussevitzky Music Foundation in the Library of Congress. Now was the time to undertake it. Congratulating Britten on his appointment to the Order of Merit ('no one could deserve it more than you'), Walton wrote on 25 March 1965: 'Incidentally I am seriously thinking of embarking on "The Bear" which Peter suggested I might do for the English Opera Group. But it falls into that rather awkward category of a "curtain raiser" and I don't know if it would fit in with something else. I don't think it could last more than 45 mins, if that. But what kind of an orchestra should I do it for?' He began by writing his own libretto, but soon decided he needed help. He had met the writer Paul Dehn in Ischia and liked him, and, when Dehn's name was suggested, he jumped at the idea. The libretto was begun in the spring of 1965 and a draft was left with Walton in November after Dehn had worked for a fortnight with him in Ischia. Walton began to compose immediately. As was his custom, he sent to London for certain scores to study, in this case Britten's *Albert Herring* and Stravinsky's *Mavra*.

In December 1965 he went to London to prepare to conduct the Westminster Abbey performance of the orchestral version of *The Twelve*. He had had influenza, which had left him with a persistent cough. Before the Abbey concert he went for a drink with a friend who was a doctor. She noticed that, after climbing a long flight of stairs, Walton was alarmingly out of breath and she arranged for him to see a heart specialist. There was nothing wrong with the heart, but the specialist noticed a shadow on his left lung. A second opinion diagnosed cancer, the result of years of smoking twenty pipes a day. (He at once instructed his wife to throw away all his pipes and tobacco and he never smoked again.) An operation was successfully performed at the London Clinic on 10 January 1966. Among his many visitors was Peter Heyworth. After extra treatment on a health farm, Walton returned to Ischia on 21 February and resumed work on *The Bear*.

By coincidence, in February Britten underwent an operation for diverticulitis. From hospital he wrote to 'My very dear William and Sue [*sic*].'

I have just awakened from a very doped sleep & they are coming to fetch me for the 'theatre' (I wish it were Covent Garden) in an hour or so; but before disappearing into oblivion, I wanted to let you know how touched I was by the present from you which arrived late last night. What with this lovely champagne, the approaching Spring (?), the glorious news that you, William, have made such a fabulous recovery, I feel that once to-day is over there is real hope & joy once more in the world. These weeks of waiting have been rather trying, but they are over now, & now the job is to get well—& you've no idea how that champagne will help . . .

He wrote again from his home on 9 March to say that the champagne

was the perfect 'pick me up' & 'appetite-giver', still in fact is the only thing (except lemon-barley!) I can drink with pleasure! . . . I have heard by devious sources that you William have been feeling steadily better—I do hope this is true & that you haven't had too many of the (I fear) quite inevitable downs which go along with the ups. Do please be careful & don't start work too soon—I find any concentration quite out of the question at the moment—I can't even read a novel! . . . I totter around the garden, go for gentle drives, play 'Scrabble', & return to bed quite exhausted. All the same I begin to feel almost human . . . I long to hear that you are both forgetting that horrid experience very quickly, & that William is completely restored to health. This comes with warmest love . . .

In his reply Walton told Britten that Paul Dehn was going to Ischia at the end of April 'so we will have a chance to re-polish the end of the "Bear" which at the moment I don't feel is very satisfactory.'

Walton wrote to Alan Frank on 7 March:

I'm just beginning to perk up a bit, though the actual part operated on still nags a good deal. One can't say one's in great pain but all the same one's conscious of it—but of course it will pass. Will you get me a record of Boulez's *Pli selon pli* being broadcast on March 16 I think. André Previn is firm in his intention to record Symph. I in London in August. I'm not sure I know who he is! Ought I to?

Previn, of course, became one of his finest interpreters and a firm friend. At the end of April he was telling Peter Heyworth he was 'beginning to feel more like my old self and have even looked a piece of MS paper in the face—not with very promising results however. But it's wonderful to have an excuse not to work!' He urged his 'tormentor-in-chief' to visit him in Ischia 'where you would find yourself very welcome.'

After a few months, Walton began to have severe pains in his left shoulder. A check-up revealed return of the cancer, which had then to be treated by 11 weeks of cobalt radiation in the Middlesex Hospital. This appeared to have no effect and Walton became progressively weaker. He was also haunted by memories of Alice Wimborne's long and painful death. The specialist doubled the strength of the treatment. This proved successful, although it left Walton easily prone to colds and bronchitis, and, as the rays had burnt his oesophagus, his food had to be cut up finely and he had temporarily to give up drinking wine.

Walton made a remarkable recovery and went to England for the Aldeburgh Festival.

I can't tell you how impressed I was and how much I enjoyed the Burning F.F. [Fiery Furnace] [he wrote to Britten on 21 July]. Besides its beautiful & dramatic form, it was so intensely moving and convincing and achieved with such small forces. It didn't surprise me at all that you had found it so difficult to do, but it sounds so spontaneous that one would never guess the agony it must have cost you—you brought it off triumphantly. Though the performance and production was excellent I would like to have heard it with Peter and hope to do so some time. We shall be in London at the end of Sept. for yet another check-up & it is just a possibility that I will bring the "Bear" with me.

He also went to the 1966 Salzburg Festival for Henze's opera *The Bassarids*. He continued work on *The Bear* for the rest of 1966 and wrote to Alan Frank on 11 December: 'The B. moves slowly but steadily, and if my next check-up is satisfactory I shall doubtless finish it in time. But I find a lot of what I did this time a year ago needs drastically re-writing.' In February 1967 he reported: 'Gloomy symptoms have begun to appear in the last few weeks,' but all was well. About this time two recordings of the First Symphony were issued, one conducted by Previn, the other by Sargent for EMI, on the sleeve of which was emblazoned Walton's highly complimentary endorsement of the interpretation. Roy Douglas, who knew and shared Walton's low opinion of Sargent, was surprised by this fit of generosity and mentioned it in a letter to the composer, who replied (on 6 March 1967):

Thank you for your kind words about S.1. Previn's recording is from all points of view the better of the two. I was let in for doing that blurb thro' circumstances and could hardly refuse despite my only having attended two 1/2 sessions of the EMI recording, and, at that, feeling like death just having had a dose of gamma rays (which incidentally seem to have done the trick and I'm cured, at least so the doctors say – but they did say that before!) and I hadn't heard the P. recording – so it is just a bit irritating.

About the 'Bear'. I'm finishing it off in Pfte score . . . I agree about the Alto Fl. I've only used it as it appears to be a permanent instrument in BB's scoring for Aldeburgh.

He finished *The Bear* on 30 April, writing to Dehn: 'Finished with B.B., not what you think. [He meant 'Bloody Bear'.] I don't know what to think of it. It has taken far longer than necessary. It is I think sufficiently in time.' To Roy Douglas, who had worked on the piano score and parts, he wrote on 17 May: 'I thought for a moment I should never finish it, but however I got through somehow and hope it isn't quite so awful as I was inclined to think. A.F. [Alan Frank] seems to think he has a winner on his hands.' The first performance was given at the Aldeburgh Festival on 3 June, in the Jubilee Hall, as a double bill with Lennox Berkeley's *Castaway*, also to a Dehn libretto. Monica Sinclair sang the widow Popova and John Shaw sang Smirnov, the 'Bear' of the title. James Lockhart conducted the English Chamber Orchestra and Colin Graham produced. A London performance followed on 12 July at Sadler's Wells and a recording was made in August, after the score had been revised. The one-act opera was warmly received, Andrew Porter going so far as to describe it in the *Musical Times* as 'one of the strongest and most brilliant things Walton has written.'[2] To friends Walton wrote:

I have become so used to being slated by those critics that I felt there must be something wrong when the worms turned on some praise. However, you will hear the record soon; pretty good.[3] The music is light and humorous. Of course *Façade* crops up frequently in the critics' notices, but it has no real connection except that I happen to have composed both. The so-called jokes are not in quotations, but *à la manière* of Russian music. In fact by now I would be hard put to tell you which they are. The odd thing is that it has all, so to speak, sunk into the landscape and all sounds as if it were me, which in fact it is.

The plot of *The Bear* concerns an attractive young widow, Yeliena Popova, attended by her old servant Luka and still aggressively faithful to her far from faithful husband. She is confronted by the bearish, boorish Grigory Smirnov, one of Popov's creditors. They quarrel and each aims a loaded pistol at

[2] *Musical Times*, July 1967, p. 632.
[3] Walton revised this opinion later. He found the performance of Popova too 'heavy', as indeed it is, and came to dislike the recording.

the other. Neither fires, because they have fallen in love. The Chekhov vaudeville was a riotous success in Moscow in 1888 and both Dehn and Walton perceived its susceptibility as a vehicle for parody. Walton was especially attracted by the caricatures of the three characters (he defined caricature as 'so accurate an exaggeration of the real thing as to be funny'): the seemingly respectable but inwardly passionate widow, the seemingly gruff but inwardly sentimental suitor, and the seemingly obsequious but inwardly rebellious old servant. Walton subtitled the opera 'an extravaganza.' Dehn wrote:

Walton, with a muttered 'It's no good pretending it's *Three Sisters*', sat down fancifully to compose it in a music-room which is sound-proofed, partly to shut out the stentorian tenor-talent of his head gardener, Antonio, and partly to shut in the creative noises made by Walton himself who admits that, among composers, he is the world's worst pianist and the world's second-worst singer.[4]

The success of *The Bear* must cause regret that, instead of spending so many years on the hollow heroics of *Troilus and Cressida*, Walton had not pursued a subject for a full-length comic opera (such as *The Importance of Being Earnest*), for which he had all the gifts. The scoring is deliciously light, in spite of the exotic array of percussion. It was originally written for five string players, but a note on the full score states that, whenever possible, the composer would prefer a larger number to be used. Trying to identify the composers parodied makes a fascinating game, but one can never be sure of accurate solutions. What is certain is that they are good-natured, affectionate parodies; and surely the first Aldeburgh audience must have identified the very opening of the opera with the opening of Britten's *A Midsummer Night's Dream*, with its portamento strings? Smirnov's 'Madame, je vous prie' (Ex. 19) is a parody of the Russian fondness for speaking French as a mark of refinement.

When he urges Popova to 'unveil, as did Salome', there is a hint of Strauss in the accompaniment. (Walton worried how much Boosey and Hawkes might charge!) Popova's enchanting 'I was a constant, faithful wife' could be by several composers; and so

---

[4] Sleeve-note of EMI recording of *The Bear*, SAN 192, issued 1967.

Ex. 19

one could fruitlessly persevere. It is one of the best of one-act operas and the highest compliment one can pay it is to apply to its sparkle, wit, and musicality the adjective Rossinian. Walton, a twentieth-century soulmate in many respects of Poulenc and Ibert, once told a surprised André Previn that he regarded Rossini as the finest composer, and he was not jesting. Walton was pleased in 1971 when *The Bear* and *Façade* were performed in the same programme—'the ideal combination', he said. On 4 January 1968 Walton wrote to thank Roy Douglas

for all your work on getting the score of the 'Bear' into order. I am delighted you like the record – so do I, with perhaps a few more reservations than you. On my machine, at any rate, I find the recording rather harsh and strident and *ff* almost throughout . . . Of course I suffered by never knowing how many strings I was supposed to be allowed, 1 of each I understood till I discovered too late that L. B. [Lennox Berkeley] was using 4,3,2,2,1, little enough but it at least allowed for dividing the strings on occasion. Double that amount would have made all the difference to the quality – a failing that Walter L. [egge] wouldn't have allowed to occur in a recording.

I do feel it is 2 or 3 minutes too long; being a bit finicky, and I think a cut from (in the printed libretto) *Luka* 'M. is indisposed and will see no one' to *Smir* 'What's that! etc' would be a help. There's too much of Smir's anger. But I've not really gone into it. And, as you suggest, the 5th verse of the 'ditty' ['I was a constant, faithful wife'] could go easily. Why didn't I think of it! That 5th verse is a bit too high for M. to get the words out clearly or easily. However there it is and it is not too bad as it stands . . . The production for the New Year is scarce and not up to standard![5]

---

[5] When the score of *The Bear* was published in 1968, Walton provided an appendix giving four optional cuts. Two of them were the passages to which he referred in this letter.

Dehn died in 1976, a greater loss potentially to Walton than Hassall, for their next collaboration would have been founded on a previous success.

In the autumn of 1967 the Queen's Private Secretary wrote to Walton to enquire if he would accept membership of the Order of Merit. When he read the letter he burst into tears, saying: 'To think that the Queen actually knows I'm alive!' The appointment was announced in November. Replying to congratulations from Roy Douglas, he wrote: 'Bless you and thank you . . . What a life you lead owing to me, and I can't thank you enough, nor I'm sure could that other O. M. whose script was worse than mine!' (a reference to Vaughan Williams, to whose manuscript scores Douglas had supplied a similar service— 'washing their face', was Vaughan Williams's term). A congratulatory letter from Szell contained pleasing news: 'The *Hindemith Variations* made a deep impression, not only in Edinburgh but also in Lucerne on our festival tour last summer. I intend to do them in Vienna, with the Philharmonic, next December—if I can get enough rehearsals and the agreement of that ultra-reactionary gang!'

It was about this time that Diana Rix, young, pretty, efficient, and with a sense of humour to match Walton's, began to work for his agent, Harold Holt Ltd (of which Ian Hunter was chairman and managing director). They soon established an easy relationship over his bookings to conduct and other business affairs. One of Walton's engaging traits was that all his correspondence, business and personal, was handwritten and this encouraged him to be informal. His letters to Diana Rix are peppered with playful amorous sallies, tongue-in-cheek, which it obviously amused him to write and her to receive. Not many agents of composers or conductors can have received letters beginning or ending: 'Darling, I'm in bed (which makes me think of you) ill . . .' 'Are you my sweetheart or aren't you? Make up your mind . . .' 'Love and a long lecherous kiss . . .' 'Sweetheart, I've only to begin writing to you for me to be filled with delicious, luscious and libidinous thoughts about you—but to business! How boring!' Both Walton and Susana were devoted to her and she to them.

An insouciant sparkle, similar to that pervading *The Bear*, irradiates Walton's next composition, an orchestral work commissioned by the New York Philharmonic Orchestra for its

125th anniversary. Yet though *Capriccio Burlesco* sounds as if it was the inspiration of a moment, Walton told a reporter of the *Brighton Evening Argus* in April 1968, when he made his first public appearance in England for two years at the Brighton Festival: 'I've done it about four times and thrown it away four times. But it's getting better, although it hasn't got a title'. On the manuscript score the original title is shown as *Philharmonic Overture NY '68*, but Walton replaced this prosaic label, thank goodness, in June 1968. The final version was begun in May 1968 and completed on 5 September. Walton went to the United States for the first performance in the new Philharmonic (now Avery Fisher) Hall in the Lincoln Center on 7 December, when the orchestra was conducted by André Kostelanetz, to whom the piece is dedicated. Lasting about six minutes, the *Capriccio Burlesco* was well described by an American critic in the *New York Post* as 'sly musical gesticulation, and weaving of saucy ideas, scampering in and around musical corners and in syncopations.' With a coruscating glitter in the scoring, Walton keeps this display of high spirits spinning like a top. Inconsequential it may be, but it is the product of a sovereign command of technique and resource, the ultimate sophistication of a Walton scherzo.

A less pleasing episode was the making of a television documentary about Walton for showing on BBC. He was shown an early version of the film and was infuriated. He wrote to the television head of music, John Culshaw, on 5 May 1968:

I was led to believe that this film was going to be a pleasing tribute, but to my surprise and discomfiture what do I see and hear but John Warrack stating that the works I have written since the war are hardly worthwhile mentioning . . . Not a word about *Troilus and Cressida*, the Cello Concerto (except in a disparaging and condescending way), nor the 2nd Symphony or the *Hindemith Variations*, in fact nothing about any work I've written since the war since 'we' - who are 'we'?—do not consider them worthy of consideration.

Culshaw admitted 'a false emphasis' and ordered

various modifications, the most radical of which is the elimination of Warrack's summing-up and assessment which I have asked him to reconsider, re-write and submit for my approval. I have asked him to include a clear reference to your later works and to put the emphasis on the international acclaim they have received irrespective of the opinions of some English critics.

Walton was pacified.

He went to Texas in January 1969 as guest conductor of the Houston Symphony Orchestra. While there he informed a representative of the *Houston Chronicle* that 'six projects near to his heart' were 'urgently pending'. He itemized some of these as 'major pieces' for the New York Philharmonic and San Francisco Symphony Orchestras, a double concerto for Heifetz and Piatigorsky, and 'something' for Szell. 'And that double bass player, Gary Karr, he's on my track.' Only one of these materialized.

He did not mention a commission to which he had agreed in June 1968 and which was to have strange and unhappy consequences. During that summer the film producer Harry Saltzman, who at one time had rented Walton's former London house, Lowndes Cottage, invited him to compose the music for a feature film about the Battle of Britain in 1940. Walton agreed, especially when he heard that his friend Laurence Olivier was to play Air Chief Marshal Dowding. Saltzman and the director, Guy Hamilton, sent the script to Ischia and when he was in London Walton was shown a rough-cut of the film. He was impressed and immediately began to write his score. He wrote to Malcolm Arnold on 8 September:

I have accepted, perhaps rashly, to write the music for the film of the Battle of Britain. I say rashly since I've done nothing of the kind since Richard III, some many years ago. Would you by any good chance be willing to conduct the sessions? There will be about twelve, with any orchestra we like to have, the recording would be in May–June of next year. May be a bit earlier. I do hope that you will be able to do this. I should feel much more confident with you about if things should go wrong. I saw a very very 'rough-cut' last time I was over, but having now completed the film the next rough-cut should be much better.

Walton and Arnold saw the second rough-cut in October. The score was completed fairly quickly early in 1969, Arnold orchestrating several episodes including some of the *Battle in the Air* and other sequences. An undated letter from Walton to Arnold, beginning 'I have taken you at your word', refers to his having sent the top lines (perhaps a little more) of two pieces almost identical . . . so you must try and make them sound quite different! I think they should be fairly loud (I've put in no expression marks) as I imagine there's a lot of background airplane noises to overcome. On the other hand it is probably better the other way round. I'm getting in a bit of a panic, but I'm aiming to finish all the small bits by the end of the week so I've time to devote to the big stuff. No sign of a tune! Every time I think of one I find I've written it before.

The music was recorded at Denham on 21 February and 10 April 1969 with Arnold conducting (except the March, which Walton conducted). Edward Greenfield was present for the *Guardian* at the second session.

The music fizzes [he wrote of the *Battle in the Air*]. This is very much the pre-war Walton of the First Symphony, uninhibited, youthful, exciting. The ominous opening tremolo (Trevor Howard as Air Vice-Marshal Park pacing the control-room) has an echo of Agincourt, but then to vapour trails in the sky Walton creates an utterly original sound. I fail to identify even which instruments are playing. Sir William points to the score where violas and cellos are instructed to play arpeggios in glissandi harmonics on the fourth string, a nice lesson in orchestration.

Greenfield then described the recording of the 'grand super-dambusting' March: 'With an outrageous whooping on the horns comes a grand patriotic tune to out-type and out-glory any that Sir William has yet written, whether for films or coronations. The first three notes may be a blatant crib from the opening of Elgar's Second Symphony, but they turn at once into the purplest Walton.'[6]

Walton returned to Ischia content. Some weeks later he was shocked and amazed when he was telephoned by a *Daily Express* reporter, who enquired why Ron Goodwin had been asked to supply a new score for the film. This was the first Walton knew of this development. After Walton had left London, Arnold and David Picker, the brothers who were in charge of United Artists, distributors of the film, demanded to have the music tracks sent to them in New York. These were sent unaccompanied by any film. The Pickers were baffled by what they heard and demanded music by a composer they knew. But before that there was a complaint to Walton that there was not

---

[6] *The Guardian*, 29 Apr. 1969, p. 10.

enough music to fill an LP recording which it was intended to issue. On 21 May he was asked for 'a 16-bar middle section' to add to the March. He told his agent Ian Hunter on 1 June: 'I fail to see what more is wanted. If they want another "Land of ho. and glo." it's not on the cards—for it is one already on a slightly smaller scale in that it doesn't last quite as long.'

The next move by United Artists was to approach John Barry for a new score, but he declined. The offer to Goodwin followed. Walton was angry and hurt. 'It's bloody cheek,' he told the *Sunday Telegraph* in June, 'all these Americans want is money and I had never even heard of Ron Goodwin'.[7] Olivier was equally angry and threatened to have his name removed from the credits if at least some of Walton's music was not used. When the film was shown on 15 September 1969, Walton's *Battle in the Air* sequence was retained, but nothing more.

On 28 August Walton wrote to Arnold: 'I am now in touch with Ron G. He is terribly upset (quite why it is a bit difficult to ascertain) about the whole business and has got *his* solicitors on to U.A. Anyhow we seem now to console one another! A bit late. Lord Goodman appeared from a yacht and was duly incensed about it all and sent cables all round, but I fear with little effect.' An amusing postscript was a letter to Walton from Alan Frank on 24 September: 'I had a notification from EMI today that in a record to be released next November your Spitfire Prelude and Fugue appears, performed by—guess who—Ron Goodwin and his Orchestra.' Walton scribbled on this letter 'Talk about turning the other cheek!'

Meanwhile United Artists claimed that Walton's score was their property and refused to release it, although Stewart Craggs, with Walton's help, bought a recording and copies of the score from them in May 1971. For Walton's seventieth birthday, the British Prime Minister, Edward Heath, arranged the release of the score from United Artists and it was restored to Oxford University Press on 7 April 1972. A concert suite from the film score was arranged after Walton's death by Colin Matthews, and Carl Davis conducted the first performance in Bristol at a Bournemouth Symphony Orchestra concert there on 10 May 1985. Davis later recorded it. The *Battle in the Air*

[7] *Sunday Telegraph*, 22 June 1969, p. 2.

recaptures the accuracy with which Walton put the Spitfire into music in *The First of the Few* nearly thirty years earlier. The March is vintage *Crown Imperial* Walton; its trio uses music associated earlier in the score with the 'Young Siegfrieds' of the Luftwaffe and based on Siegfried's horn-call from Wagner's *Siegfried*.

23

Britten and Bagatelles, 1969–72

NOTWITHSTANDING his chagrin over *The Battle of Britain*, Walton was busy completing another orchestral commission between August and November 1969. This was his work for the San Francisco Symphony Orchestra, commissioned by its benefactor, the scientist Dr Ralph Dorfman, in memory of his first wife Adeline Smith Dorfman. Choice of composer had been left to the orchestra's conductor, Josef Krips, and Walton was approached (and accepted) in the autumn of 1967. He began work during 1968. On 9 September he wrote to Britten:

> I hope that you will not think that I am making a too strange request—namely that you will allow me to attempt to write Variations (orchestral) on the theme of the 3rd movement of your Piano Concerto. I realise that you have used it as a passacaglia, but not strictly speaking as a theme and variations—I hope very much you will let me have a try. It is not a new idea of mine but one which I've been thinking about for some time.

To Alan Frank on 22 November he wrote: 'I could hardly have chosen a more infantile theme, no need for a pill with this one!'

Remembering the success of his *Hindemith Variations*, Walton had decided to repeat his experiment of entering another composer's world and transforming it into his own. Britten's concerto was written in 1938, but in 1945 he withdrew the original slow movement, Recitative and Aria, and substituted the Impromptu. The new movement was a passacaglia, a favourite form with Britten, on the nine-bar theme Walton chose (Ex. 20).

It is a stately, melancholy theme constructed from a descending phrase of augmented seconds, minor thirds, and semitones, with intervals of major and minor thirds and semitones. Before

Ex. 20

it is fully stated by Walton, he prefaces it with a brief
introduction, with high strings suggesting the East Anglian
sky. In the concerto the theme is first played by the piano and
then by the strings with the piano supplying chordal harmonies. Walton gives it to the clarinet and the harmonies to the
harp. The first variation, or improvisation, flows gently in
unisons and octaves. The textures are as spare as in a Britten
score. The ascending intervals of the theme principally occupy
Walton's attention, and he again suggests the Dawn interlude
from *Peter Grimes* with his writing for strings. Next comes a
variation marked 'vivo' and 'martellato', brassier, with dotted
rhythms and the brass twirling like Puck's trumpet gyrations
in *A Midsummer Night's Dream*. A drumming rhythm on the
strings ('col legno') begins the third variation, working to a
massive climax with staccato brass and a clatter of percussion.
Here it is the descending intervals which are under scrutiny.
The lyricism of Walton's Cello Concerto invades the fourth
variation, with romantic strings caressing the theme's rising
intervals over a pizzicato ostinato. Harp and horns bring it to an
end. For the fifth and last variation the mood is 'scherzando'.
The theme returns in the lower strings, the pace increases, and
the end is all syncopated animation.

It was impossible in 1969 that Walton could have foreseen
that in 1976 Britten would be dead, yet if this work had

appeared in 1977 it would have been accepted as a memorial tribute, for much of it is sombre and touchingly valedictory in tone and mood. It is no surprise, therefore, to discover that its first title was *Elegiac Variations on a Theme of Benjamin Britten*. It was only because Dr Dorfman expressed a wish that his wife should be commemorated by music with a joyful rather than a mournful conclusion, that Walton in November 1969 changed the work's name to *Improvisations on an Impromptu of Benjamin Britten*—and one wonders if the 'joyful' coda was therefore a necessary afterthought. It is a hauntingly beautiful composition, as elegant and graceful a homage to Britten as the earlier set of variations was to Hindemith. Krips conducted the first performance in San Francisco on 14 January 1970. Walton's view of the work was characteristically low-key. On 29 August 1969 he had written to Alan Frank: 'May have the Vars. finished by Sept. 12th when we come to London for 10 days or at any rate well on the way. I can't say I'm mad about them—rather Capriccio B. level, I fear.' But he wrote to Britten on 19 January 1970: 'I've just received the following cable, "Your work enthusiastically received, an excellent composition. Josef Krips" - which is all to the good as I was rather frightened that it wasn't going to be up to standard. I understand that it has also been released for Aldeburgh. I hope that you will conduct it.'

Even though he had declared after the *Battle of Britain* fiasco that he would never write another note for the cinema, Walton could not resist a request from Laurence Olivier for music for his film version of the National Theatre production of *Three Sisters*. The score was composed between mid-November and Christmas 1969. The recording sessions were at Shepperton on 29 December and Walton flew from Ischia to attend them. For a ballroom scene he used the waltz he had composed in 1935 for the Cochran ballet *The First Shoot*. There was a near-miss with another film early in 1970. This was the story of the origins, building, and history of the Basilica of St Peter's, Rome, to be called *Upon this Rock*, with Edith Evans, Ralph Richardson, Orson Welles, and Dirk Bogarde in the cast. At first Walton was interested, especially for a fee of £10,000. He wrote to Diana Rix on 29 January 1970: 'I'm all for this new film, apart from anything else it is an appealing kind of subject, not that I'm not a bloody heathen—I'll do it all the better for that.' By 4 February he had decided he could not do it. 'I'm no longer what I was since my operation and can no longer work hours on end which this film would need.' Six days later he had changed his mind and agreed in principle, estimating he would need to write sixty minutes of music. He then saw a rough-cut of the film again and re-read the script. He cabled to Harold Holt that it would be 'disastrous for me'. He amplified this to Diana Rix on 18 February: 'Most uninspiring' was his final verdict on the film.

He was also exercised about potential soloists for the Cello Concerto.

Peter Pears writes à propos of the *Improvisations* that they are including the Prokofiev Sinfonia Concertante (for vlc & orch.) [he wrote to Diana on 29 January 1970], and it's being played by Slava Ros. Does that name mean anything to you—it doesn't to me, but if he can play the Prokofiev he must be first class, so you might find out about him. [Walton apparently did not recognize Mstislav Rostropovich from the diminutive form of his name used by his Aldeburgh friends!] He's playing at Aldebugger in July and I demeaned myself by asking P.P. to say a good word for my Vlc. Con.—probably a mistaken tactic. The *Improvisations on an Impromptu by B.B.* (what a title!) open the programme. The work seems to have been a big success in San Francisco—bro. Alec has sent me some notices—v. good—headings such as 'W's superb improvisations' and continuing in a similar vein. My 70th birthday had better be put off as I very much doubt if I shall ever reach it. But if I should, I'm quite definitely not going to write any piece for it. First the NPO [New Philharmonia Orchestra] want it, then the LSO want it, then the LPO and the BBC so I can't squabble with them all so I had better not write anything at all.

In the summer of 1970 Walton began to transcribe the guitar accompaniment of his song-cycle *Anon. in Love* for orchestra. In a letter to Malcolm Arnold on 8 June 1970 he said he was being 'very dilatory' because 'I don't know what to do with the 'armonies which are so often implied on the "box" - whether to fill them in or not.' He added: 'Very sorry to hear about the gout. Osbert Sitwell had a theory that only geniuses suffered from it—a theory certainly borne out in your case. Larry O. is also a fellow-sufferer, but he has discovered some pills or other that have completely cured him. I will find out about them (It must have been the lobster, it can't have been the booze, as

George Robey used to sing).' He went to England a few days later for a television performance of *The Bear* and to hear the first performance in Britain of the *Britten Improvisations* by the Royal Liverpool Philharmonic Orchestra conducted by Sir Charles Groves in the Snape Maltings on 27 June as part of the Aldeburgh Festival. The televised *The Bear* was an unhappy experience. Walton did not take easily to the special conditions of recording music for television. He attended the annual lunch of the Performing Right Society on 1 July at which the Prince of Wales was guest of honour. Each composer-member of the Society's Council was invited to write a piece of music for inclusion in an album to be presented to the Prince, who was a cellist. Walton contributed a *Theme (for Variations)* for cello solo.[1] At the end of July he was saddened by the death of Georg Szell and John Barbirolli. 'It's obviously very tricky this time of life,' he wrote to Diana Rix. Towards the end of this year Ian Hunter tried to interest him in setting a Cecil Day Lewis poem commemorating the 150th anniversary of the death of John Keats, to be performed at a Brighton Festival concert also containing Britten's *Serenade*. 'I've got a better idea than competing with B.B. at setting a Keats sonnet for Brighton', he replied on 11 December, 'and that is that you should get Mr Tear or some tenor to give the first performance of *Anon. in Love* arranged by the composer for str. orch., harp and 1 perc.'

While staying at the Ritz, his favourite London 'haunt' in his later years, in 1970, Walton was asked by the conductor Neville Marriner if he would write a work for the chamber orchestra Marriner had founded, the Academy of St Martin-in-the-Fields. Walton declined, but Marriner had another idea up his sleeve: would he make a full-strings version of the A minor String Quartet? This appealed to Walton. His main concern at this time, however, was a commission from Julian Bream, who had long wanted some solo guitar pieces by Walton because he had been so impressed by the guitar accompaniment to the *Anon. in Love* song-cycle. 'I managed to write some rather pretty

---

[1] The other contributors were Lennox Berkeley, Bliss, Ronald Binge, Vivian Ellis, John Gardner, Joseph Horovitz, Mitch Murray, Steve Race, Ernest Tomlinson, Guy Warrack, and Brian Willey. Two Welsh composers, Grace Williams and David Wynne, were also asked to contribute.

pieces for him', Walton said in a television interview, 'except that the first six notes of the first piece all need to be played on open strings. So when he begins to play, the audience will probably think he's tuning the bloody thing up.' On 21 May 1971 Walton wrote to Malcolm Arnold: 'We've just returned from staying with Hans W. H. [Henze] to hear his new piece.[2] I don't quite know what to think, except that I wish he'd take to writing music again and not indulge in electronic (very good) noises . . . May I dedicate my guitar pieces to you?' The pieces now became *Five Bagatelles* (edited by Julian Bream). By 1 June he was experiencing his usual doubts: 'I'm beginning to have cold feet about dedicating the *Bagatelles* to you—I don't think they are good enough or worthy enough for you—from which you may gather they are not going at all well. I shall dedicate something else if these aren't up to the mark . . . Ischia is not much fun at the moment, very cold and wet and no "crumpets". I expect it will cheer up as soon as I leave for London.' On 30 August he wrote again: 'I've finished the guitar pieces (5) with dedication to yourself. I'm inclined to think they are rather good, but Julian will let you know about them. He's coming to inspect them on Sept. 1st!' A few days later he reported: 'In a few days I shall be sending you a copy of the *Bagatelles*. Julian on the other hand when he saw the dedication[3] was very pleased. After all, what is important to him is his rake-off as Editor!'

The *Five Bagatelles* are among Walton's most piquant and delightful miniatures. He exploits the guitar's resources to the full and the music always sounds Waltonian. The outer movements are Allegro and Con slancio (impetuously) enclosing a slow waltz, a seductive movement marked 'Alla Cubana' and a slow movement that is mainly an essay in rapid tremolando. Bream played No. 2 on its own at the Queen Elizabeth Hall on 13 February 1972 and Nos. 1 and 3 in a television programme relayed on 29 March, Walton's seventieth birthday. The complete set was first played at the Bath Festival on 27 May, with a London performance on 21 January 1973.

---

[2] *Der langwierige Weg in die Wohnung der Natascha Ungeheuer.*

[3] The dedication is 'To Malcolm Arnold, with admiration and affection for his 50th birthday'.

On 6 July 1971 Walton enlisted Arnold's aid in the string quartet transcription:

I hope this reaches you before a letter from Neville Marriner or Alan Frank as I should like to be the first to ask you about the proposed project of doing my String Quartet for String Orch. (i.e. with double-bass). During our discussion about it and about who was to do it, naturally enough your name cropped up. I said I couldn't think of anyone I'd prefer more, but I said that I thought it was too much of an imposition to ask you to do it especially as I know you are very busy with conducting dates and going to Vancouver and eventually to Ireland. They wanted me to 'cut' it. I said I'd do that (this was about 10 days ago), but I've been through it time and time again (it's the 1st mov. that they were interested in cutting) and I've found it impossible to do without its sounding as if it had been castrated, had its stomach out, with hysterectomy thrown in. I must say I rather agree with them that it is a bit long, that is if it is played by the Allegri Quartet, whose performance is a bit sluggish and lifeless—it is I think the record I once played to you. However, the record of the Hollywood Quartet is altogether a different affair. I played it this morning and was pleased with it. N.M. asked that it should be 'cut' to 23 mins, and lo and behold without any cutting the Hollywood 4tet play it in exactly that length of time, so the cutting element is eliminated, except for the repeat in the 2nd mov. which is easily removed and is better for it.

You have doubtless perceived what all this long preamble is about—that is to ask you to undertake the scoring for St. Orch. I know it is a lot to ask, but I'm too involved with it myself to do it really well, otherwise I'd do it or rather try to do it. You needn't do it all—just those parts which you think would gain by more volume of strings and it might have a solo string quartet in it, as in the Intro. and Allegro of Elgar . . . It is cold, windy and wet—a most peculiar July for here. Also there's Sir Robert Mayer in the offing!

Arnold agreed to this request, but in September Walton wrote: 'N. Marriner seems to want the string orchestra version of the 4tet by the end of the year. In which case as you are very busy till then and I seem to have eff-all to do, I might as well pull myself together and do my own dirty work! However I would dearly like to consult you and have your advice about it.'

Walton worked on what was to be known as *Sonata for Strings* throughout September and October. Towards the end of October, Arnold went to Ischia to complete a work of his own. 'We are absolutely delighted and looking forward to your arrival on Oct. 25,' Walton wrote on 26 September. 'By then I shall hope to have got this boring 4tet out of the way and all you will have to do is to point out where it can be improved, to the accompaniment of popping corks. I know that critics and others will ask why he can't write something new instead of rehashing an old 4tet—in fact, I'd like to know too. But I'm not going to!' When Arnold arrived in Ischia, he found that Walton had completed the first two movements and was half-way through the slow movement. Arnold worked on the last movement under Walton's supervision—the manuscript of the Finale is all in Arnold's hand. So the work was delivered on time and was first performed by the Academy of St Martin-in-the-Fields at the Festival of Perth, Western Australia, on 2 March 1972. The Academy gave the British première at the Bath Festival on 27 May and the first London performance at the Mansion House on 11 July.

The only significant differences between the sonata and the quartet occur in the first movement. The remaining three are straight transcriptions (one bar is omitted from the Scherzo and one is added to the Lento). Walton heavily reworked the first movement, changing keys, cutting, compressing, and redistributing the balance of textures. Altogether about thirty-three bars were cut, with consequent adjustments and changes. The sonata was hailed as worthy to rank with the best of English works for string orchestra and it is certainly very effective, but this writer has a stubborn preference for the music in its original form. Walton made some alterations after he had heard the work performed, but these were too late for a recording. As he told Marriner in a letter in 1975: 'The present record as you know is not really very satisfactory, having been done without my having been able to make the emendations which are now in the printed score. It is sometimes a mistake to rush into a recording before the work has been properly assimilated. I've done it myself to my great regret with *The Bear*—a glaring instance of stupid impetuosity on my part.'[4]

Towards the end of 1971 Walton also composed a short (three-minute) *Jubilate Deo*, written as a present for Lina Lalandi, director of the English Bach Festival, after they had had 'a very happy meeting' at the Ritz. She had gone to see him

[4] Walton, however, did not conduct the recording of *The Bear*.

to commission a work with which the festival could celebrate his seventieth birthday. When he returned to Ischia he began to regret his promise. 'And why can't I, I'd like to know,' he wrote to Alan Frank on 2 September, 'slide out of the *Jubilate*? I'm a very slippery customer, as you should well know by now ... The Jubilate is not the most inspiring bit of nonsense—in fact the only thing to be said for it is its brevity.' Frank mentioned rumours of another Auden setting for Oxford. 'There is *nothing* to it,' Walton replied on 1 January 1972. 'If there's to be anything it will be a *Jubilate* and a pretty smart one.'

24
Seventieth Birthday, 1972

BRITAIN likes to make a fuss of its famous men and women when they become septuagenarians and octogenarians. But Walton approached his seventieth anniversary with mixed feelings. He would enjoy the celebrations, but he was still smarting from the wounds inflicted on him by his critics over the previous twenty years, and would not a prolonged celebration involving many performances of his music reopen barely healed scars? 'Thank you so much for your telegram of good wishes,' he wrote to Peter Heyworth. 'I shall need them, especially from you.' But he was encouraged by Malcolm Arnold and even more by the knowledge that Edward Heath, the Prime Minister, was to hold a dinner party for him at No. 10 Downing Street on the evening of his birthday. 'I'm glad you were impressed by Heath's rendering of *Cockaigne*,' he wrote to Arnold on 24 January. 'No other P.M. could do it nor think of giving me a 70th birthday party, from which I propose to be carried out as I shall have been on the wagon from March 6th.'[1]

The year began with the award to Walton of the Benjamin Franklin Medal of the Royal Society of Arts 'for his work for Anglo-American understanding.' The regular performance of Walton's music in the United States, the citation stated, had 'made a notable contribution to Anglo-American culture'. There were a few disturbing signs of physical decline. He wrote to Angus Morrison on 9 February: 'Bar being troubled with my legs—I can barely walk at times and no one seems to be able to diagnose what's wrong—it's really old age—I am very well, as I hope you are.' Walton and his wife went to England during April where one of the first birthday celebrations was conferment of honorary membership of the Royal

---

[1] The idea for the party came from John Peyton, Minister of Transport in Mr Heath's Government, who with his wife had often visited the Waltons in Ischia. Mr Heath insisted on making all the arrangements.

Manchester College of Music, where his father had been a student eighty years earlier. On 22 April he undertook another nostalgic journey, to Oxford, where Simon Preston conducted the Christ Church Cathedral Choir in *The Twelve* and the first performance of the *Jubilate Deo*. Lina Lalandi remembers his being 'deeply moved, I believe to tears.' William Mann in *The Times* of 24 April, after noting the new work's 'two enchanting quiet sections, typical in their rhythmic vitality, the second involving treble and alto solo', continued:

Between these recent works the choir sang 'Drop, drop slow tears' which Walton set at the age of 15, while he was a chorister of Christ Church. The technical fluency of the choral writing is explicable though still remarkable; the creative boldness and poetic feeling of the music are much more astounding. It is a real piece of music, no student exercise, and 55 years later it provided a genuine moving experience even in the company of mature Walton, J. S. Bach and Taverner, who was Master of Choristers in the 1520s when it was still called Cardinal College, after the founder, Wolsey.

On the eve of his birthday Walton went to the Festival Hall, where Previn conducted the London Symphony Orchestra in the *Britten Improvisations* and *Belshazzar's Feast*. Walton conducted the Viola Concerto, with Menuhin as soloist. Ronald Crichton, in the *Financial Times*, caught the flavour of the evening in his notice on 30 March. Menuhin's viola, he wrote, had such a penetrating tone that

one wished that just this once they had gone back to the old scoring with triple wind and without harp—no doubt the revisions make life easier for the solo, but the smoothing and streamlining tone down an acerbity that was very much part of the music, while the harp brings it nearer the Tennysonian euphony of Ischia and the later period, very beautiful, yet different . . . *Belshazzar* is something of a marvel. Year in, year out, it comes back for punishment, in performances brilliant or stodgy. Even critics who have loved it for 40 years must be forgiven for sometimes wishing it might for a season or so be allowed to rest. Yet its resilience is phenomenal. In all but the dimmest performances some page or other is sure to catch the glow of the old fire.

The birthday surprise came in the form of six minute-long improvisations on 'Happy Birthday to You', in each of which a Walton quotation was embedded, commissioned for the occasion by Previn and the London Symphony Orchestra. Each composer conducted his or her own tribute, except Richard Rodney Bennett, who played the piano part in his *Intrada* while Previn conducted. (There had been a project in 1969, which alas came to nothing, for Bennett to be soloist in Walton's re-recording of the *Sinfonia Concertante*.) Thea Musgrave pleased the critic of *The Times* most: 'an absolutely characteristic piece, with real harmonic tensions, and colourful, too.' Malcolm Arnold quoted from 'Popular Song', while Nicholas Maw also used part of *Façade* in a mock waltz. Robert Simpson's—'easily the wittiest', Stanley Sadie wrote—'started as if to be an hour-long symphony jointly composed by Beethoven, Bruckner, Sibelius, and Nielsen, but ended promptly, the theme having been symphonically developed and pared to a delicate little motif.' The theme was a parody of the opening of Walton's First Symphony. Finally—'a real tribute from the avant-garde to the almost dead and buried', Walton said—Maxwell Davies converted the *Crown Imperial* march into a foxtrot.

Various 'Walton at 70' articles appeared in the Press.

Why, more than other composers of unproblematical idiom, does his standing fluctuate so widely, not only in the world of music in general, but even in the estimation of a single listener at different times? [asked Desmond Shawe-Taylor in the *Sunday Times*]. . . . In all his music there is something that attracts an ambivalent response: it is so lively, so pointed, so energetic, often so lyrical, always so beautifully made; and yet so limited in emotional scope and variety. What we feel about it at any given moment is apt to depend upon our awareness of these limitations. The listener whose first response to some new Walton piece was an ungracious mutter about 'the mixture as before' may find to his shame, on returning to the same work some years later, that he is captivated by invention so shapely and workmanship so fine. If it is a mark of the great composer to be instantly recognizable, then Walton is a great composer.

Gillian Widdicombe, in the *Financial Times*, squared up to the alleged dichotomy between pre- and post-war Walton:

What has he done with his fifties and sixties; and what may we wish for him in his seventies? Both decades have been coloured by his residence in Ischia, but the sense of isolation has I think been only superficial. It has definitely undermined the popularity of his works, since this has been left to a few faithfuls instead of being constantly promoted, like Britten's, by the composer in person. But Ischia has not prevented Walton from keeping his ear close to the pulse of contemporary music. Walton's writing for orchestra has definitely been the keynote to his development during the past 20 years, after the completion of *Troilus* in 1954 ... These [works] have followed a stylistic arch, at its most taut and brilliant in the Second Symphony, at its most relaxed and broad in the humorous good nature of the *Britten Improvisations*. This arch is in no way a decline but a change. Walton now feels that the orchestral writing of his early works was too difficult; that the same effect could have been achieved with less struggle.

On the birthday itself, 29 March, was the Downing Street dinner. Queen Elizabeth the Queen Mother was present with many of Walton's friends—Olivier, Frederick Ashton, Lionel Tertis, Henry Moore, Kenneth Clark, Arnold Goodman, and Benjamin Britten. His brother Alec and the doctor who had treated Walton for lung cancer were there too. Herbert Howells composed a grace and Paul Dehn and Arthur Bliss collaborated on 'An Ode for William Walton' which contained lines like these:

> Concertos written by this fertile fellow
> For violin, viola and cello.
> *Troilus and Cressida*—sad, pursuant pair—
> Pursued, a decade later, by *The Bear*.
> Prince Hamlet, Richard Crookback, Henry Five
> (Graced by his music-track) became alive.
> And still his Overtures and his *Façade*
> Quicken the feet of those who promenade.

This, with Walton's 'Set me as a seal upon thine heart', was sung by the Martin Neary Singers. David Atherton conducted the London Sinfonietta, with Alvar Lidell as reciter, in ten items from *Façade*. The climax of the evening was described by Mr Heath in a book:

As midnight approached, I recalled to everyone there how I heard William Walton asked, in an interview on his sixtieth birthday, whether there was any piece of music he would have liked to have composed himself. Without pausing for a moment, he had replied 'Yes, Schubert's B flat Trio.' And so, at midnight, with the Queen Mother and the Waltons sitting on the sofa, and the rest of us, including the Blisses, Herbert Howells and Ben of the older generation, Malcolm Arnold and Richard Rodney Bennett of the younger composers; Lionel Tertis, the greatest of viola players, then well over ninety, with his wife; the Soltis, Fred Ashton from the ballet,

Laurence Olivier, Bryan Forbes, Nanette Newman and those whom many would term the "Arts Establishment", Lord Clark, Arnold Goodman, Jennie Lee and the Droghedas, with many other friends, sitting around the floor, we heard John Georgiadis and Douglas Cummings, the leader of the London Symphony Orchestra and the first cellist, together with the pianist John Lill, play the Schubert B flat Trio.[2]

The celebration continued for almost six months throughout Britain. At Bath the *Five Bagatelles* had their first complete performance, and the Leicestershire Schools Symphony Orchestra impressed Walton with *Belshazzar's Feast*. His fellow-composers gave him a stereo tape-recorder and record-player with loudspeakers to replace his primitive equipment in Ischia. Oxford University Press gave a party and presented him with a de luxe edition of *Façade*.

It was all a triumphant progress and Walton wrote on 2 May to the composer-friend who had finally persuaded him it would be:

Dear Malcolm, Dining the other night I asked about my letter which should have reached you with the vino—but no sign of it, or perhaps it's in the boot of your car! However this is more or less what it said as it was a spontaneous burst of thanks from the heart: Dear Malcolm. This is a small token of my affection, esteem and regard and thanks for your help and support both practical and moral, for I was often at the point of cancelling before we started as I was feeling very low about everything etc. etc. I can't remember the rest of it, but it was in a similar strain and I just want you to know in writing how I feel about it all . . . as ever, William.

Another letter to Arnold, who was living in Ireland, on 8 August described some of the other birthday events:

We are just back (till Sept. 7th when we go to London again!) from King's Lynn. An enjoyable festival including, I need hardly say, *Façade*, not very well done, and the Sonata for St. orch. Good. Marriner now seems to have really got it going and it sounded extremely well. Then on to Aldeburgh for a W.W. evening consisting of *The Bear, Siesta* (a new ballet by Fred. A.), *Façade* (in its original form), poems recited by P.P. and the original scoring—very good and effective. Ben incidentally extremely amiable about you. Lilias [Mrs Robin Sheep-shanks] managed to persuade the Prince of Wales to attend the

[2]　E. Heath, *Music – A Joy for Life* (London, 1976), pp. 143–4.

opening night which added to the glamour. What a charmer—but he must be a mixed-up kid what with studying the workings of a submarine—flying a jet—playing the cello etc. etc. He turned up with several girls and it would seem that he alone of the whole lot had seen an opera before. As it was very well sung, played and produced,[3] it went down with a bang.

To Britten, Walton wrote that 'it was really very noble of you to provide such a handsome 70th birthday present. The B. couldn't have gone better and as for Peter—he was superb. He must record it, for there isn't, in spite of the number made, a really vaguely good one amongst the lot, in fact one has to go back to the one he did with Edith years ago.' The two friends later in the year exchanged recordings. 'The Improvisations sound fine', Britten wrote, '& make much more sense to me than they did at the Maltings last year; although whether that little idea can take all that I am still not sure! I haven't (being much away) yet had time to play marvellous old Belshazzar, but Previn does you so magnificently & John S-Q is so remarkable that I can guess it will be a knock-out.' And Walton wrote to Britten and Pears: 'It was most kind of you to send me the quite superlative recording of *Gerontius*. It has almost overcome my antipathy for it. At any rate my Protestant hackles didn't rise quite as much as they usually do when I listen to the work.'

He was pleased, too, that Previn and the London Symphony Orchestra had played the Second Symphony in London.

It seems to have worked out very well [he wrote to the conductor on 18 November], judging from the press, which I thought on the whole was quite good, much better than the work usually gets. But it is, I know, very difficult to play—perhaps more so than the 1st? Anyhow it is not so familiar. I will come over if you are going to record it. And if you record *Henry V*, try and make them use the original version, much more exciting than the rather tame version that Muir Mathieson arranged in the Suite. E.M.I. could give you an old recording of it with Larry thrown in! . . . S.III is not progressing—in fact, I think I must start all over again.

---

[3] Patricia Kern and Thomas Hemsley sang Popova and Smirnov. The English Opera Group Orchestra was conducted by David Taylor and the producer was Colin Graham.

The birthday year ended with a nostalgic gesture. On 28
December he asked Diana Rix to arrange for the Christ Church
Cathedral Choir's recording of his church music to be sent to
the Dowager Viscountess Wimborne, Alice's daughter-in-law
and a widow since 1967, for whose wedding he had composed
'Set me as a seal upon thine heart'.

25

Roaring Fanfares, 1972–7

WALTON's return to London in September 1972 was to attend a London Symphony Orchestra–Previn Promenade Concert which not only included *Scapino* and *Belshazzar's Feast* but the revised Act II of *Troilus and Cressida* with Richard Lewis and Jill Gomez. Back in Ischia he wrote on 4 October to Malcolm Arnold, apologizing for failing to telephone him, 'but life was a bit too hectic what with the performance and recording of the Violin Concerto with Previn and Kyung-Wha Chung. What a girl! She has to be heard to be believed. In addition she's very easy on the eye. I will have the record sent to you when it is out. Did you receive the copy of B's F?' This query referred to Previn's recording, which he had also sent to Britten. Walton discussed it in a letter to Arnold dated 15 October:

I've just been playing (for the first time) the record of B's F. How right you are! Conductors never seem to realize that in B's F. there is no need to add to the excitement—on the contrary, it should be kept on a very tight rein, otherwise it becomes a shambles, as unfortunately this recording often does. It just shows how necessary it is for the composer to be present for a recording. Owing to the P.M.'s party I couldn't be there, but I just had the time to instruct him about 'the trumpeters and pipers' and the very end which had been completely wrong at the perf. the previous night. These bits are O.K. but the speed in other places completely defeats its object. However it can't be helped—it's too late now. It's so irritating, because he's so quick and sensitive to act on a hint about pace or anything, and if only I had been there it would have been splendid, I feel.

Walton, as this letter shows, admired Previn's conducting and had said he would compose a third symphony for him. He mentioned this to Arnold in October. 'Unfortunately I can't tell you that my Symph. III is progressing and I don't suppose it will if I continue like this! However I hope something will turn up fairly soon.' Something did, for he refused an invitation to

the 1973 Hong Kong Festival, telling Diana Rix on 9 January:
'S. 3 is getting under way and I fear it would be fatal for it to
break off and then have all the agony of re-starting again.' To
Alan Frank on 7 February 1973 he wrote: 'S. 3 is I fear going to
be like Sib.'s 8. Alright for him, he'd already done seven.' In
March he wrote to Arnold: 'What marvellous progress about
your Symph. I wish I could say the same for mine.' When
Diana Rix returned from Hong Kong, he wrote to her on 30
March: 'It was just as well that I didn't try and come with you
for in Rome we went to hear Schippers do my 2nd S.
Marvellous orch. and perf. a grande [sic] success both with
public and press. However I found I could hardly walk and
hobbled to the platform with the aid of a stick to take my bows
. . . You must tell and demonstrate to me all the Chinese
positions.'

    Understandably he had composed virtually nothing during
1972 except for an eight-bar eightieth Birthday Greeting to
Herbert Howells, on the manuscript of which he wrote: 'This I
fear is a very inadequate return for your beautiful "grace" at
No. 10 for my 70th!' Work on the third symphony occupied the
first half of 1973, except for an Anniversary Fanfare for EMI's
75th anniversary concert in London on 29 November. This was
an improvisation on 'Happy Birthday to You' designed to lead
without a break into *Orb and Sceptre*. But on 9 August he
mentioned a new project to Malcolm Arnold: 'We are coming to
Ireland at the end of April next year. Cork Univ. desires to
doctor me (I'm not sure I've anything left!) and I have to write a
little piece for it. Being, I presume, a Catholic Univ. I thought of
setting "Cantico del Sole" by St Francis of Assisi. Anything to
put off the evil day of tackling a third S.!'

    Cork's invitation to Walton was for an unaccompanied
partsong for the Cork International Festival's 25th anniversary.
Boris Blacher, Roman Vlad, and Brian Boydell had also been
asked for compositions. Walton's was commissioned by Sir
Robert Mayer's wife Dorothy. The Senate of the National
University of Ireland had agreed in principle to award
honorary doctorates of music to the four composers. A stipu-
lation by the festival was that the work was preferably to be
sung by an amateur choir. This caused a hitch and in July 1973
Walton said he could not accept the commission. Professor

Aloys Fleischmann asked him to reconsider his decision and as a compromise suggested he should accept the degree *in absentia* and send some early unpublished partsong. Walton then said he would write a new piece, which he finished early in 1974. 'The "Cantico" has worked out as a deplorably dull and unexciting piece', he wrote to Arnold, 'and one for which I need doctoring in the other sense.' Yet on 5 November 1973 he had written to Alan Frank: 'I am having a bout of chronic depression and have destroyed what there was of S. 3. [He told André Previn on 10 November: 'I consigned it to the flames the other day']. The *Cantico* is good but needs I think the addition of a small orch.' The first performance of the work was given in Cork on 25 April 1974 by the BBC Northern Singers conducted by Stephen Wilkinson. A critic singled out 'a new, almost Latin sensuousness about the harmony, particularly at the work's radiant climax.'

Walton went to London in the summer of 1973 for the fiftieth anniversary on 12 June of the first public performance of *Façade*. This was to be the last time he conducted his own music. The reciters at the Aeolian Hall were Mary Thomas and Derek Hammond-Stroud. He had told Diana Rix: 'I think I can manage to conduct it alright especially if I can wear a pullover and sit on a high stool as does Neville Marriner.' He worked on his revision of the 1918–21 Piano Quartet during 1973–4: 'It was written when I was a drooling baby,' he wrote to Diana Rix, 'but since my "2nd childhood" I've revised it a little and it is a very attractive piece (like you).' But the possibility of a Covent Garden revival of *Troilus and Cressida* conducted by Previn principally occupied his mind. The subject was raised to Arnold on 21 December 1973:

Just back from London. What a relief to be in one piece! And the gloom there—just like the beginning of the late great wars. The only thing to be said for it is the comparative lack of bombs which deficiency the IRA are trying to rectify. Here owing to the lack of benzine it is strangely quiet—hardly any lorries and not many cars . . . Cov. Gar. keep procrastinating over T. & C. so I've persuaded the Coliseum to take an interest, which they are doing, but there's the bother of getting the dates right. Janet B. was marvellous in *Maria Stuarda* and I formed a very high opinion of the whole company. It is much more of a unit than Cov. Gar.

He was still keeping in close touch with Previn over Covent Garden's casting problems. On 10 November 1973 he wrote:

I was on the point of writing to ask you to have supper after the B.B. beano on the 25th [concert to celebrate Britten's 60th birthday]—that is if you aren't too exhausted—we could go to the Hungry Horse which is fairly near to the A.H. [Albert Hall] and open on Sun. I intend to hear the *Spring S.* [Spring Symphony] which I'm v. partial to. The usual Cov. Gar. capers. Tooley rang me to say that John Alexander could not sing Tro., and should we postpone the perf. till someone is found . . .

In January 1974 he started anew on the third symphony. He sent Previn a page of score, dated 3 January, inscribed: 'Here are a few, very few, bars which I hope will eventually turn out to be Symphony III—we'll see!' Later in the month he went to Rome for a concert of his music organized by Jack Buckley, arts officer of the British Council, and the Radio Audizione Italiana Orchestra. Previn conducted, Kyung-Wha Chung played the Violin Concerto, and John Shirley-Quirk and ninety members of the London Symphony Chorus flew out to take part in *Belshazzar's Feast*. 'Rome was quite a success from all points of view,' Walton wrote to Arnold on 2 February. 'And about time too, say I, having lived here for more than 20 years. But how much I should have disliked being drawn into Italian musical life, so it is just as well to live in quiet obscurity!'

On 18 January Walton had written to Previn:

Thank you for the marvellous performances the other night. The press has up to now been unusually ecstatic . . . Actually Siciliani [RAI director of music] said to me that we must now see about *Troilus*. RAI want it in English with Janet. In fact it may be the solution of our troubles, except we shall have to work with the RAI Orchestra—not too great a hardship, especially as T & C is not so difficult as the works it has already played very excellently and I think it would work out for a record and I'm writing to Peter Andry [of EMI] about it. It seems to me to be our best hope.

Previn was then ill with measles, though at first it was thought to be mumps. 'We are devastated . . .' Walton wrote to him on 6 February. 'Most unpleasant—do be careful and keep your balls in a bag (a large one—they swell! as you doubtless know by now) though doubtless your doctor will have impressed this precaution on you. I once suffered from mumps, but I was only about 30, but I'm not sure I have ever been the same since—but I managed, so cheer up.' He wrote again on 17 February, after a recording of the Radio Audizione Italiana concert had been broadcast: 'The Concerto was superb, but the balance [by the radio engineers] of B's Feast was not at all good. The chorus often obliterated the orch. and vice-versa, and Shirley-Quirk sounded like the Almighty arriving for Judgement Day. In fact they made a balls of it if I dare mention those sacred objects at this delicate moment . . .' On 22 March he reported that his talk with Siciliani had gone 'reasonably well. Jack Buckley told me the rather alarming news that recording in Italy was more expensive than in England which is a bit of a blow. Anyway Janet can't do it before '76.' The plan for a studio recording in Milan was still extant in April 1976, seven months before the opera was at last performed at Covent Garden. Lord Goodman told Walton then that EMI was willing to finance a studio recording and that Janet Baker would be available in September 1977. However, when Previn's illness caused him to withdraw from the Covent Garden performances, the scheme was abandoned and EMI recorded the opera 'live' in the Royal Opera House, with Baker as Cressida.

Revisions of *Troilus* and providing a new ending to the Cello Concerto were the main creative endeavours of 1974. He was looking forward to Piatigorsky's return to the concert platform after illness and when this was cancelled (and before he knew of his old friend's cancer) he wrote to Diana Rix: 'It won't really surprise me about G.P. I did warn him that after 70 it needed more than a cello between one's legs to set one off.' He travelled to hear other composers' music, as two letters to Arnold show. On 28 September: 'Just returned from a marvellous week in Venice and happened to hear Mehta with the L.A. [Los Angeles] orch. He's a splendid conductor and the orch. first-class. Quite a revelation for the Italians. Music is rapidly disintegrating as is everything else for that matter. The Fenice is occupied by the orch. as a protest at not having been paid for 3 months and the opera in Rome is shut.' On 27 October:

I went to London last week for *The Bassarids* which he [Henze] conducted. It is a most wonderful piece as done by him. I had seen it twice before in Salzburg and at La Scala in Milan—both inadequate performances so much so that I had had no high regard for the opera and I am so glad that I went to hear it in the Coliseum. It is a first-rate company. Excellent soloists, chorus and orch. (the least good part but very adequate).[1] It just shows one how devastating a bad performance can be. I heard also his new piece—*Tristan*—which I didn't enjoy so much—spoilt for me by an electronic tape which confused the whole sound.

Arnold sent Walton a regular supply of cuttings of press criticisms which had amused, enraged, or mystified him. To one of these Walton replied (19 November): 'As it happens I've just been reading a notice of a new piece by that ghastly man Stockhausen which bears out everything you say about him. It's a Japanese religious piece called *Inori*. Keep away from it.' Walton attended a Stockhausen concert on one occasion in Rome. 'Mother's milk as far as I was concerned. I wasn't at all impressed by it as being something very advanced. It seemed to me rather old-fashioned.' He thought that 'all those people like Hába' of the early 1920s were 'far more advanced than what's taking place today. In actual disagreeable noises, there was nothing to exceed them.'[2]

Earlier in the year, 12 July, he had written: 'My No.3 is a non-starter! I'm doing Mag. & Nunc. for Chichester—about all I can manage if I can even manage that.' This had been commissioned by the Dean of Chichester, the Rt. Revd. Walter Hussey, for the cathedral's 900th anniversary. Hussey was a noted patron of the arts who, when vicar of St Matthew's Church, Northampton, had asked Walton for a work to mark the golden jubilee of the church in 1943. Walton had declined and Hussey asked Britten instead, with the result that *Rejoice in the Lamb* was written. Walton went to the first performance of this cantata on 21 September 1943, when a Henry Moore statue was also unveiled. He later rather regretted he had not written something and belatedly made it up to Hussey thirty-two years later. 'How I dislike the words of Mag. & Nunc.', he complained to Alan Frank on 10 July 1974, 'most uninspiring. But as the queer dean has been very generous, I feel I must try to do

---

[1] The cast of the 1974 ENO production was Josephine Barstow, Katherine Pring, Kenneth Collins, Gregory Dempsey, Norman Welsby, Tom McDonnell, and Dennis Wicks. Henze produced as well as conducted his opera.

[2] BBC *Music Weekly*, 28 Mar. 1982.

something at least respectably good.' (The commission fee was £500.) He completed the Magnificat and Nunc Dimittis in November 1974. They were performed at Chichester, conducted by John Birch, on 14 June 1975. Walton revised both settings before publication. He was by now finding composition increasingly difficult and arduous. He refused a request for a work for the 250th Three Choirs Festival in 1977 but accepted a commission in January 1974 from the Greater London Council to contribute to a concert to mark the Festival Hall's 25th anniversary in 1976. The idea was that Herbert von Karajan would conduct the Berlin Philharmonic Orchestra in the new work. Walton soon had misgivings. On 28 July 1974 he wrote to Alan Frank:

I've a feeling that I've taken on more than I can do and feel inclined to get out of the Fest. Hall commission especially as I've had no details and there is no guarantee about Karajan. He's the most dreadful shit. I heard from Walter Legge the other day and he was reminding me about the extraordinary perf. he did years ago of B.'s F. But has he done it since? Certainly not and he easily could have countless times—in fact it was only the presence of the British Army that forced him to do it in Wien! After all there are others, Malcolm A. and Lennox B. and I'm sick of writing ceremonial balls so I'm very inclined to cry off. What else am I supposed to be writing?—I've forgotten so you might try and remember for me. I know there's a symphony but it makes no headway at all. Help, help!

Four months later, on 28 November, he wrote again:

I'm still vacillating about the R.F.H. piece. I wish I knew what it is supposed to be. I couldn't be more uncertain about it or for that matter about myself. All my ideas seem to be so out of date and not 'with it'. With what! I should like to know? But anyhow I feel very out of touch and don't know what the piece should be like—shall I call it R.F.H. Celebration '76? Festival something or other seems and is so trite. It's awfully difficult to go on celebrating for 10 mins, too long in my old age!

Another four months passed and he was still uncertain. He wrote to Malcolm Arnold on 26 March 1975: 'I've let myself in for this fucking piece for the R.F.H. but I've still got a bit of time to get out of it if I make up my mind quickly. Ideas are sparse, bare and ugly so "per forza" I think I shall have to give it up. In fact I must face it that it is highly probable that I shall never write a note again. It doesn't depress me too much—nor I imagine anyone else.' Arnold told him not to talk nonsense and Walton replied on 24 May:

Thank you enormously for your encouraging letter and I believe it has had some effect for I believe I've started, but I'm not too sure and will have to wait to see how it goes on. But I think it's going to be alright . . . Unfortunately as well as having trouble with my muse I have been having a lot of bother with my blood pressure—up and down like a see-saw—most unpleasant. Luckily there's a very good young German doctor here. Of course the first thing he's done is to knock me off all alcohol, which sounds a bit of a bore but I'm sorry to say that I think I feel a bit better for it.

Five months later he wrote:

I too am disappointed that the wrong Malcolm should have been chosen as the new M of Q's M. [Master of the Queen's Music].[3] I did my best but nowadays cementing the cracks in the Commonwealth obviously takes precedence. I know nothing about it and for all I know it may be a good appointment. I've even considered trying to shove off onto him this dreadful R.F.H. piece which is ruining my life and doesn't progress at all. And to add to all my horrors I'm at the moment afflicted with a chronic bout of sciatica so all in all I'm feeling pretty low.

To Diana Rix on 18 September he wrote: 'I'm alright but my muse ain't. She's like you, showing no sign of giving in. Too late now! Only yesterday I was 70 and tomorrow I shall be 75 and looking and feeling it. What a life—what a death!'

The Festival Hall piece, which was to have been called *Adagio ed Allegro Festivo*, was finally abandoned when Karajan declined the GLC's request that he should conduct the new work; and on 4 January 1976 Walton told Arnold that the third symphony had been 'abandoned, or rather not yet started. Perhaps it will be sometime, but there are very few signs.' Instead of a new work for the Festival Hall anniversary, he agreed to provide a reworking of a not-very-old one by transcribing the guitar *Bagatelles* for full orchestra. First mention of this task is in a letter to Previn on 7 October 1975:

They are working out well, but not easy to do. What to add and what

---

[3] Sir Arthur Bliss died in 1975 and was succeeded by the Australian composer Malcolm Williamson.

to leave out, those are the questions! I thought of asking you to do them, but I realized it was not fair to ask you especially as you are so occupied. Messiaen, I see—that will keep you busy! I don't know it at all, only that it is longer than Mahler! [*Turangalîla-Symphonie*]. I don't know whether John Denison [manager of the Royal Festival Hall] will want to do my transcriptions. They are called *Vari Caprici* [*sic*] in their new garb, by the way. But it will be a first performance in their new garb—and much better than the pompous bore I was turning out for that occasion. It doesn't matter when, only that you will do them sometime.

By 16 April 1976, when he wrote to Malcolm Arnold, he had completed the task:

I've been much occupied and pre-occupied by scoring those 5 *Bagatelles* for a large orchestra and they appear or rather re-appear under the name *Varii Capricci*. They are to be done at the F.H. on May 4th at its 25th anniversary concert. I fear I've not done them very well, in fact I should have asked you to do them, but having let myself in for something for the F.H. I thought these would be easy. I couldn't have been more mistaken. I found them full of pitfalls especially the last one which is musically very much changed and hurriedly scored and full of wrong notes! I'd left it to the last minute as the OUP were panicking about part-copying. The only thing that is intact is the dedication . . . We shall be at the Ritz if it's still there. Luckily the Labour Gov. put it on the scheduled list so it should be alright for the future, tho' I'm sure the inside will be fairly buggered up.

Fastidious and conscientious as ever, after the first perform-ance of *Varii Capricci* (by the London Symphony Orchestra conducted by Previn) Walton rewrote the last movement. The first performance of the revised version was at a BBC Welsh Symphony Orchestra studio concert in Cardiff on 28 January 1981, conducted by Owain Arwel Hughes. The music was later used for a ballet by Frederick Ashton. The final revisions were sent from Ischia to his old friend two days before Walton's death in 1983 and the ballet, dedicated 'to the memory of my lifelong friend', was first performed in the Metropolitan Opera House, New York, on 19 April 1983 and at Covent Garden on 20 July of the same year.

Having returned to Ischia from the Festival Hall première of *Varii Capricci*, Walton in May 1976 wrote another of those brief fanfares which had qualified him, since Elgar's death in 1934, to be unofficial master of the monarch's music no matter who held the title officially. This was for the inauguration of the new Lion Terraces at the London Zoo in Regent's Park and was written at the request of Walton's friend Solly Zuckerman, president of the Zoological Society of London. It was played by the trumpeters of the Royal Military School of Music, Kneller Hall, on 3 June, when the Queen arrived to open the new terraces and when she attended an evening reception to mark the Society's 150th anniversary. The *Roaring Fanfare* was dedicated to 'Solly Z. that Lion of Lions.' He had also, in October 1974, written a fanfare for a London Weekend Television programme to mark the opening of the National Theatre on the South Bank. Because of delays in completing the theatre, this was not recorded until April 1976, when it was decided to record a revised version. Several versions were then produced in different styles and were used throughout the programme when it was transmitted on 21 August 1976.

But most of 1976 was occupied with revisions of *Troilus and Cressida* and anxieties about its revival at Covent Garden in November. On 30 January he wrote to Diana Rix: 'C.G. has still not found a tenor for T. & C. What about Placido Domingo? He speaks and sings perfect English—knows the piece, tho' he's not sung it. He'd fill the house and C.G. would not lose so much if anything. But Tooley, I suspect, is frightened of flying high!' In a letter to Malcolm Arnold on 1 September he wrote:

I am trying, but not succeeding, to work—nothing more irritating—a state of life which I ought to have got used to by now. I'm supposed to be writing an anthem for Rochester U.S. and another for the Queen's Jubilee next year on a poem by Paul Dehn. It's not a very inspiring poem. How can it be when poor Paul is fading away with lung cancer and I hate to chivvy him too much. And various other fritteries of the same ilk . . . We are both well really except I get very depressed . . . We shall alas be no longer staying at the Ritz; it's now prohibitively expensive, so we have to make do with the Waldorf in Aldwych, which has its advantages in that it is within walking distance of Cov. Gar.

When Walton arrived in London for *Troilus and Cressida*, Alan Blyth interviewed him for *The Times*. He spoke about future plans and present difficulties.

When you reach 70 everything seems to go downhill. I've had another symphony in mind for some time but I haven't yet put down very

much on paper . . . Another idea I have is for a clarinet concerto . . .
One of my best-sellers at the moment is *The Bear*. It will have 25
performances this season in Germany. I would very much like to write
a companion-piece for that . . . I still work regular hours. It's the only
way of getting anything done. I usually work near the piano just to
make sure I'm on the right notes, but I work out most of the music in
my head. Composing has never come easily to me, and the older I get
the more difficult it seems to become. If one's not careful one tends to
become repetitious; an idea comes into your head and you find it's the
same one you had ten years ago. Then I see the headline: 'Walton
makes no progress.' I do believe that a composer must stay true to
himself. I don't believe in trying to keep up with every fashion as
Stravinsky tried to do. It's like having your face lifted.[4]

The Waltons had gone to London in good time to supervise
rehearsals of the opera. In the days before the performance it
became obvious he was not well. He seemed to be 'somewhere
else' and twice fell in his hotel after losing his sense of balance.
After the final performance on 30 November he collapsed
during a celebratory dinner at the Garrick Club and could stand
and walk only with difficulty. A slight stroke was suspected
and a brain scan was carried out. He had not had a stroke but
his cerebral circulation was severely restricted. Lady Walton
took him back to Ischia, where sleeping-pills which he had
been given in London caused him to have hallucinations that
he was being burned alive in an aircraft or eaten by insects.
After the pills had been discontinued, he slept for most of the
next few weeks, unaware that his old friendly rival Britten had
died, aged sixty-three, on 4 December. On 23 December he
wrote to Peter Pears: 'My thoughts have been with both of you
frequently, and with the tragedy that such a genius as Britten
died so young. I'm relieved to hear that at least for Ben it was
peaceful, gradual, and not too painful. My heart goes out to
you . . .'

Su Walton wrote to Malcolm Arnold on 5 February 1977:

He has been very ill . . . and among other things quite convinced his
beautiful opera was a flop after which he fainted on the stage of the
opera house and cannot remember anything further. Luckily lots of
letters telling him how lovely T. & C. was have arrived and we have
told him repeatedly that he never got on the stage or fainted so he is a
little happier. I am overjoyed to say that he is enormously recovered,

[4] *The Times*, 13 Nov. 1976, p. 9.

he is walking again without help and though still getting very tired after the treatments I think he will be able to enjoy his birthday concerts . . . We hope to attend the Solti/LPO concert at the RFH on March 1st, then retire for diet and treatment to Shrubland Hall nr. Ipswich and return to London March 23rd for Previn/LSO on the 27th . . . This visit we shall stay at the Savoy.

A course of acupuncture brought about a dramatic improvement in Walton's health. He had to abandon some commissions, among them a work for the London Sinfonietta and a choral piece intended for St Alban's Cathedral. But he completed the *Antiphon* (George Herbert's 'Let all the world in every corner sing') for the 150th anniversary of St Paul's Church, Rochester, New York. 'I'm just finishing *Antiphon* for Rochester (Kodak),' he told Alan Frank on 13 June. 'Of course it has taken me a lot of bother. How stupid I am to worry, but I find I've a great antipathy for the organ and don't know how to write for the bloody thing . . . And for only $1,000. I must be mad.' *Antiphon* was first sung in St Paul's Church on 20 November 1977. Walton also wrote a carol, 'King Herod and the Cock', for the King's College, Cambridge, Festival of Nine Lessons and Carols conducted by Philip Ledger on Christmas Eve. And he managed thirty seconds of title music for the BBC's televised series of all Shakespeare's plays.

26

The Haunted End, 1977–83

ALTHOUGH confined to a wheelchair, Walton attended his seventy-fifth birthday concerts in London in March 1977. He still felt neglected, although not as much as fifteen years earlier. He had semi-seriously written to Alan Frank on 9 February 1976: 'And now I think I ought to be given a prize for my 75th, the Aspen–Ravel what have you. But I think with a little Lockheeding something could be arranged. B.B.'s had the lot and I think I might be allowed one! Approach who shall be approached. B. and H. obviously know who.' He naughtily broached the same subject to Diana Rix:

It has just struck me after reading a wretched mag. called *The Composer* containing an article on prizes, most of which, some of considerable sums, have been awarded to a certain B.B., now I feel having reached the venerable age of my 75th year that one of these might be awarded to me. Not moving in those circles I'd never heard of them except in a vague way—the Aspen or the Ravel. Not having a B. & H. of a publisher to do a little spadework, perhaps H.H. [Harold Holt] would know the right approach . . . I think I qualify, don't you?

At the Fishmongers' Hall on 22 March Walton was presented with the Incorporated Society of Musicians' 'Musician of the Year' award. Three days later, at a concert organized by Lina Lalandi in the Plaisterers' Hall, some of the suppressed numbers from the original *Façade* were revived, with Richard Baker reciting and Charles Mackerras conducting the English Bach Festival Ensemble. Listeners were taken aback by the avant-garde nature of the pieces compared with the *Façade* they knew so well. Walton agreed to their publication, but, while checking the proofs during 1978, he decided to reject three of the items, replace them by others, and to rework and reorder the music. The result was known as *Façade 2*. The first performance was at the Aldeburgh Festival on 19 June 1979, when Peter Pears recited and Steuart Bedford conducted the ensemble. Bedford conducted a new recording of *Façade* and *Façade 2* later in the year, when the superb reciters were Cathy Berberian and Robert Tear. Walton dedicated *Façade 2* to Berberian and a facsimile of the manuscript score was issued with the recording. He had heard her at Siena in August 1978 when there was a performance to mark the fiftieth anniversary of the ISCM *Façade*.

It went well [he wrote to Alan Frank on 2 September], except for many mistakes in the parts (I wasn't at a rehearsal) and Cathy Berberian was quite excellent, partnered by Jack Buckley of the Brit. Coun. in Rome. Enthusiastic reception, to me rather surprisingly. It will probably be my only work to be heard on its 100th anniversary! I've been revising some of the unused ones at OUP's insistence. Leftovers rehashed. What a mistake!

He was always wary of 'rehashes' of early works. When Britten revised some boyhood songs as *Tit for Tat*, Walton remarked: 'If you ask me, it's a case of *Tat for Tit*'.

Walton's first letter to Malcolm Arnold after his illness was written on 4 June 1977. He was delighted to hear of the success of his friend's piano concerto but added:

I'm sorry you have to endure the boredom and horror of losing so much weight—it is like losing that useful tool! But take care about the weight, it was while Su was trying to do the same that I started my collapse and I think if she hadn't discovered the acupuncturist I should have been completely paralysed by now—as it is, I am fairly well, if only I hadn't let myself in for composing some piddling music! Fucking fool—or rather non-effing fool that I am.

It disappointed him that he did not write a new work for the Queen's Silver Jubilee in 1977, but he released an unpublished work from 1962 for the Young Musicians' Symphony Orchestra concert in St John's, Smith Square, on 25 June. This was the original orchestral version of the *Prelude for Orchestra*, the prelude written for Granada Television, which the company had handed over to Gilbert Vinter to convert into a march for wind band. He was well enough—and determined to be well enough—to visit London in November for a lunch given by the Queen at Buckingham Palace for members of the Order of Merit to celebrate the 75th anniversary of the creation of the Order by King Edward VII. Five of Walton's closest friends were members, Laurence Olivier, Frederick Ashton, Henry Moore, Kenneth Clark, and Solly Zuckerman. Walton was made especially welcome by the Savoy Hotel on his last visits to London. He had a nostalgia for English food and drink—steak-and-kidney pudding and bitter beer. An obliging waiter would go into the Strand to fetch him a pint of his favourite brew.

Rumours of a third symphony persisted. In the February 1978 issue of *Hi-Fi News and Record Review*, Previn, referring to the page of score Walton had sent him in January 1974, told Andrew Keener: 'There's a Third Symphony on the way ... I've had the first page, with a very kind inscription and dedication, so I've duly framed it, but I'm still waiting for further pages.' But writing to Roy Douglas on 17 February, Walton said:

I've rather foolishly said I was writing a 3rd Symph. Not a hope! though I have done a few sketches lately. Still more stupidly I said I'd do some piece for the American wind bands before I realized what scoring for it would be like, let alone thinking about the music. Also, more disastrously, a piece for the Brass Band Competition. I had to get out of that too! Not having tried to compose, except for an odd anthem or two, I find it more difficult than ever and, as you know only too well how difficult I found it always to get going, even I myself am now completely horrified at my impotence, there's no other word for it, and I am afraid it won't pass and that I shall never be able to put pen to paper again. I would much like to do a Brass Band piece for they are such wonderful players, especially the Grimthorpe [*sic*] and Bess [*sic*] o' the Barn bands. There is a splendid record with Elgar Howarth conducting them in various pieces by such composers as Henze and Birtwistle. If you've not already got it, get it at once, since you have gone brass bandy too! You'll be very surprised and taken by it, it's fantastic. On the whole I'm much better than I was, especially now that I've come to face that I've no real necessity to write anything any more. After all, better composers than myself stopped, Verdi for instance soon after 70 and didn't start again till he was about 78 or so ... Susana is well (I don't know what I'd do without her) and sends regards. Write to me sometime when you have nothing better to do.

On 17 April he wrote to Alan Frank: 'I feel very out of things and work doesn't progress—far from it—tho' I'm getting the shape of Sinf.III in my head, but not on paper (allegretto—

scherzo diabolico—slow finale). Bound to be a flop with the slow mov. at the end.'

The American wind band piece had been commissioned by Professor Robert Reynolds, Director of University of Michigan Bands, in 1977, on behalf of the 'Big Ten Band Conductors' Association', an organization of ten mid-western universities.[1]

No music was written during 1978. The summertime visit to London was spent seeing doctors—'I for my old oesophagus,' he told Malcolm Arnold, 'for which has been found a cure. It seemed to be going to be most unpleasant. It wasn't at all and has had a most beneficial effect and I can now eat anything without choking and, what's more important, I can drink almost anything with moderation!' But, as he told Tony Palmer, 'I loathe growing old. Getting gaga and incapacitated, I don't like it. I don't see how anyone could.' To Alan Frank on 17 November he wrote:

I seem to be alright, even regarding my eyes, in fact the only thing that eludes me is my apparent inability to write a note of music down on paper, in spite of the tempting 'carrots' offered in the form of my doing pieces for brass band (you'd think I could manage that one), for American wind bands (too many instruments and they are paying me in dollars, so until it recovers its health it holds no temptation for me to proceed!). I keep hoping that the Sinf.III will somehow materialise, but it shows little sign of doing so.

In 1979 he produced another fanfare, a *Salute to Sir Robert Mayer on his 100th Birthday*, which was played by twelve trumpeters at Sir Robert's centenary concert in the Festival Hall on 5 June 1979. After amendment and revision it was published as *Introduction to the National Anthem*. He was determined to provide something for the Grimethorpe Colliery Band, as he told Roy Douglas in his last letter to his old friend and helper, written on 21 April 1979:

Sorry to have been a little dawdling about thanking you for your kind (both encouraging and depressing) letter for my birthday. I would like to be able to say it is because I have been seized by a frenzy of inspired work, but I fear that that is not quite true. However I have done about 3 mins. of a piece for a Bra. Ba. competition, which in all probability will be sent to you by OUP to discover if I'm still in my right mind! I

---

[1] Michigan, Ohio State, Northwestern, Wisconsin, Purdue, Minnesota, Indiana, Iowa, Michigan State, and Illinois.

know only too well about commissioned works; the only thing I can say about them is if I'd not had a commission I should have produced nothing at all, which would probably be just as well. I shall proceed with the dotty Br. Ba. piece as I'm getting rather intrigued by it. Anyway it won't frighten the Bands, for though the piece is slightly Henze-ish, it is not at all Birtwistle-ish. Which leads me to Elgar Howarth—a remarkable young man who I suspect will turn out to be a remarkable composer. You know of course *Fireworks*[2] for Bra. Ba. without which I shouldn't be attempting to compose this piece. It is interesting for me to learn that he is the conductor of the Tun. Wells S.O.[3] What excellent experience for both parties and for you too, filling in the missing instruments. The kind of experience I've missed and miss now more than ever . . . Give my regards to Gary H. [Elgar Howarth]. And I promise to get down to work on this rather silly piece.

Whether he thought the piece too silly, it was never completed and instead the band received an arrangement of his 1935 Cochran ballet score *The First Shoot*, for which he had always had a soft spot. He had twice before used its waltz in films. On the score he thanked Elgar Howarth for 'expert advice' and dedicated it 'In mem. C. B. Cochran and his Young Ladies.' Howarth conducted a performance in Goldsmiths' College, London, on 19 December 1980, which was filmed and partially used in Tony Palmer's moving television profile of Walton, *At the Haunted End of the Day*, first shown on 19 April 1981. The first public performance of the piece was at a Promenade Concert in the Royal Albert Hall on 7 September 1981. 'It will, I hope, give a good boost to the somewhat dilapidated state of the W.W. fortune,' he wrote to Alan Frank. The Palmer film, containing archive material and also a long interview with Walton, with memorable shots of Susana and him in their home on Ischia, was a sensitive and revealing portrait. He wrote on 25 April to Peter Heyworth:

Thank you for your kind words about that TV biography. I am glad that that 'crutch' is at last safely buried! I may add that your articles in *The Observer*, which is occasionally obtainable, are the only bits of musical criticism that are worth paying attention to. Come and see us

---

[2] *Fireworks*, composed by Elgar Howarth, 1975, is a modern classic of the band repertoire.
[3] Roy Douglas lived in Tunbridge Wells and played percussion etc. in the orchestra there.

whenever you feel like it. It could be a good place for you to finish
your book. Love from us both.

He was a jovial and generous host to friends who stayed with
them. Sunbathers by the pool were liable to find themselves
sprayed with cooling water from the hose, especially if they
were Diana Rix or Gillian Widdicombe, when Walton would
gleefully appear from behind an exotic shrub, 'like Alberich
with the Rhinemaidens', as a victim said.

Walton's mood at the start of 1980 can be gauged from a letter
written to Alan Frank on 27 February. He had always enjoyed
Alan's letters and was extremely glad that he continued to
receive them after Alan had retired from Oxford University
Press—they kept him in touch with gossip.

It has been a pretty ghastly winter [Walton wrote], cold, wet and grey,
an exceptionally bad one as far as I can recall, but except for some bad
days in March (there are always bad patches during that month) it
should get better. But things are in a pretty bad way here, especially
politically and one can't ever open a paper without reading about
some awful shoot-up and 'terroristi' goings-on. The authorities seem
to be quite powerless to cope with them. Luckily being on an island it
hasn't as yet started here, but I expect it will any minute. All the same
I don't think we shall be visiting England this year. What with our
rebuilding programme [on the five houses they owned to rent to
visitors] we are in quite appalling debt—a nightmare—but if we can
survive the next year or two, things should be quite okay, in fact v.
good—but for the time being, not so good. But between the PRS and
the OUP I think we shall get on. After all I get a good, even fat, income
from each (so does OUP from me, if it earns double what I do!) and the
houses have been taken over by an English company called Villa
Venture. Our first year with it started on May 1st and if successful it
will continue for five years at a good rental, rising with the cost of
living. Su, in fact, is a marvellous businesswoman, and it will be a
great thing for her having all the administration problems taken out of
her hands—so wish us luck. I have got the Stravinsky book, but as yet
have only just dipped into it, but I shall go on dipping for some time.
Actually he and I always got on—it was B.B. that for some unknown
reason he was against and went out of his way to be as offensive as
possible. In fact it was rather a compliment to Ben, I think, though he
didn't intend it as such, that I'm sure. Vera S. must have been a
strikingly good-looking woman—still was when I last saw her not so
long ago . . . I'm supposed to be doing an unaccompanied solo piece
for Rostropovich, which I very stupidly let myself in for and I now

find it to be an extremely difficult thing to bring off. But there it is—
I'm always doing this kind of thing without thinking how difficult it
would be beforehand.

To Previn he confided on 30 April that

> my projected Symphony has not responded to treatment and is where
> it was the last time I wrote you, which may be disappointing for you—
> it certainly is for me. At the moment I am writing a *Passacaglia* for solo
> cello for Rostropovich. I find it not very inspiring, on the other hand it
> avoids the necessity of scoring, which I find so arduous to tackle in my
> old age. The fact is that I've found my creative spirit to be flagging, and
> anything I now try to write I find out that I've done it much better
> before, so I don't proceed.

On 5 February a year later he reported completion of the
Passacaglia to Alan Frank. 'I don't know (or for that matter care)
when he plays it. It's not really a piece for public perf. I don't
think the Bach pieces for solo vl. or vlc. are either, marvellous
tho' they may be.' He added: 'Health is getting pretty
precarious, especially this business of walking. Very boring
and incapacitating being hardly able to walk even from the pfte
to my desk. So it makes work even slower than ever! In fact
better give it up. I'm deteriorating rapidly & hope I shall reach
my 80th!'

He had met Mstislav Rostropovich at an Aldeburgh Festival.
Walton asked him when he would play his concerto. 'You
write me new work, and I will play new work and old work,'
was the reply. Reporting this to Diana Rix in May 1975, Walton
added: 'He obviously doesn't know me!' Nevertheless he
composed the *Passacaglia*. Rostropovich meanwhile had asked
various composer-friends, including Walton, for works for the
National Symphony Orchestra of Washington of which he had
been appointed principal conductor in 1977, the year before he
was deprived of his Soviet citizenship. Composition was even
more laborious than usual for Walton because of cataracts of
both eyes—'there's no fool like an old fool, how could I have
been so stupid to think that I could do it' was the familiar
plaint to Alan Frank, but he worked on what was to become
the *Prologo e Fantasia* throughout 1981. 'I've done about four and
a half minutes', he wrote again to Alan Frank, 'but as usual find
myself bogged down—lack of a good tune. May be one will
turn up. I doubt it.' This was the constant burden of Walton's

letters. 'I'm desperate to find a tune—if I do the OUP must pay me double royalties.' In his waves of pessimistic depression, he told Alan Frank, he was kept going by Susana's optimistic outlook.

He went to London for the first performance, on 20 February 1982, of the *Prologo e Fantasia*, which was given while the National Symphony Orchestra of Washington was on a European tour. It is a short work, beginning with rarified passages for the string sections in turn. These build up to a climax, after which the trombone introduces the Fantasia, in Walton's 'scherzando' vein and including a 'make-believe' fugal passage. The work may be slight, but the prologue has a strange other-worldly air that is curiously haunting. Its perfunctory ending, just when one is hoping for an extended coda, is frustrating—and from it we may infer the tragic bewilderment pervading Walton's creative struggle at this period. Three weeks later, on 16 March, Rostropovich gave the first performance of the *Passacaglia* during a London Philharmonic concert in the Festival Hall. Immediately afterwards he gave the second performance as an eightieth birthday greeting to the composer. There are ten variations on the eight-bar theme. The first three explore the instrument's lower registers. In the fourth the higher register is reached with the theme extended. The tempo is faster in the fifth and sixth variations. In the seventh, 'tranquillo espressivo', the cello's flowing tune has plucked chords as accompaniment. The eighth and ninth variations run together as a quicksilver 'scherzo' and the Finale is a flurry of repeated notes.

London now had a new concert-hall, the Barbican, and there were Walton eightieth birthday concerts there and at the Royal Festival Hall on 29 March. Walton went to the rehearsal of the London Symphony Orchestra at the Barbican, when Nobuko Imai played the Viola Concerto, and in the evening heard Previn conduct the Philharmonia in the Violin Concerto, with Kyung-Wha Chung, and *Belshazzar's Feast*, with Thomas Allen and the Philharmonia Chorus. The audience's reception for *Belshazzar* moved Walton to tears—and there were tears, too, in the eyes of those who saw the frail, white-haired, gaunt-faced old man and remembered the debonair figure of the 'white hope of English music' when it seemed he would never grow old. But the concert he loved most was at Westminster Abbey, when Simon Preston conducted the Coronation Te Deum. After a short stay in Suffolk, the Waltons returned to the Savoy, where he was taken ill again and rushed to the intensive care unit of St Thomas's Hospital. He had nearly died and again had alarming nightmares. Yet his sense of humour later made capital of the experience. Had he experienced anything of 'another life', he was asked. Yes, he had distinctly heard a fanfare—by Arthur Bliss, of course. (Even if this was the invention of Gillian Widdicombe, it had the authentic Walton touch.)

By the end of April he had recovered sufficiently to return to Ischia, where a local celebration of his birthday had been arranged in the castle. Tony Palmer's film was shown and there was a photographic exhibition. The Koenig Ensemble, conducted by Jan Latham-Koenig, performed *Façade*, with Marghanita Laski and Jack Buckley as the reciters. This was the last 'live' performance of his music that Walton heard. He made a bizarre appearance in Palmer's nine-hour film *Wagner*. Walton and Susana were filmed on 22 July in front of a backdrop on the terrace of their home, as the King and Queen of Bavaria. In August Walton went to London for removal of cataracts. The first operation went well, but after the second he again had to be taken into intensive care. On 2 October, in Ischia, he wrote four bars of *Duettino* for oboe and violin 'to continue as an exercise in music theory.' This little 'andante espressivo' was composed for the young son and daughter of his Köchel, Stewart Craggs, and it was the last music he wrote. But not the last music he thought about.

I'm having a go at a motet [he wrote to Alan Frank on 15 January 1983], studying Palestrina like mad, with little effect. It reminds me of my choirboy days when Pal. seemed to be the most boring composer there ever was. I'm now much taken with his music, especially 'The Song of Songs'. There is an excellent recording by Michael Howard and his group. Really splendid. I'd like to think that I could write something even 100th part as good. We'll see!

And there was something else. On a visit to England in 1975 he had greatly admired a performance of *Belshazzar's Feast* conducted by Owain Arwel Hughes. Some years later Hughes became conductor of the Huddersfield Choral Society and wrote to Walton asking him to compose a work for the choir's 150th anniversary in 1986 as he had done for the 125th anniversary in 1961 (the Gloria). Any kind of work, Hughes stressed, short or long, but he himself hoped Walton might compose a Stabat Mater. There was no immediate response, but late on the evening of 7 March 1983 his telephone rang in London. It was Walton, from Ischia, to say that he had become quite excited about the idea and would write a Stabat Mater. Hughes must come out to Ischia to advise him as it progressed. Early the next morning, 8 March, exactly three weeks short of his eighty-first birthday, Walton awoke feeling unwell and having difficulty in breathing. A doctor gave him a booster for his heart. While Susana propped him up on pillows, Walton said: 'Don't leave me, please don't leave me,' a last poignant manifestation of the fear of loneliness that had perhaps begun in the train on the way to Oxford in 1912, or even earlier. As the doctor was writing a prescription for an oxygen cylinder to be sent from the hospital, William Walton died. His body lay covered by a silk brocade bedspread which had belonged to Alice Wimborne.

The journey which had begun in Oldham in the year after Edward VII had ascended the throne ended in a Florence crematorium in bizarre circumstances of which the only redeeming feature was that they would have appealed to Walton's sardonic sense of humour. The eight mourners had difficulty finding the crematorium. The zinc-lined coffin arrived on a dustcart, brought by workmen in overalls. Susana Walton, in *Behind the Façade*, says that one of the mourners—a woman—insisted that the coffin should be opened so that she could take a last look. They then adjourned to a tavern while waiting to collect the ashes.

On 20 July 1983 a memorial stone, donated by the Performing Right Society, was unveiled in the North Aisle of Westminster Abbey, near to the memorial stones for Elgar and Britten and the graves of Purcell and Vaughan Williams. His ashes are in a rock in Ischia above the garden which Susana and he created and loved. Pointing out the site he had chosen to a friend, he had made a characteristic jest of the matter. 'For my ashes,' he said. 'Like cricket, you know.'

27
Epilogue

ROY CAMPBELL's comment is the chief clue. Even as an Oxford undergraduate, Walton showed him 'how a man may live for his art.' However nonchalant he appeared—as someone said of him, 'he intrigues interest by appearing not to be very interested'—and whatever he said in deprecation of his gifts, the fact remains that Walton was totally committed to the art of composition for sixty-seven years. He composed something every day to the last day of his life, even if it was only two bars which he later threw away. 'He kept office hours,' Julian Bream observed in 1981. Walton was a supremely professional, dedicated composer, or, as he put it, 'I was so damned stupid all I could do was to write music.' He admitted that composing was difficult for him; and perhaps he lacked the resourcefulness of Henze and Britten. 'Fluency's not a question of actual writing the notes down. It's thought, the actual creative business. Creativity and facility aren't quite the same thing. When it's creating something like a symphony, it's not like writing film music. One can probably write two hours of film music in a couple of weeks and it'll take you a couple of years to write half an hour of a symphony.'[1] He was profoundly self-critical and claimed that the most important instrument in his repertoire was an india-rubber. 'Without an india-rubber, I was absolutely sunk. So I surrounded myself with them, and I seem to have spent my entire life rubbing out what I've written.'[2]

He was not by temperament a composer who wanted to change the world by his art. He was not propelled by a moral, political, and social conscience like Michael Tippett, nor did he seek some kind of spiritual redemption for himself and for mankind through his music, like Mahler. He came nearer to the aesthetic of Richard Strauss. His music deals with human

[1] BBC *Music Weekly*, 28 Mar. 1982.
[2] *At the Haunted End of the Day.*

emotions, not with spiritual values and perceptions. 'Involved but detached' might be his motto. *Belshazzar's Feast* is a human drama, not a religious experience. His other works on religious subjects are secular in mood. Even his most tempestuous music, such as the First Symphony, is, like Strauss's, presented with supreme craftsmanship, but owes more to the Gallic example of Ravel and Roussel and to the rhythmic excitement of Stravinsky than to German models. But something deeper than the obvious emotion being expressed throbs beneath the surface of all his music—anger, pain, frustration, a sense of loss, one cannot be sure exactly what it is, but one senses it and it is disturbing and often uncomfortable. What one does not sense, except once or twice, is that consoling compassion which makes Elgar's music a source of solace.

Inevitably, criticism of his music has tended to divide it into two sections, pre–1940 and post–1945. It is not such an unfair division, because there is a distinct change, not so much of style as of tone, in the post–1945 works. It is reasonable to suppose that he underwent some kind of crisis of creativity after the Violin Concerto, which he declared at the time to be the worst music he had written. One may ascribe that remark to the despondency and insecurity which were part of his make-up and which came to a head when a composition was giving him problems, but the fact that he also said that he would then compose chamber music to teach himself composition lends substance to the theory. Although the incursion of film music during the war delayed this re-educative process, it is significant that, at the first opportunity, in 1945, he began work on a string quartet followed by a violin sonata. It is as if he had suddenly become painfully aware of the gap in his musical education caused by the absence of any strict academic training and disciplines.

But it is unfair, and unperceptive, to maintain that all the eggs in the pre-war basket were good and that the post-war basket contained the same eggs, only addled. The question is whether one believes that he failed to develop within his established idiom and that his manner often became manner-ism, that he effectively reworked his idiom but did not develop or enliven it, or that he consistently displayed what William Glock called a 'strenuous musical morality' by remaining true to himself and to an acute self-knowledge of his range and limitations such as few artists have displayed. The assumption that a composer's development must be related to the rate of change in his idiom bedevilled much criticism during the mid–twentieth century. Few composers, having reached an artistic maturity, explore wholly new ground. It was said that Stravinsky had always 'something new' to say, but it is now clear that he brilliantly showed new aspects of the old Stravinsky. Admirers of Vaughan Williams spoke of his continual 'exploration' but after *Job* in 1930 he illumined (magnificently) aspects of an already formed personality. Elgar, Mahler, Strauss, Prokofiev, and Sibelius are others of whom the same may be said; and Britten, whom—ironically in relation to Walton—some now regard as having 'fallen away' after *Billy Budd* in 1951.

Walton, like most composers, knew only one way to compose and it would have been disastrous for him to court the approval of avant-garde critics by adopting serialism or other techniques which were foreign to his nature in every way. Besides, he had himself forged what many a lesser composer would envy, a distinctive, immediately recognizable style and tone of voice, conservative in essence and outlook (though sometimes overlaid with dissonance) and based on classical models. He thus was able to convey his sanguine musical temperament, in which wit and vitality were accompanied by an understated vein of poetic melancholy. This was achieved technically by melodies which are compounded of wide intervals (often sevenths), use of conjunct motion, the alternation of held notes and *gruppetti*, and falling—drooping, rather—cadences. His harmony is mainly diatonic, with the additions of clashes of adjacent notes and major–minor modality (the famed 'bitter–sweet' quality). Rhythm is jagged, gritty, and strewn with the syncopations of jazz. His music has persistent and vital energy, springing perhaps from the ruthlessness that was part of his nature. For structural form he prefers the classical patterns but varies them in emphasis and perspective and favours a motto-theme, which, as in the concertos, often means that a work's end is in its beginning. As an orchestrator, he is in the highest class, with the clarity of Ravel, the richness of Elgar, the precision of Stravinsky. His use of solo woodwind and his distinctive writing for strings and brass are immediately recognizable. 'There sometimes seem to be too many notes at a first glance,' Previn remarked, 'the string parts are diabolical. But every semiquaver is necessary to make that sound.' With these tools, Walton wrought one of the finest symphonies and three of the best concertos of the first half of the twentieth century, an achievement never to be underestimated.

Then why is it that the widely held opinion persists that Walton 'went off' after the war, that none of his post–1945 works (with the exception of the Cello Concerto) has yet achieved a popularity comparable with that of, say, the First Symphony, and that the 'something' that Harty believed should happen to Walton's 'soul' did not? One simple answer is that people, including many critics, know the pre–1940 works so much better because they are played so much more often. This does not necessarily imply that they are, *ipso facto*, better works. The reluctance of conductors and concert-managers to substitute a lesser-known, and therefore riskier, work for a sure-fire favourite is a regrettable but inescapable feature of musical life. It was Walton's fortune, but also his misfortune, to compose some early 'hits'. *Façade*, initially as a ballet suite, the First Symphony, *Belshazzar's Feast*, and the Viola and Violin Concertos were five masterpieces which made Walton's name, attracted immense publicity, and became rapidly established. Just as Strauss suffered because his most easily appreciated and sensational works were composed up to 1911, the year of *Der Rosenkavalier*, and he too was (until only quite recently) considered to have fallen away for the next thirty years until his so-called Indian Summer, so Walton suffered—and as it seems Britten may also, when one considers that the post–1950 works are heard much less frequently than the early masterpieces which made his name.

There is also much more of Walton's post–1945 music from which to choose. It is said that he was a slow composer, but he was not unprolific. A glance at Stewart Craggs's catalogue of his works is sufficient to show that the works up to 1940 occupy only just over a third of the space, forty entries compared with fifty-nine after 1941, of which one is a large-scale opera which occupied Walton for the best part of eight years. As far as the general musical public is concerned, the pre–1940 Walton is easier to grasp at a first hearing. The First Symphony's emotional impetus is so great that he could not have repeated it if he had wished—and he would not have wished to repeat the personal experience that communicates itself so powerfully to the listener to the music. *Belshazzar's Feast* too is a swift kick in the teeth to the audience, delivered with a touch of showmanship that is as effective as it is unsubtle. In the post–1945 works there is a mature synthesis of mood, feeling, and means of expression that is musically an advance on the early works. The *Hindemith Variations* is a consummate achievement, but it requires more from a listener than, say, *Portsmouth Point*. The most overtly emotional of the later works is the Cello Concerto, which is why it alone has rivalled the popularity of the pre–1940 compositions. Walton, like Elgar, was in touch with audiences and he believed that, if he was not 'modern' enough for the critics, he was still too modern for many audiences—and for many orchestras, which found his music difficult to play. This is perhaps why lighter works like the *Partita* (1957) and the *Capriccio Burlesco* (1968) have not become as popular as *Scapino*. They are music devoted in gloomy times to 'the great cause of cheering us all up' and to neglect them as we do is to render the world of music much drearier. But to perform them really well, that is another matter.

The theory is advanced, too, that, by 'exiling' himself in Ischia, Walton cut himself off from contact with the wider world of music. He certainly missed the competitive element on which he had thrived. Friends who visited him noticed how eagerly he went to collect the mail to see if there was a letter from some friend in London. But no one can say with certainty that Walton would have composed differently or more or fewer works if he had remained in London after 1948. It is just as easy to cut oneself off from the world of music at large in Aldeburgh or Worcester or Orkney as it is in Ischia. Moreover, he lived in an age of easy communication and of the radio, the long-playing record, and tape-recording. All his life Walton had assiduously studied other composers' music. He continued to do so in Ischia. He frequently flew to London to see operas or to hear an important new work in the concert-hall. Oxford

University Press kept him constantly supplied with scores and recordings, whether it was of the Beatles, Mahler symphonies, or Stockhausen. He knew as much as, probably more than, most of his contemporaries about the music of Stravinsky, Ravel, Bartók, Villa-Lobos, Henze, Nono, and Schoenberg, not to mention Hindemith, Britten, Vaughan Williams, Shostakovich, Prokofiev, and Arnold. Among works he asked to be sent to him in Ischia were Berio's *Sinfonia*, several by Henze and Nono, all Shostakovich's string quartets, and Mahler's Sixth and Seventh Symphonies. He admired Koechlin's *Les Bandar-Log*. He studied Vaughan Williams's later works, particularly enjoying the Eighth Symphony ('what a surprise!' he commented). In 1972 he asked for Gerhard scores. 'I've only just lately cottoned on to him through the Dorati recording and the N. Del Mar. At first hearing one thinks, here goes, another Stock.—Cage—Foss—Xenakis, but no! he is far better and an individuality emerges.' At the same time he had been 'doing a little reassessment of S.'s [Stravinsky's] music and I'm disappointed to find most of it wears very thin—the exceptions being Firebird, Sacre, S. of P. [Symphony of Psalms], but it gets thinner and thinner as R.C.'s [Robert Craft's] influence extends with his 12 tones! Maybe I'm prejudiced and wrong. But he's surely not the great figure that they all make out?'

In December 1964 he became very enthusiastic about the music of Ronald Stevenson, who had sent him a recording of his piano work *Passacaglia on DSCH*. Walton told Alan Frank:

I was slightly put off, since the work lasts 1 hr. 20 mins. Nothing daunted I put it on thinking I'd take it off within 5 mins, but not at all, I found it absolutely riveting, full of continuous invention and became more enthralled as it progressed and I'm inclined to suspect that it is a really 1st class work of its kind. When I say of its kind, it is in the Liszt–Busoni tradition up to a point, more in the pianistic writing than in the actual content. The music is contemporary, though not influenced by Strav. or Schoenberg or Britten or Boulez, but something quite of his own. In fact it is a very exciting work, lucid in spite of a certain amount of complication, and he's a splendid pianist. I see from the 'sleeve' that he's written over 100 works some of which I feel must be good if they don't all last 80 mins!

When Tippett had his sixtieth birthday, Walton sent for recordings and scores 'to see what all the fuss is about.' In later years he confessed to being 'bored blue' by the recording of

King Priam. In 1978 he told Alan Frank he was playing records of Sibelius symphonies—'marvellous recordings by Karajan of 4, 5, 6 and 7. I've not played a note by him for years. What a composer! Constant and Cecil Gray were very right after all. It seems to me to make what I read about the Pompidou Centre in Paris[3] seem a bit fatuous but of course, not having heard any of it, I'm probably quite wrong.' He was delighted when Alan Frank 'discovered' Spohr—'a very underrated composer. I only know comparatively well, needless to say, his church music, full if I remember rightly of pre-Wagneriana which the old sod took full advantage of and why not.'

When he first chose his *Desert Island Discs* in 1965, he included Schoenberg's *Variations*, which he regarded as one of the century's masterpieces. He once listed the five greatest composers of the twentieth century as Stravinsky, Schoenberg, Debussy, Sibelius, and Mahler, with Hindemith and Britten added if seven names were required. His friendship with Henze grew from an interest in his music, and, although he was not interested in electronic developments, he listened to them. 'I think he's the only proper composer going today,' he said of Henze. 'I hate to say it, but it's no good mentioning Britten or Tippett or myself in the same breath as Hans. The only thing I've got against him is that he takes on so much.'[4] He liked Berio's music, and he admired Birtwistle, recognizing a fellow-spirit in a Lancashire man from the cotton towns, who had kept firmly to his own path in the face of every kind of initial discouragement. Henze recalled a train journey in Italy when Walton first introduced him to white truffles and then insisted on going to hear contemporary works in Bologna, where he spent his time with Luigi Nono. 'I tried to cheer him up when he was depressed over his own music,' Henze said. 'I tried to show him that people wanted his music and would wait for it. I also tried to show him how a tone-row was made, but it wasn't his kind of thing.' ('Can't get my mathematics right', said Walton.) Henze discovered that Walton knew more about art and literature than people realized. 'He would find the most wonderful terms to describe a beautiful painting he had seen in

---

[3] The Institut de recherche et de co-ordination acoustique musique (IRCAM) is an electronic studios and laboratory in the Georges Pompidou Centre, Paris, inaugurated in 1977 under the direction of Pierre Boulez.

[4] *The Guardian*, 30 July 1976, p. 10.

Siena or somewhere. At those times you could see he had incredible sensitivity, great pride and a great generous heart.'[5]

Residence in Ischia cannot be advanced as a reason for any 'isolation' in Walton's outlook. If he was 'behind the times', like most British composers, it was not because he was out of touch. The truth is that Walton measured himself by the highest standards. He was plagued by misgivings and doubts. He could not believe he was as highly regarded as he was. Like Elgar and Britten, he was a prey to insecurity, which was intensified by adverse criticism. He was aware that everything he wrote would be judged against the Viola Concerto and he was in constant fear of not living up to his own self-critical yardstick. Eventually this made him feel creatively impotent, and towards the end of his life he said: 'I think I could have done it better than I have done. I'm a disappointment to myself, if you really want to know. I don't say that the works are disappointing, but I could have done better if I'd thought about it. Rather sad, really.'[6]

From a more personal standpoint, the criticism was unfair that marriage to a woman twenty years his junior would mean that he was diverted from the serious business of composition by the need to entertain her and keep her amused. Those who said this did not understand that he lived for his art, to the self-centred exclusion of much else. In his youth in England, he would dissipate time with Constant Lambert in pubs or by going to night clubs. He enjoyed relaxing by flirting and *la dolce vita*. By choosing to live in Ischia to complete his opera without interruptions or diversions, he knew he was leaving that life behind. It was, as it happened, a life Susana Walton would not have enjoyed and he grew to love the joy of creating a unique garden with her, bringing back plants from Australia where once he might have thought only of bringing back wines and tobacco. The garden and the property on Ischia brought him also a load of debts, and these were met by the commissions from American orchestras and elsewhere.

Walton's letters to his publishers and his agent show very clearly that he had an acute business sense. His compositions and the performance of them were his livelihood, and he was

[5] BBC *Music Weekly*, 28 Mar. 1982.
[6] BBC *Kaleidoscope*, 8 Mar. 1983.

as assiduous about their sale and promotion as any industrial or commercial manufacturer. If he heard a conductor or soloist of whom he approved, he would see that his scores were sent to him in the hope that performances and perhaps recordings would follow. He was keenly interested in the sales of his recordings throughout the world. 'New Year resolutions for you for '76', he wrote to Diana Rix. '1. Daniel [Barenboim] and Pinky [Pinchas Zukerman] and a cellist (?) to record my Pfte Quartet. Also my Vl. Sonata. 2. Get Solti to record Syms. I and II.' Where his royalties and performing rights were concerned, he was aware of every rise and fall and wanted to know the reasons and was always ready with suggestions how to halt a decline. In this respect he was a wily, hard-headed northern businessman and never left Oldham far behind. (John Ireland described Walton to Alan Rawsthorne as 'the most mercenary-minded composer I have ever met'.) To give many examples from his letters on financial matters would be tedious, but one quotation alone will reveal his acumen:

27 December 1963 . . . In spite of my royalties from the OUP being up last year and the Ascap 'first times' supposedly being in force, my P.R.S. [Performing Right Society] warrant is over £1,000 less than last year. Last year's was £3,962–0–8 & this year £2,922–7–6. I wonder why—you might, as it concerns both OUP and myself, have a discreet scout around. Not that the P.R.S. aren't strictly accurate, but it might be interesting to know where there have been fewer perfs. than last year. I don't know whether the OUP has already started to take its whack on the film royalties—if it has, that could account for the drop or part of it as last year the royalties on H.V and Ham. (I don't know about Rich.III) totalled £543. This unfortunate droop makes my borrowing £2,000 all the more imperative . . .

Bitterness about the period of neglect suffered in the 1960s spills out in some letters. For example, on 21 January 1965, in a letter to Alan Frank:

Glad to hear of the old faithfuls George H. [Hurst] and Norman D.M. [Del Mar] especially as I see I'm not mentioned in the BBC's American tour. I suspect I am being squeezed out between B. and T. [Britten and Tippett]. You might see what Glock and Dorati have to say. Or is Du Pré doing the Vlc. Concerto? No, she's doing B's Vlc. Sym. is the answer, I suspect, God help her. They might at least include P.P. [*Portsmouth Point*] if not the Con. or S.2 or the H.Vars.

Britten's *Cello Symphony* did not earn as much admiration from Walton as most of his other works:

Not up to standard I feel [he wrote to Alan Frank on 2 January 1965]. Rather boring and lacking in personality, strangely enough. It is not in the same class as Prokofiev's Concerto Sinfonico with Rostro.[povich] nor for that matter, I'm unpopularly inclined to think, is it as good as mine. True, it's more full of tricks, but what of 'em? In fact the Vlc. Sonata is much much better having at any rate 2 splendid movs. and the other 3 not too bad.

He was saddened that Jacqueline du Pré's proposed performance of his concerto (in Guildford on 2 June 1973) was cancelled because of the onset of her multiple sclerosis—she last played in public in February of that year (the Elgar concerto conducted by Zubin Mehta).

On 14 July 1978 he wrote to André Previn:

I see in *Time* that your leading cellist in Pittsburgh has won a prize in Russia and the thought strikes me that, if egged on by you, he might be persuaded to play my cello concerto. It was, as you may know, commissioned by Piatigorsky, his teacher, and I think one of my better works, but hardly ever played, as the cello, as you know, is not a popular 'draw' however good the player, and conductors (you are the exception, I hope) don't like having to learn a new work which does not show off themselves but a mere cellist. But as he is at the moment in the public eye, perhaps the Pittsburgh might be willing to risk it. There is a new ending as yet unperformed. It was going to be played by Piatigorsky in his 'come-back' in London, but death intervened. Have a look at it and see. You have always been such a champion of my music, for which my gratitude is unbounded, and I believe you could do for the Cello Concerto what you did for my Violin Concerto, by persuading Rosen to perform it.

From 1955 to 1975 a continuous preoccupation was the possibility of a revival of *Troilus and Cressida*. Every conductor or producer whom he met was sounded out and given a score. Time and again promises were made and broken but he was never entirely discouraged. He listened to every soprano and tenor with Cressida and Troilus in mind. He sent for the recording of Tippett's *The Midsummer Marriage* solely so that he could get an idea of the voices used by Covent Garden for an English opera. 'Up to now I can only manage the first Act,' he told Alan Frank on 30 November 1971. 'How long—how loud—how (if I may say so) boring.' During the 1972 discussions, mentioned in an earlier chapter, of a collaboration between EMI and RAI over a recording of *Troilus and Cressida*, he wrote to Alan Frank: 'I certainly think that EMI–RAI may need some topping up and myself think the Brit. Coun. should oblige. I see Arse-over-tip-it has not only Philips but the Brit. Coun., the Arts Council and the Gulb. [Gulbenkian Foundation] backing Mid. Mar. Downright favouritism, but as it paid off they all might think it would not be a bad idea to do the same for T. & C.'

He remains, as a man, an enigma. The assumed diffidence was a mask. The boy from Oldham no less than the old man in Ischia was ruthless in getting his own way. He had a will of steel and he deliberately escaped from situations which he knew would be dead-ends for him, whether they were the parental home in Lancashire, the stifling world of the Sitwells (as it became) or the rat-race of the English post–1945 musical scene. By charm and elegant detachment, he pursued the only goal he wished to reach—to write good music. In *Façade* he created a work unique of its kind which no one would dare to emulate if they could; he wrote an opera in the grand tradition, as he intended; *Belshazzar's Feast* is among the classics of English choral music; he wrote orchestral works which are, or should be, indispensable to the repertoire of the world's virtuoso orchestras; he enriched the repertoire of international string soloists; he inherited for the nation Elgar's pomp and circumstance; he belonged to no exclusive coterie, eschewed pretentiousness, and never admitted a shoddy note into his scores. His music, too, is ultimately an enigma, because it does not tell us the whole truth about William Walton. There was an area of his psyche which he kept shrouded even from himself. But Guido Adler, the musical scholar who was Mahler's friend, heard the Viola Concerto in the 1930s when he was over seventy-five. He approached an English colleague and, with tears in his eyes, said: 'Here's the real thing—at last.' The real thing—it is an epitaph Walton would have liked, even if he might not wholly have believed it.

Classified list of works

STAGE

Operas

Troilus and Cressida
Opera in 3 Acts. Libretto by Christopher Hassall.
Composed 1947–54.Revised 1963, 1972–6.
F.p. London, Royal Opera House, Covent Garden, 3 Dec. 1954.
   Richard Lewis (ten.), Magda Laszlò (sop.), Peter Pears (ten.), Otakar
   Kraus (bar.), Frederick Dalberg (bass), cond. Sir Malcolm Sargent.
F.p. first revised version, London, Royal Opera House, Covent Garden,
   23 Apr. 1963. André Turp (ten.), Marie Collier (sop.), John Lanigan
   (ten.), Otakar Kraus (bar.), Forbes Robinson (bass), cond. Sir
   Malcolm Sargent.
F.p. second revision, London, Royal Opera House, Covent Garden, 12
   Nov. 1976. Richard Cassilly (ten.), Janet Baker (mez.-sop.), Gerald
   English (ten.), Benjamin Luxon (bar.), Richard Van Allan (bass),
   cond. Lawrence Foster.
See also **Orchestral (Suites)**.

The Bear
Extravaganza in 1 Act. Libretto by Paul Dehn and William Walton,
   after Anton Chekhov.
Composed 1965–7.
F.p. Aldeburgh (20th Aldeburgh Festival), Jubilee Hall, 3 June 1967.
   Monica Sinclair (mez.-sop.), John Shaw (bar.), Norman Lumsden
   (ten.), English Chamber Orchestra, cond. James Lockhart.
F. London p. Sadler's Wells Theatre, 12 July 1967. Cast as above.

Ballets

Dr Syntax
Projected one-act ballet, probably based on overture (see **Orchestral
   (Overtures)**).

1 Act, scenario by Sacheverell Sitwell and P. Wyndham Lewis.
Title-page only exists.
Probably 1921.

Façade
Ballet based on *Façade, Suite No. 1* (see **Orchestral (Suites)**).
1 Act of 6 scenes ('Polka', 'Valse', 'Tango', 'Jodelling Song', 'Taran-
    tella', 'Polka').
Choreography by Günter Hess, probably 1929.
F.p. Hagen, Westphalia, 22 Sept. 1929. German Chamber Dance
    Theatre.
See also **Choral and Vocal (Voice(s) and Instrument(s))**.

Façade
Ballet based on *Façade, Suite No. 1* (see **Orchestral (Suites)**).
1 Act of 7 *divertissements* ('Scotch Rhapsody', 'Jodelling Song', 'Polka',
    'Valse', 'Popular Song', 'Tango-Pasodoble', 'Tarantella Sevillana'.)
Choreography by Frederick Ashton 1931.
F.p. London, Cambridge Theatre, 26 Apr. 1931. The Camargo Society,
    cast including Lydia Lopokova, Alicia Markova, Prudence Hyman,
    Diana Gould, Anthony Tudor, Frederick Ashton, Walter Gore.
    Cond. Constant Lambert.
'Country Dance' added 1935.
'Noche espagnole' and 'Foxtrot' added 1940.
See also **Choral and Vocal (Voice(s) and Instrument(s))**.

Façade
Ballet based on Entertainment (see **Choral and Vocal (Voice(s) and
    Instrument(s))**).
1 Act. Choreography by Frederick Ashton, reciter Peter Pears.
F.p. Snape, Suffolk, The Maltings, 28 July 1972. Cond. David Taylor.
F. London p. Sadler's Wells Theatre, 9 Oct. 1972. Cond. David Taylor.
See also **Orchestral (Suites)**.

The First Shoot
Scene 19 in Part II of C. B. Cochran revue *Follow the Sun*. Scenario,
    Osbert Sitwell.
Choreography by Frederick Ashton.
Composed 1935.
F.p. Manchester, Opera House, 23 Dec. 1935. Claire Luce, Nick Long
    Jr., and corps de ballet. Cond. Frank Collinson.
F. London p. Adelphi Theatre, 4 Feb. 1936. Cast as above.
See also **Brass Band**.

Siesta—Pas de Deux
Ballet, based on *Siesta* (see **Orchestral (Miscellaneous)**).

Choreography by Frederick Ashton, 1935.

F.p. London, Sadler's Wells Theatre, 24 Jan. 1936. Pearl Argyle and Robert Helpmann. Sadler's Wells Orchestra, cond. W.W.

As new *Pas de Deux*, choreography by Frederick Ashton, 1972.

F.p. Snape, The Maltings, 28 July 1972. Vyvyan Lorrayne and Barry McGrath. English Opera Group Orchestra, cond. David Taylor.

F. London p. Sadler's Wells Theatre, 9 Oct. 1972. Cast as above.

The Wise Virgins

Ballet in 1 Act, to music by J. S. Bach chosen and arranged by Constant Lambert and orchestrated by William Walton.

Choreography by Frederick Ashton.

Composed 1940.

F.p. London, Sadler's Wells Theatre, 24 Apr. 1940, cast headed by Margot Fonteyn and Michael Somes. Cond. Constant Lambert.

See also **Orchestral (Suites).**

The Quest

Ballet in 5 scenes. Scenario by Doris Langley Moore after *The Faerie Queene* by Edmund Spenser. Choreography by Frederick Ashton. Composed 1943.

F.p. London, New Theatre, 6 Apr. 1943. Sadler's Wells Ballet Company, cast including Margot Fonteyn, Beryl Grey, Moira Shearer, Julia Farron, Robert Helpmann, Alexis Rassine. Cond. Constant Lambert.

See also **Orchestral (Suites).**

Varii Capricci

1983 revision (new coda to Finale) of 1975–6 arr. for orchestra of *Five Bagatelles* for guitar (see **Orchestral (Miscellaneous)** and **Instrumental Solos).**

Choreography by Sir Frederick Ashton.

F.p. New York, Metropolitan Opera House, 19 Apr. 1983.

F. London p. Royal Opera House, Covent Garden, 20 July 1983. Antoinette Sibley, Anthony Dowell. Royal Opera House Orchestra, cond. Ashley Lawrence.

Incidental Music for Plays

See also **Incidental Music for Radio.**

A Son of Heaven

Tragic melodrama in 4 Acts by Lytton Strachey (1880–1932).

Composed 1924–5.

F.p. London, Scala Theatre, 12 July 1925. Cond. W.W.

The Boy David
Play in 3 Acts by J. M. Barrie (1860–1937). Overture and 5 entr'actes (pre-recorded by unknown orchestra).
Composed 1935.
F.p. Edinburgh, King's Theatre, 21 Nov. 1936.
F. London p. His Majesty's Theatre, 14 Dec. 1936.

Macbeth
Play in 3 Acts by William Shakespeare (1564–1616).
Composed 1941–2. Music pre-recorded by London Philharmonic Orchestra, cond. Ernest Irving.
F.p. Manchester, Opera House, 16 Jan. 1942.
F. London p. Piccadilly Theatre, 8 July 1942.
See also **Orchestral (Fanfares)**.

Title Music for BBC TV Shakespeare Series
Composed 1977.
F.p. London, Lime Grove, 26 Jan. 1978. English National Opera Orchestra, cond. David Lloyd-Jones.

ORCHESTRAL

Symphonies

No. 1 in B flat minor
Composed 1931–5.
F.p. (3 movements), London, Queen's Hall, 3 Dec. 1934. London Symphony Orchestra, cond. Sir Hamilton Harty.
F.p. (complete), London, Queen's Hall, 6 Nov. 1935. BBC Symphony Orchestra, cond. Sir Hamilton Harty.

No. 2
Composed 1957–60.
F.p. Edinburgh, Usher Hall (14th Edinburgh Festival), 2 Sept. 1960. Royal Liverpool Philharmonic Orchestra, cond. John Pritchard.
F. London p. Royal Festival Hall, 23 Nov. 1960. Royal Liverpool Philharmonic Orchestra, cond. John Pritchard.

Concertos

Fantasia Concertante, for 2 pianos, jazz band, and orchestra
Composed 1922–3.
Unpublished and unperformed.

Sinfonia Concertante, for orchestra with piano obbligato
Composed 1926–7. Revised 1943.
F.p. London, Queen's Hall, 5 Jan. 1928. York Bowen (piano), Orchestra of Royal Philharmonic Society, cond. Ernest Ansermet.
F.p. revised version, Liverpool, Philharmonic Hall, 9 Feb. 1944. Cyril Smith (piano), Liverpool Philharmonic Orchestra, cond. Dr Malcolm Sargent.

Concerto in A minor for Viola and Orchestra
Composed 1928–9. Revised 1936 and 1961.
F.p. London, Queen's Hall, 3 Oct. 1929. Paul Hindemith (viola), Queen's Hall Orchestra, cond. W.W.
F.p. 1961 revision, London, Royal Festival Hall, 18 Jan. 1962. John Coulling (viola), London Philharmonic Orchestra, cond. Sir Malcolm Sargent.

Concerto in B minor for Violin and Orchestra
Composed 1938–9. Revised 1943.
F.p. Cleveland, Ohio, Severance Hall, 7 Dec. 1939. Jascha Heifetz (violin), Cleveland Orchestra, cond. Artur Rodzinski.
F. London p. Royal Albert Hall, 1 Nov. 1941. Henry Holst (violin), London Philharmonic Orchestra, cond. W.W.
F.p. revised version, Wolverhampton, 17 Jan. 1944. Henry Holst (violin), Liverpool Philharmonic Orchestra, cond. Dr Malcolm Sargent.

Concerto for Violoncello and Orchestra
Composed 1955–6.
F.p. Boston, Mass., Symphony Hall, 25 Jan. 1957. Gregor Piatigorsky (violoncello), Boston Symphony Orchestra, cond. Charles Munch.
F. London p. Royal Festival Hall, 13 Feb. 1957. Gregor Piatigorsky (violoncello), BBC Symphony Orchestra, cond. Sir Malcolm Sargent.
Finale revised 1974: new 23-bar ending. Unperformed.

Overtures

Dr Syntax
Pedagogic overture.
Composed 1920–1.
Unpublished and unperformed.
(See also **Stage (Ballets)**).

Portsmouth Point
After an etching by Thomas Rowlandson (1756–1827).
Composed 1924–5.

F.p. Zurich (4th Festival of International Society for Contemporary Music), Tonhalle, 22 June 1926. Tonhalle Orchestra, cond. Volkmar Andreae.

F. London p. His Majesty's Theatre, 28 June 1926. Orchestra of Ballets Russes Company, cond. Eugene Goossens.

Scapino
Comedy overture after an etching from Callot's *Balli di Sfessania* (1622).
Composed 1940. Revised 1950.

F.p. Chicago, Ill., Orchestra Hall, 3 Apr. 1941. Chicago Symphony Orchestra, cond. Frederick Stock.

F. English p. Bedford, Corn Exchange, 12 Nov. 1941. BBC Symphony Orchestra, cond. W.W.

F. London p. Royal Albert Hall, 13 Dec. 1941. London Philharmonic Orchestra, cond. W.W.

F.p. revised version, London, Royal Albert Hall, 13 Nov. 1950. Philharmonia Orchestra, cond. W.W.

Johannesburg Festival Overture
Composed 1956.

F.p. Johannesburg, City Hall, 25 Sept. 1956. South African Broadcasting Corporation Symphony Orchestra, cond. Sir Malcolm Sargent.

F. Eng. p. Liverpool, Philharmonic Hall, 13 Nov. 1956. Liverpool Philharmonic Orchestra, cond. Efrem Kurtz.

F. London p. Royal Festival Hall, 23 Jan. 1957. BBC Symphony Orchestra, cond. Sir Malcolm Sargent.

Miscellaneous

Siesta, for small orchestra
Composed 1926. Revised 1962.

F.p. London, Aeolian Hall, 24 Nov. 1926. Aeolian Chamber Orchestra, cond. W.W.

See also **Stage (Ballets)**.

Music for Children
Orchestral version, 1940, of *Duets for Children* (see **Keyboard**).
Order of pieces changed to:

1. 'The Music Lesson'. 2. 'The Three-legged Race'. 3. 'The Silent Lake'. 4. 'Pony Trap'. 5. 'Swing-boats.' 6. Puppet's Dance. 7. 'Song at Dusk'. 8. 'Hop-Scotch.' 9. 'Ghosts.' 10. 'Trumpet Tune.'

F.p. London, Queen's Hall, 16 Feb. 1941. London Philharmonic Orchestra, cond. Basil Cameron.

'Galop Finale'

New Finale for a ballet, *Devoirs de Vacances*, in Paris on 8 Nov. 1949; not used.

Spitfire Prelude and Fugue
Arranged from film music for *The First of the Few* (see **Film Music.**)
Composed 1942.
F.p. Liverpool, Philharmonic Hall, 2 Jan. 1943. Liverpool Philharmonic Orchestra, cond. W.W.
F. London p. Royal Albert Hall, 21 Feb. 1943. London Philharmonic Orchestra and section of BBC Symphony Orchestra, cond. Dr Malcolm Sargent.

Two Pieces for Strings
From the film music for *Henry V* (see **Film Music**).
1. Passacaglia: Death of Falstaff. 2. Touch her soft lips and part.
Composed 1943–4.

Variation on an Elizabethan Theme ('Sellinger's Round')
    for strings
Composed 1953.
F.p. Aldeburgh (6th Aldeburgh Festival), Parish Church, 20 June 1953. Aldeburgh Festival Orchestra, cond. Benjamin Britten.
F. London p. Wigmore Hall, 29 May 1957. Collegium Musicum Londinii, cond. John Minchinton.

National Anthem
Arr. for full orchestra, 1953.
F.p. London, Royal Opera House, Covent Garden, 8 June 1953. Covent Garden Orchestra, cond. John Pritchard.
Arr. for full orchestra, 1955.
F.p. London, Royal Festival Hall, 18 Oct. 1955. The Philharmonia Orchestra, cond. Herbert von Karajan.

The Star-Spangled Banner
Arr. for full orchestra, 1955.
Unpublished.

Partita
Composed 1957.
F.p. Cleveland, Ohio, Severance Hall, 30 Jan. 1958. Cleveland Orchestra, cond. Georg Szell.
F. English p. Manchester, Free Trade Hall, 30 Apr. 1958. Hallé Orchestra, cond. W.W.
F. London p. Royal Festival Hall, 2 May 1958. Hallé Orchestra, cond. Sir John Barbirolli.

Granada Prelude, Call Signs and End Music
Composed 1962.
F. concert p. of *Prelude*, London, St John's, Smith Square, 25 June
   1977. Young Musicians' Symphony Orchestra, cond. James Blair.

Variations on a Theme by Hindemith
Composed 1962–3.
F.p. London, Royal Festival Hall, 8 Mar. 1963. Royal Philharmonic
   Orchestra, cond. W.W.

Hamlet and Ophelia
Poem for orchestra adapted by Muir Mathieson, 1967, from film music
   for *Hamlet* (see **Film Music**).
F.p. unknown.

Major Barbara, a Shavian Sequence
Arr. Christopher Palmer, 1987, from film music for *Major Barbara* (see
   Film Music).
F.p. London, Barbican Hall, 15 Mar. 1988. Royal Philharmonic
   Orchestra, cond. Elmer Bernstein.

Capriccio Burlesco
Composed 1968.
F.p. New York, Phiharmonic Hall (Lincoln Center), 7 Dec. 1968. New
   York Philharmonic Orchestra, cond. André Kostelanetz.
F. London p. Royal Festival Hall, 5 Feb. 1969. BBC Symphony
   Orchestra, cond. Colin Davis.

Improvisations on an Impromptu of Benjamin Britten
Composed 1968–9.
F.p. San Francisco, War Memorial Opera House, 14 Jan. 1970. San
   Francisco Symphony Orchestra, cond. Josef Krips.
F. English p. Snape (23rd Aldeburgh Festival), The Maltings, 27 June
   1970. Royal Liverpool Philharmonic Orchestra, cond. Charles
   Groves.
F. London p. Royal Festival Hall, 20 Oct. 1970. London Philharmonic
   Orchestra, cond. Josef Krips.

Sonata for Strings
Transcription of String Quartet in A minor (see **Chamber Music**).
Transcribed 1971.
F.p. Perth, W. Australia, Octagon Theatre, 2 Mar. 1972. Academy of St
   Martin-in-the-Fields, cond. Neville Marriner.
F. English p. Bath, Assembly Rooms, 27 May 1972. Academy of St
   Martin-in-the-Fields, cond. Neville Marriner.

F. London p. Mansion House, 11 July 1972. Academy of St Martin-in-the-Fields, cond. Neville Marriner.

Varii Capricci
Free transcriptions of *Five Bagatelles* for guitar (see **Instrumental Solos**).
Transcribed 1975–6.
F.p. London, Royal Festival Hall, 4 May 1976. London Symphony Orchestra, cond. André Previn.
Finale re-written 1977. F.p. Cardiff, Broadcasting House, 28 Jan. 1981. BBC Welsh Symphony Orchestra, cond. Owain Arwel Hughes.
See also **Stage (Ballets)**.

Prologo e Fantasia
Composed 1981–2.
F.p. London, Royal Festival Hall, 20 Feb. 1982. National Symphony Orchestra of Washington, D.C., cond. Mstislav Rostropovich.

Marches

Crown Imperial
Composed 1937. For Coronation of King George VI and Queen Elizabeth.
F.p. London, Kingsway Hall, 16 Apr. 1937. BBC Symphony Orchestra, cond. Sir Adrian Boult.
F. broadcast p. London, 9 May 1937. BBC Symphony Orchestra, cond. Clarence Raybould.
F. public p. London, Westminster Abbey, 12 May 1937 (Coronation of King George VI and Queen Elizabeth). The Coronation Orchestra, cond. Sir Adrian Boult.

Orb and Sceptre
Composed 1952–3. For Coronation of Queen Elizabeth II.
F.p. London, Kingsway Hall (recording), 18 Mar. 1953. The Philharmonia Orchestra, cond. W.W.
F. public p. London, Westminster Abbey, 2 June 1953 (Coronation of Queen Elizabeth II). The Coronation Orchestra, cond. Sir Adrian Boult.
F. concert p. London, Royal Festival Hall, 7 June 1953. London Symphony Orchestra, cond. Sir John Barbirolli.

A History of the English-speaking Peoples
Composed 1959. For television series based on Winston Churchill's *A History of the English-speaking Peoples*.
F.p. Elstree, 25 May 1959, London Symphony Orchestra, cond. W.W.

Funeral March
Arr. Muir Mathieson, 1963, from film music for *Hamlet* (see **Film Music**).
F.p unknown.

Fanfares

For the Red Army (2 fanfares.)
Composed 1943.
F.p. London, Royal Albert Hall, 21 Feb. 1943 (Red Army Day Celebrations). Trumpets and drums of Life Guards, Royal Horse Guards and Royal Air Force, cond. Dr Malcolm Sargent.

Memorial Fanfare for Henry Wood, for full orchestra
Composed 1945.
A revision and amplification of one of the Red Army fanfares.
F.p. London, Royal Albert Hall, 4 Mar. 1945. BBC Symphony Orchestra, London Symphony Orchestra, London Philharmonic Orchestra, cond. Sir Adrian Boult.

Fanfare for a Great Occasion
Fanfares from *Hamlet* (see **Film Music**) grouped into single piece by Sir Malcolm Sargent, 1962
F.p. Wayfield, Adelaide, S. Australia, Centennial Hall, 17 Mar. 1962. London Philharmonic Orchestra, cond. Sir Malcolm Sargent.
F. British p. Ely Cathedral, 8 July 1962. London Philharmonic Orchestra, cond. Sir Adrian Boult.

A Queen's Fanfare
Composed 1959. For the Queen's entrance at the NATO Parliamentary Conference.
F.p. London, Westminster Hall, 5 June 1959. The State Trumpeters.

Anniversary Fanfare
Composed 1973. For EMI's 75th anniversary concert.
F.p. London, Royal Festival Hall, 29 Nov. 1973. Trumpeters of the Royal Military School of Music, Kneller Hall, cond. Lt.-Col. R. B. Bashford.

Fanfare for the National
Composed 1974. Revised 1976. For television programme to mark opening of the National Theatre.
F.p. Wembley, De Lane Lea Studios, 1 Apr. 1976. Band of the Life Guards, cond. Harry Rabinowitz.

Roaring Fanfare
Composed 1976. For inauguration of Lion Terraces, London Zoo.
F.p. London, Zoological Gardens, 3 June 1976. Trumpeters of the
    Royal Military School of Music, Kneller Hall, cond. Trevor Platts.

Salute to Sir Robert Mayer on his 100th Birthday
Composed 1979.
F.p. London, Royal Festival Hall, 5 June 1979. 12 trumpeters from
    schools of Inner London Education Authority.
Later amended and revised and published as *Introduction to the National
    Anthem* for 3 trumpets and 3 trombones.

A Birthday Fanfare
Composed 1981. For seventieth birthday of Dr Karl-Friedrich Still (a
    friend and neighbour of the Waltons in Ischia).
F.p. Recklinghausen, Wengerner Mühle, 10 Oct. 1981. Trumpeters of
    the Westphalia Symphony Orchestra, cond. Karl Rickenbacher.
F. London p. Royal Albert Hall, 7 June 1982. Trumpeters of the Royal
    Military School of Music, Kneller Hall, cond. Lt.-Col. G. E. Evans.

Fanfare and March
Arr. Christopher Palmer, 1987, from music for *Macbeth* (see **Incidental
    Music for Plays**).

Suites (including those arranged by others)

Façade, Suite No.1
Arranged from the Entertainment (see **Choral and Vocal (Voice(s)
    and Instrument(s))**).
Orchestrated 1926.
1. 'Polka'. 2. 'Valse'. 3. 'Swiss Jodelling Song'. 4. 'Tango-Pasodoble'. 5.
    'Tarantella Sevillana'.
F.p. London, Lyceum Theatre, 3 Dec. 1926 (excluding No. 3). Lyceum
    Theatre Orchestra, cond. W.W.
See also **Stage (Ballets)**.

Façade, Suite No.2
Arranged from the Entertainment (see **Choral and Vocal Voice(s)
    and Instrument(s))**).
1. 'Fanfare'. 2. 'Scotch Rhapsody'. 3. 'Country Dance'. 4. 'Noche
    espagnole'. 5. 'Popular Song'. 6. 'Old Sir Faulk'.
F.p. New York, Carnegie Hall, 30 Mar. 1938. Philharmonic-Symphony
    Orchestra of New York, cond. John Barbirolli (No. 6 excluded and
    replaced by No. 4 from *Suite No.1*).

F. London p. Queen's Hall, 10 Sept. 1938. BBC Symphony Orchestra, cond. Sir Henry Wood.
See also **Stage (Ballets)**.

The Wise Virgins
Suite in 6 movements from the ballet *The Wise Virgins* (see **Stage (Ballets)**).
Arranged 1939–40.
F.p. probably recording session 24 July 1940. Sadler's Wells Orchestra, cond. W.W.

The Quest
Arr. Vilem Tausky, 1961, from the ballet *The Quest* (see **Stage (Ballets)**).
F.p. London, Royal Festival Hall, 3 June 1961. BBC Concert Orchestra, cond. Vilem Tausky.

Suite from Henry V
Arr. Malcolm Sargent, 1945, from film music for *Henry V* (see **Film Music**). For orchestra and chorus (4 movements).
F.p. London, Royal Albert Hall, 14 Sept. 1945. BBC Choral Society, Croydon Philharmonic Society, BBC Symphony Orchestra, cond. W.W.

Suite from Henry V
Arr. Muir Mathieson, 1963, from film music for *Henry V* (see **Film Music**). For orchestra (5 movements).
F.p. unknown.

A Shakespeare Suite from Richard III
Arr. Muir Mathieson, 1963, from film music for *Richard III* (6 movements) (see **Film Music**).
F.p. unknown.
See also **Keyboard**.

Prelude, Richard III
Adapted by Muir Mathieson, 1963, from film music for *Richard III* (see **Film Music**).
F.p. unknown.

The Battle of Britain
Suite adapted by Colin Matthews, 1984–5, from film music for *The Battle of Britain* (see **Film Music**).
F.p. Bristol, Colston Hall, 10 May 1985. Bournemouth Symphony Orchestra, cond. Carl Davis.

F. London p. Barbican Hall, 10 August 1985. London Symphony
   Orchestra, cond. Carl Davis.

Christopher Columbus
Suite, arr. Christopher Palmer, 1987, from music for play (see
   Incidental Music for Radio).
1. Fiesta. 2. Romanza. 3. Gloria.
See also **Choral and Vocal (Chorus and Orchestra)**.

Troilus and Cressida
Symphonic suite, arr. Christopher Palmer, 1987, from opera (see
   Stage (Operas)).
1. Prelude and Seascape. 2. Scherzo. 3. The Lovers. 4. Finale.
F.p. London, Royal Albert Hall, 3 Aug. 1988. BBC Welsh Symphony
   Orchestra, cond. Bryden Thomson.

CHORAL AND VOCAL

Chorus and Orchestra

The Forsaken Merman
Cantata for soprano, tenor, double women's chorus, and orchestra.
   Text by Matthew Arnold (1822–88).
Composed 1916.
Unpublished and unperformed.

Belshazzar's Feast
Cantata for baritone, mixed chorus, and orchestra. Text selected from
   the Old Testament by Osbert Sitwell (1892–1969).
Composed 1930–1. Revised 1931, 1948, and 1957.
F.p. Leeds, Town Hall (Leeds Triennial Festival), 8 Oct. 1931. Dennis
   Noble (bar.), Festival Chorus, London Symphony Orchestra, cond.
   Dr Malcolm Sargent.
F. London p. Queen's Hall, 25 Nov. 1931. Stuart Robertson (bar.),
   National Chorus, BBC Symphony Orchestra, cond. Dr Adrian Boult.
F.p. 1948 revision, London, Royal Albert Hall, 8 Mar. 1950. Dennis
   Noble (bar.), BBC Choral Society and Goldsmiths' Choral Union,
   BBC Symphony Orchestra, cond. Sir Malcolm Sargent.

In Honour of the City of London
Cantata for mixed chorus and orchestra. Text by William Dunbar
   (1465–1520).
Composed 1937.

F.p. Leeds, Town Hall (Leeds Triennial Festival), 6 Oct. 1937. Festival Chorus, London Philharmonic Orchestra, cond. Dr Malcolm Sargent.

F. London p. Queen's Hall, 1 Dec. 1937. BBC Choral Society, BBC Symphony Orchestra, cond. W.W.

Coronation Te Deum
For 2 mixed choruses, 2 semi-choruses, boys' voices, organ, orchestra and military brass.
Composed 1952–3.
F.p. London, Westminster Abbey, 2 June 1953 (Coronation of Queen Elizabeth II). Coronation Choir and Orchestra, trumpeters of Royal Military School of Music, Kneller Hall, Osborne Peasgood (organ), cond. Sir William McKie.
F. concert p. London, Royal Albert Hall, 27 July 1953. BBC Choral Society, Goldsmiths' Choral Union, BBC Symphony Orchestra, Arnold Greir (organ), cond. Sir Malcolm Sargent.

Gloria
For contralto, tenor, and bass soloists, mixed chorus, and orchestra. Text from modern Roman Missal.
Composed 1960–1.
F.p. Huddersfield, Town Hall, 24 Nov. 1961. Marjorie Thomas (cont.), Richard Lewis (ten.), John Cameron (bar.), Huddersfield Choral Society, Royal Liverpool Philharmonic Orchestra, cond. Sir Malcolm Sargent.
F. London p. Royal Festival Hall, 18 Jan. 1962. Marjorie Thomas (cont.), Ronald Dowd (ten.), Owen Brannigan (bass), London Philharmonic Choir and Orchestra, cond. Sir Malcolm Sargent.

Henry V, a Shakespearean Scenario
For speaker, chorus, boys' choir, and orchestra.
Arr. Christopher Palmer, 1988, from film music for *Henry V* (see **Film Music**).
1. Prologue. 2. Interlude—At the Boar's Head. 3. Embarkation. 4. Interlude—Touch her soft lips and part. 5. (a) Harfleur. (b) The Night Watch. 6. Agincourt. 7. Interlude—At the French Court. 8. Epilogue.

Christopher Columbus
Cantata for contralto, tenor, baritone, mixed chorus, and orchestra, devised by Carl Davis, 1988, from music for play.
See **Incidental Music for Radio** and **Orchestral (Suites)**.

Voice(s) and Instrument(s)

Tell me where is fancy bred?
Song for soprano and tenor voices, 3 violins, and piano. Text by
   William Shakespeare, *The Merchant of Venice*.
Composed 1916 (2 July).
Unpublished.

The Winds
Song for voice and piano. Text by A. C. Swinburne (1837–1909).
Composed 1918.
F.p. Probably 1920 and 11 Dec. 1921 (at Lady Glenconner's home, 34
   Queen Anne's Gate, London).
F. documented p. London, Aeolian Hall, 30 Oct. 1929. Odette de Foras
   (sop.), Gordon Bryan (piano).

Tritons
Song for voice and piano. Text by William Drummond (1585–1649).
Composed 1920.
F.p. London, Aeolian Hall, 30 Oct. 1929. Odette de Foras (sop.),
   Gordon Bryan (piano).

The Passionate Shepherd
Song for tenor and 10 solo instruments. Text by Christopher Marlowe
   (1564–93).
Composed 1920.
Unpublished.

Façade
Entertainment for reciter and 14 instruments (6 players). Poems by
   Edith Sitwell (1887–1964).
Composed 1921–8, with further revisions thereafter.
Definitive published version (1951): Fanfare. I. 1. 'Hornpipe.' 2. 'En
   famille'. 3. 'Mariner Man'. II. 4. 'Long Steel Grass'. 5. 'Through
   Gilded Trellises'. 6. 'Tango-pasodoble' ('I do like to be beside the
   seaside'). III. 7. 'Lullaby for Jumbo'. 8. 'Black Mrs Behemoth'. 9.
   'Tarantella'. IV. 10. 'A Man from a Far Countree'. 11. 'By the Lake'.
   12. 'Country Dance'. V. 13. 'Polka'. 14. 'Four in the Morning' ('in
   collaboration with C.L.'). 15. 'Something Lies beyond the Scene'.
   VI. 16. 'Valse'. 17. 'Jodelling Song'. 18. 'Scotch Rhapsody'. VII. 19.
   'Popular Song'. 20. 'Foxtrot' ('Old Sir Faulk'). 21. 'Sir Beelzebub'.
F.p. London, 2 Carlyle Square, 24 Jan. 1922. Edith Sitwell (reciter),
   cond. W.W.

F. public p. London, Aeolian Hall, 12 June 1923. Edith Sitwell (reciter), cond. W.W.

F.p. of 1926 revision: London, New Chenil Galleries (Chelsea), 27 Apr. 1926. Neil Porter (reciter), cond. W.W.

F.p. of 1926 revision with some different items: London, New Chenil Galleries (Chelsea), 29 June 1926. Constant Lambert (reciter), cond. W.W.

F.p. of 1928 revision: Siena (6th ISCM Festival), Rozzi Theatre, 14 Sept. 1928. Constant Lambert (reciter), cond. W.W.

F.p. of definitive 1942 version (published 1951): London, Aeolian Hall, 29 May 1942. Constant Lambert (reciter), cond. W.W.

See also **Stage (Ballets)** and **Orchestral (Suites)**.

Façade 2: A Further Entertainment

1977 revision of eight discarded items from original *Façade*, originally called *Façade Revived*. Revised 1977–8.

Flourish. 1. 'Came the Great Popinjay'. 2. 'Aubade'. 3. 'March'. 4. 'Madam Mouse Trots'. 5. 'The Octogenarian'. 6. 'Gardener Janus Catches a Naiad'. 7. 'Water Party'. 8. 'Said King Pompey'.

F.p. London, Plaisterers' Hall, 25 Mar. 1977. Richard Baker (reciter), English Bach Festival Ensemble, cond. Charles Mackerras.

F.p. revised version, Snape, The Maltings (32nd Aldeburgh Festival), 19 June 1979. Peter Pears (reciter), cond. Steuart Bedford.

Bucolic Comedies

5 songs from *Façade* to words by Edith Sitwell, with accompaniment for 6 instruments.

Composed 1923–4.

Unpublished.

(Three of the songs were revised as *Three Songs* (q.v.) of 1931–2.)

Three Songs

Songs for voice and piano. Texts by Edith Sitwell.

Composed 1931–2.

1. 'Daphne', 2. 'Through Gilded Trellises'. 3. 'Old Sir Faulk'.

F.p. London, Wigmore Hall, 10 Oct. 1932. Dora Stevens (Dora Foss) (sop.), Hubert Foss (piano).

'Under the Greenwood Tree'

Song for voice and piano.

Composed 1936.

See also **Unaccompanied Voices**

'Beatriz's Song'

Song for voice and strings or guitar (adapted for guitar by Hector Quine). Text by Louis MacNeice (1907-63).

From radio-drama *Christopher Columbus* (see **Incidental Music for Radio**).
Composed 1942.
F.p. Bedford, 12 October 1942. Joan Lennard (sop.), George Elliott (guitar).

Three Solo Songs from Troilus and Cressida (see also **Stage (Operas)**).
Texts by Christopher Hassall. Prepared for concert performance by W.W., with special endings for Nos. 1 and 2.
Possibly 1970.
1. 'Slowly it all comes back'. 2. 'How can I sleep?' 3. 'Diomede!'

Anon. in Love
6 songs for tenor and guitar (ed. Julian Bream). Texts from anonymous sixteenth- and seventeenth-century poems selected by Christopher Hassall from *The English Galaxy of Shorter Poems* (ed. Gerald Bullett).
Composed 1959.
1. 'Fain would I change that note'. 2. 'O stay, sweet love'. 3. 'Lady, when I behold'. 4. 'My Love in her Attire'. 5. 'I Gave her Cakes'. 6. 'To Couple is a Custom'.
F.p. Claydon, Ipswich, Shrubland Park Hall (13th Aldeburgh Festival), 21 June 1960. Peter Pears (ten.), Julian Bream (guitar).
Guitar accompaniment transcribed for piano by Christopher Palmer, 1988.
Rescored for tenor and small orchestra, 1970–1.
F.p. London, The Mansion House (Festival of the City of London), 21 June 1971. Robert Tear (ten.), London Mozart Players, cond. Harry Blech.

A Song for the Lord Mayor's Table
6 songs for soprano and piano. Texts selected by Christopher Hassall from poems by William Blake (1757–1827), Thomas Jordan (?1612–85), Charles Morris (1745–1838), William Wordsworth (1770–1850), and 2 anonymous eighteenth-century poets.
Composed 1962.
1. 'The Lord Mayor's Table'. 2. 'Glide gently'. 3. 'Wapping Old Stairs'. 4. 'Holy Thursday'. 5. 'The Contrast'. 6. 'Rhyme'.
F.p. London, Goldsmiths' Hall, 18 July 1962. Elisabeth Schwarzkopf (sop.), Gerald Moore (piano).
Rescored for soprano and orchestra, 1970.
F.p. London, The Mansion House (Festival of the City of London), 7 July 1970. Janet Baker (mez.-sop.), English Chamber Orchestra, cond. George Malcolm.

Unaccompanied voices

A Litany ('Drop, Drop, Slow Tears')
Motet for SATB.
Text by Phineas Fletcher (1582–1650).
Composed 1916, revised for publication 1930.

'Make we joy now in this fest'
Carol for SATB.
Composed 1931.

'Under the Greenwood Tree'
Arr. of song for unison voices, 1937.
See also **Choral and Vocal (Voice(s) and Instrument(s))**.

'Set me as a seal upon thine heart'
Anthem for SATB. Text from *Song of Solomon*.
Composed 1938.
F.p. London, St Mary Abbots Church, Kensington, 22 Nov. 1938. St
  Mary's church choir, dir. F. G. Shuttleworth.

'Where does the uttered music go?'
Setting for SATB of 'Sir Henry Wood' by John Masefield (1878–1967).
Composed 1946.
F.p. London, St Sepulchre's Church, Holborn, 26 Apr. 1946. BBC
  Chorus and Theatre Revue Chorus, cond. Leslie Woodgate.

'Put off the serpent girdle'
Song for three-part women's voices (SSA). Text by Christopher
  Hassall and Paul Dehn. Verse 1 taken from *Troilus and Cressida* (see
  Stage (Operas)) but discarded in 1976 version. Verse 2 added by
  Dehn.

'What Cheer?'
Carol for SATB. Text from *Richard Hill's Commonplace Book*, sixteenth
  century.
Composed 1960.

'All this Time'
Carol for SATB. Text anon., sixteenth century.
Composed 1970.

Cantico del Sole
Motet for SATB. Text by St Francis of Assisi (1182–1226).
Composed 1973–4.

F.p. Cork, University College, 25 Apr. 1974. BBC Northern Singers, cond. Stephen Wilkinson.

F. British p. Manchester, Milton Hall, 14 Sept. 1974. BBC Northern Singers, cond. Stephen Wilkinson.

F. London p. Queen Elizabeth Hall, 23 Nov. 1974. Louis Halsey Singers, cond. Louis Halsey.

'King Herod and the Cock'
Carol for SATB. Text traditional.
Composed 1977.
F.p. Cambridge, King's College, 24 Dec. 1977. King's College Choir, cond. Philip Ledger.

Voices with Organ

The Twelve
Anthem for SATB and organ. Text by W. H. Auden (1907–73).
Composed 1964–5.
F.p. Oxford, Christ Church Cathedral, 16 May 1965. Christ Church Cathedral Choir, cond. Dr Sydney Watson, with Robert Bottone (organ).
Organ part arranged for orchestra 1965.
F.p. London, Westminster Abbey, 2 Jan. 1966. Ann Dowdall (sop.), Shirley Minty (cont.), Robert Tear (ten.), M. Wakeham (bar.), London Philharmonic Choir and Orchestra, cond. W.W.

Missa brevis
For double mixed chorus (with organ in Gloria only).
Composed 1965–6.
F. broadcast p. BBC Radio 3, 29 Mar. 1967. BBC Chorus, cond. Alan G. Melville, with Simon Preston (organ).

Jubilate Deo
For double mixed chorus and organ.
Composed 1971–2.
F.p. Oxford, Christ Church Cathedral (English Bach Festival), 22 Apr. 1972. Christ Church Cathedral Choir, cond. Simon Preston, with Stephen Darlington (organ).

Magnificat and Nunc Dimittis
For SATB and organ.
Composed 1974. Revised 1975.
F.p. Chichester, Chichester Cathedral, 14 June 1975. Chichester Cathedral Choir, cond. John Birch, with Ian Fox (organ).

Antiphon
For SATB and organ. Text by George Herbert (1593–1632).
Composed 1977.
F.p. Rochester, NY, St Paul's Church, 20 Nov. 1977. St Paul's Church
    Choir, cond. David Fetler, with David Craighead (organ).

CHAMBER MUSIC

Quartet for Piano and Strings
Composed 1918–21. Revised 1973–4.
F.p. (surmised), London, Aeolian Hall, 30 Oct. 1929. Gordon Bryan
    (piano), Pierre Tas or William Primrose (violin), James Lockyer
    (viola), John Gabalfa (cello).

String Quartet
Composed 1919 in two movements. Revised 1921–2, with a third
    movement (Scherzo) inserted second.
F.p. London, 19 Berners Street, 4 Mar. 1921. Pennington String
    Quartet.
F.p. of revised quartet, London, Royal College of Music, 5 July 1923.
    McCullagh String Quartet.

Toccata in A minor, for violin and piano
Composed 1922–3.
F.p. London, 6 Queen Square, 12 May 1925. K. Goldsmith (violin),
    Angus Morrison (piano).
Withdrawn.

String Quartet in A minor
Composed 1945–7.
F.p. London, Broadcasting House (BBC Third Programme), 4 May
    1947. Blech String Quartet.
F. public p. London, Broadcasting House, Concert Hall, 5 May 1947.
    Blech String Quartet.
Revised 1971 as *Sonata for Strings* (see **Orchestral (Miscellaneous)**).

Sonata for Violin and Piano
Composed 1947–8. Revised 1949–50.
F.p. Zurich, Tonhalle, 30 Sept. 1949. Yehudi Menuhin (violin), Louis
    Kentner (piano).
F.p. revised version, London, Theatre Royal, Drury Lane, 5 Feb. 1950.
    Yehudi Menuhin (violin), Louis Kentner (piano).

Two Pieces, for violin and piano
Composed 1948–50.

1. Canzonetta. 2. Scherzetto.
F.p. London, Broadcasting House (BBC Third Programme), 27 Sept.
   1950. Frederick Grinke (violin), Ernest Lush (piano).

Duettino, for oboe and violin
Composed 2 Oct. 1982.
Unpublished.

KEYBOARD (ORGAN AND PIANO)

Choral Prelude on 'Wheatley', for organ
Composed 1916 (16 Aug.).
Unpublished.

Valse in C minor, for piano
Composed 1917 (2 Feb.).
Unpublished.

Valse from Façade
Arr. for solo piano by W.W. (see also under **Choral and Vocal
   (Voice(s) and Instrument(s)))**.
Published 1928.

Portsmouth Point
Arr. for piano duet by W.W., 1925 (see **Orchestral (Overtures)**).

Siesta
Arr. for piano (4 hands) by W.W., 1928 (see also under **Orchestral
   (Miscellaneous)**).

Sinfonia Concertante
Arr. for 2 pianos by W.W., 1928 (see **Orchestral (Concertos)**).

Choral Prelude, Herzlich thut mich verlangen (J. S. Bach)
Freely arr. for piano by W.W., 1931. Contribution to *A Bach Book for
   Harriet Cohen.*
F.p. London, Queen's Hall, 17 Oct. 1932. Harriet Cohen.

Ballet Music from Escape Me Never
Arr. for solo piano by W.W., 1935 (see **Film Music**).

Crown Imperial
Arr. for solo piano by W.W., 1937 (see **Orchestral (Marches)**).

Duets for Children, for piano duet
Composed 1940.
1. 'The Music Lesson'. 2. 'The Three-legged Race'. 3. 'The Silent Lake'.
   4. 'Pony Trap'. 5. 'Ghosts'. 6. 'Hop-scotch'. 7. 'Swing-boats'. 8. 'Song
   at Dusk'. 9. 'Puppet's Dance'. 10. 'Trumpet Tune'.
F.p. (probably) London, recording studio for Columbia Graphophone
   Company, 7 May 1940. Ilona Kabos, Louis Kentner.
See also *Music for Children* under **Orchestral (Miscellaneous)**.

Three Pieces for Organ
From the film *Richard III* (see **Film Music**).
Composed 1955.
1. March. 2. Elegy. 3. Scherzetto.
F.p. on film sound-track (recorded on organ of Denham church).
See also *A Shakespeare Suite*, movements 6, 5, and 2, under **Orchestral
   (Suites)**.

INSTRUMENTAL SOLOS

Theme (for Variations), for cello solo
Composed 1970. Contribution to *Music for a Prince*.
F.p. Rome, Villa Wolkonsky, 29 Apr. 1985. Antonio Lysy.

Five Bagatelles, for guitar
(ed. Julian Bream)
Composed 1970–1.
F.p. No. 2, London, Queen Elizabeth Hall, 13 Feb. 1972. Julian Bream;
   Nos. 1 and 3, London, BBC Television Centre, 29 Mar. 1972. Julian
   Bream; Complete, Bath, Assembly Rooms, 27 May 1972. Julian
   Bream.
F. London p. Queen Elizabeth Hall, 21 Jan. 1973. Julian Bream.
See also *Varii Capricci* under **Orchestral (Miscellaneous)** and **Stage
   (Ballets)**.

Passacaglia, for cello solo
Composed 1979–80.
F.p. London, Royal Festival Hall, 16 Mar. 1982. Mstislav Rostro-
   povich.

BRASS BAND

The First Shoot
Arrangement of ballet for Cochran revue *Follow the Sun* (see under
   Stage (Ballets)).

Arr. 1979–80 for brass band by W.W.

F.p. London, Goldsmiths' College (television recording), 19 Dec. 1980. Grimethorpe Colliery Band, cond. Elgar Howarth.

F. public p. London, Royal Albert Hall, 7 Sept. 1981. Grimethorpe Colliery Band, cond. Elgar Howarth.

Arr. for orchestra, 1987, by Christopher Palmer.

FILM MUSIC

Escape Me Never
Composed 1934.
Film first shown: London Pavilion, 14 Oct. 1935.
See also **Keyboard.**

As You Like It
Composed 1936.
Film first shown: London, Carlton, Haymarket, 3 Sept. 1936.

Dreaming Lips
Composed 1937.
Film first shown: London Pavilion, 11 Oct. 1937.

Stolen Life
Composed 1938.
Film first shown: London, Plaza, 18 Jan. 1939.

Major Barbara
Composed 1940.
Film first shown: Nassau, Bahamas, 21 Mar. 1941; London, Odeon, Leicester Square, 4 Aug. 1941.
See also **Orchestral (Miscellaneous).**

Next of Kin
Composed 1941.
Film first shown: London, Curzon Theatre, Jan. 1942 (special audiences only). Public showing: London Pavilion, 15 May 1942.

The Foreman Went to France
Composed 1941–2.
Film first shown: London Pavilion, 13 Apr. 1942.

The First of the Few
Composed 1942.
Film first shown: London, Leicester Square Theatre, 20 Aug. 1942.
See *Spitfire Prelude and Fugue* under **Orchestral (Miscellaneous).**

Went the Day Well?
Composed 1942.
Film first shown: London Pavilion, 1 Nov. 1942.

Henry V
Composed 1943–4.
Film first shown: London, Carlton, Haymarket, 22 Nov. 1944.
See also *Two Pieces for Strings* under **Orchestral (Miscellaneous)**, *Suite*
under **Orchestral (Suites)**, and **Choral and Vocal (Chorus and Orchestra)**.

Hamlet
Composed 1947.
Film first shown: London, Odeon, Leicester Square, 6 May 1948.
See also **Orchestral (Miscellaneous)** and **(Marches)**.

Richard III
Composed 1955.
Film first shown: London, Leicester Square Theatre, 13 Dec. 1955.
See also **Orchestral (Suites)** and **Keyboard.**

The Battle of Britain
Composed 1969.
Film first shown: London, Dominion Cinema, 15 Sept. 1969.
See also **Orchestral (Suites).**

Three Sisters
Composed 1969.
Film first shown: Venice, Sala Volpi, 26 Aug. 1970; London, Cameo-
Poly, 2 Nov. 1970.

INCIDENTAL MUSIC FOR RADIO

Christopher Columbus
Play by Louis MacNeice (1907–63).
Composed 1942.
F.p. Bedford, BBC broadcast, 12 Oct. 1942. Joan Lennard (sop.),
Bradbridge White (ten.), BBC Chorus, BBC Symphony Orchestra,
cond. Sir Adrian Boult.
See also **Orchestral (Suites)**, and **Choral and Vocal (Chorus and
Orchestra)** and 'Beatriz's Song' under **Choral and Vocal (Voice(s)
and Instrument(s))**.

APPENDIX II

Troilus and Cressida

In his preface to the libretto of this Trojan War opera, the late Christopher Hassall made it clear that he did not base the plot on Shakespeare's *Troilus and Cressida*, in which Cressida is depicted as little more than a tart, but on Chaucer's poem *Troylus and Criseide*. Chaucer took a more charitable view of the heroine, a view expounded, as Hassall explained, by C. S. Lewis in *The Allegory of Love* when he analysed Cressida's ruling passion as fear—'fear of loneliness, of old age, of death, of love and of hostility . . . and from this fear springs the only positive passion which can be permanent in such a nature, the pitiable longing, more childlike than womanly, for *protection*, for some strong and stable thing that will hide her away and take the burden from her shoulder.'

In the *Iliad* itself Troilus and Cressida are barely mentioned. There is a reference by Priam to 'Troilus that happy charioteer', while Cressida (Chryseis) was a prisoner given to Agamennon and is in no way connected with Troilus. The legend of their love story began to evolve in the *chansons* of the medieval troubadours. The character of Pandarus, which is so brilliantly developed by Hassall and Walton, was invented by Boccaccio in *Il filostrato* in the fourteenth century and borrowed by Chaucer, who turned him into a light-hearted, middle-aged schemer. Hassall explained that the Chaucerian conception of 'courtly love' would be too remote from present-day audiences so 'as I lifted the story out of the Middle Ages and retold it in a setting of legendary Troy, all that was essentially Chaucerian fell away . . . The rules of courtly love required a go-between in the ordinary course of events. Transfer the story to another period and Pandarus must have a new justification or lose his office.' The Hassall Pandarus is a well-meaning but irresponsible entrepreneur, something of a dilettante, doing rather well out of the war, to judge by his style of living, and cynically ill-equipped to cope with the tragic forces he sets in motion. 'To ease the world's despair was never worth the trying', he says in Act III.

Hassall's aim in his libretto was to set the private tragedy against a strong background of the public tumult of the Trojan War. The traitorous High Priest, Calkas, occurs briefly in Chaucer, described as 'in science so expert was' that he 'knew well that Troy should destroyed be.' Later, 'to the Grekes ost ful prively he stol.' Hassall invented the episode of Calkas's attempt to persuade the Trojans to sue for peace, and he invented the character of Cressida's servant Evadne, who (unlike Desdemona's Emilia) is the deliberate cause of the lovers' downfall. He shows the characters in the grip of a fate they cannot control, thus establishing a link between the Greek notion of tragedy and contemporary attitudes.

SYNOPSIS

Act I. The Citadel of Troy, before the Temple of Pallas at dawn

The curtain rises after a short prelude to reveal a number of Trojans kneeling outside the Temple of Pallas, and chanting part of the invocation which the priests and priestesses are singing. Some bystanders remark on the senselessness of this activity after ten years of siege and want to drag away the worshippers. Calkas, the High Priest, whose voice has been heard leading the prayers, enters to rebuke them. He tells them that they have the means of deliverance in their own hands—the Delphic Oracle, he says, has advised the start of peace talks with the Greeks: surrender, in a word. The crowd find this hard to believe and so does Antenor, a young army officer just about to lead his men on a patrol. He demands to see the messenger who has brought this advice from Delphi. Calkas fudges the issue, and Antenor accuses him of being in the pay of the Greeks and of forging the oracle's message. The crowd turn on Calkas and are about to attack him just before he takes refuge in the temple when Prince Troilus, son of King Priam of Troy, arrives and, with drawn sword, drives them back. Troilus rebukes his friend Antenor and assures him of Calkas's good faith. But Antenor thinks he knows another reason for Troilus's defence of the High Priest—has he not seen him frequenting the temple lately? Could it be interest in the lovely novice, Calkas's daughter Cressida, who keeps the altar candles burning?

Left alone, Troilus admits that Antenor is right. He sings the recitative and aria which declare his love: 'Is Cressida a slave that she must trim those guttering candles? . . . Child of the wine-dark wave mantled in beauty.' When the aria ends, Cressida comes from the temple, a crimson scarf round her neck and carrying flowers for the altar. In her first beautiful aria, 'Morning and evening', she expresses to Troilus some of her fears and why she cannot return his love. Life has left her numbed: she is a war widow and she knows her father is preparing to desert to the Greeks. She finds peace only in tending the goddess's altar.

Cressida returns to the temple. But the last part of her conversation with Troilus has been overheard by her dandyish uncle, Calkas's brother Pandarus, who has been carried on to the stage in a silk-curtained litter befitting its owner, who is rich, rather effeminate and a skilled intriguer. It has surprised him to hear Troilus in amorous mood—'quite out of character', he says, for he thinks of the prince as a rugged warrior. Pandarus, delighted by the chance, offers to intercede with Cressida for him ('I've a genius for this game'). He is about to enter the temple to look for her when Calkas, dressed for a journey, comes out with Cressida and her servant Evadne. They vainly plead with him to stay in Troy. Pandarus follows Calkas; and Cressida sings her second big aria, 'Slowly it all comes back—out of my childhood', in which she recalls that, even when she was a child, Calkas seemed to her an inconstant shadowy figure.

Pandarus and Evadne return. It is evident to them all that Calkas has gone over to the Greeks. Who will protect us now, Cressida asks. This is the chance for Pandarus to point out to her that her best protection is the fact that a member of the Trojan royal family is in love with her. He is interrupted by the coincidental return of Troilus and of Antenor's soldiers who tell how they were ambushed and Antenor himself taken prisoner. Troilus declares that he will persuade his father to buy Antenor back by an exchange of prisoners; better still, he decides, they will get him back by force and Calkas shall first bless their swords. But the temple priest tells him that Calkas is no longer there nor at his home. Troilus orders a search of the city and himself goes into the temple. Pandarus takes this opportunity to invite Cressida to supper next day. He persuades her against her better judgement to give her scarf to Troilus as a token to comfort him in his daring exploit. When Troilus returns stricken by the realization that Cressida's father is a traitor, Pandarus gives him the scarf and is delighted to observe the effect it has.

Act II. A room in the upper floor of Pandarus's house

Scene i

Supper is over in Pandarus's luxurious home. Horaste, a young beau who has been helping Pandarus as host, is beating Cressida at chess, watched by Evadne. Pandarus is pacing, or mincing, restlessly about. The weather has turned nasty, he says (which is just what he wants). Surreptitiously he sends a servant to fetch Troilus. A thunderstorm begins and Cressida rises to leave. Pandarus persuades her to stay the night and wishes her sweet dreams. Unable to sleep, Cressida admits to herself that she has been thinking of nothing but Troilus and that, though she never wanted it to happen, she is in love again.

She is disturbed by Pandarus who tells her that by 'marvellous accident' Troilus has called at the house at this late hour because he is racked by jealousy. Troilus bursts into the room, denouncing Pandarus's stupid gossip. Pandarus admits he set it all up and leaves the lovers together, with the inevitable result—after a lyrical duet. Now, in an entr'acte in the manner of Berlioz rather than that of Richard Strauss, the orchestra describes the storm that is taking place both outside and inside the room.

Scene ii

As the passionate tumult subsides into slow, contented chords, Cressida opens the shutters to watch the dawn. The lovers' rapture is shattered by the agitated arrival of Pandarus. Greek soldiers have come to the house and he urges Troilus and Cressida to hide. Prince Diomede, heir to the throne of Calydon and Argos, a soldier with 'a certain hearty charm', to quote the librettist, has come with news of Calkas—the price of the services he has rendered to the Greeks is that Cressida should join him. We learn that Troilus's attempt to recapture Antenor failed and Diomede has come to arrange an exchange of prisoners, with Cressida-for-Antenor top of the list. Pandarus's protest is silenced when Diomede produces the agreement bearing the royal seals of Greece and Troy. Diomede searches the room and finds Cressida. He is taken aback by her beauty, and orders her to be ready to leave shortly. He departs with his escort. Troilus returns in anguish, vowing to dissolve the bargain, but, when Cressida pleads with him not to let her go, he replies 'We cannot fight both gods and men.' He says he will bribe the Greek sentries so that they can meet and will smuggle messages to her. They sing a slow, soft duet and Troilus gives her back her crimson scarf. Vowing eternal love, Cressida is escorted away.

Act III. The Greek camp outside Troy, early evening ten weeks later

A slow prelude, marked 'lugubre', precedes the rise of the curtain. Sentries are calling 'All's well'. Cressida comes from the ornate pavilion allotted to Calkas, who is obviously being treated as a VIP defector. She is waiting for Evadne to return from the palisade with a message from Troilus, but there is none, nor has there been for ten weeks. 'Love cannot feed on air', she says, but asks Evadne to go back once more. Evadne advises her to forget Troilus, who, she says, has obviously forgotten her, and to encourage Diomede. Cressida is furious with her, but relents and pleads with her to go to the palisade for the last time. Alone, Cressida gives way to her despair and implores Persephone, the queen of the dead, to help her find Troilus.

Calkas joins her and rebukes her for 'nursing a nameless sorrow', adding that her twilight wanderings are causing gossip. Troy is doomed, he tells her, but Diomede can save them both if she will love him. He roughly orders her to accept Diomede that evening when the Greek prince comes for her final answer. Alone again, Cressida sings of her dilemma—she is tempted by Diomede, forsaken by Troilus, her father, and the gods. She cries out passionately for Troilus. Her soliloquy is ended by the unexpected arrival of Diomede. Cressida yields to his patient wooing and he asks for a token from her—her scarf. She offers him a ring instead, but he insists on the scarf so that he can wear it on his helmet in battle. At that moment Evadne returns and indicates 'no news from Troilus.' This is the final blow and impulsively she thrusts the scarf into Diomede's hand, saying, 'Take all you ask of me and let it be for ever.' They embrace and depart, she to Calkas's pavilion and a jubilant Diomede to arrange for proclamation of their marriage.

Evadne has overheard some of this and takes a scroll to the sentries' brazier. It is Troilus's latest message, the last of the many from him that she has been burning for ten weeks. She feels remorse about her deceit, but is convinced she has been acting for the best—life will obviously be preferable with Diomede. She is warming her hands at the fire when Troilus and Pandarus arrive, asking for Cressida's tent and explaining that there is an hour's truce for talks. Troilus explains that in his latest message he had warned Cressida to be ready to return to Troy. Evadne's glum reception alerts Pandarus's suspicions. Cressida comes out of Calkas's tent but has not seen Troilus and Pandarus—she wants to know if Evadne thinks Diomede will like her hair and dress. Troilus calls to her. Cressida, demented, asks why he sent no word to her. He sent message after message, he explains, the sentries must have tricked him; now he has persuaded his father Priam to bargain for her release. Cressida is telling him to leave and to forget her, he has come too late, when trumpets sound and voices are heard hailing her as bride of Calydon and Argos. Greeks, including Diomede and Calkas, come to pay her homage. Troilus sees that Diomede has Cressida's scarf wound round his helmet and assumes that Diomede has plundered it. He claims Cressida as his own. Diomede calls on Cressida to denounce Troilus, but she remains silent and goes to kneel by Troilus. The crowd yell 'False Cressida' and Diomede begins the sextet 'Troy, false of heart, yet fair', in which each of the main characters comments on the situation. Diomede likens Cressida to Troy itself, a rose with a fatal canker; Troilus rejects Cressida as a 'woman at wanton play'; Cressida pleads that her suffering shows her love for Troilus; Pandarus decides that man prefers war to anything else; Calkas regrets his treachery; but Evadne urges Calkas to let Diomede kill Troilus, then everything might turn out well again.

Diomede throws the scarf to the ground and tramples on it. Troilus attacks him but is stabbed in the back by Calkas. Diomede is horrified by this treacherous (Siegfried-like) murder and orders Calkas back to Troy. But he insists that Cressida, 'that whore', shall stay. Cressida picks up Troilus's sword—'one part of Troilus shall still be mine.' She sings a tender farewell to her lover and, after a final rapturous outburst, kills herself.

Walton conducts Walton: *A list of the composer's recordings of his own works.*

Façade, an Entertainment
11 items, with Edith Sitwell and Constant Lambert as reciters, and instrumental ensemble. E. Sitwell recites 'Black Mrs Behemoth', 'Jodelling Song', 'Long Steel Grass', and 'A Man from a Far Countree'; C. Lambert in 'Polka', 'Foxtrot' ('Old Sir Faulk'), 'Tango-Pasodoble', 'Scotch Rhapsody', 'Tarantella', 'Valse', and 'Popular Song'. Recorded in Chenil Galleries, London, 28 Nov. 1929.
Decca T. 124–5 (78 r.p.m.) Reissued in 1972 at 45 r.p.m. as OUP 110 (mono), part of de luxe edition of score of *Façade.*
Complete 1951 version, with Peggy Ashcroft and Paul Scofield as reciters, and London Sinfonietta. Recorded in Decca Studio No. 3, 156 Broadhurst Gardens, London, 5, 6, and 7 May 1969 (music was recorded first, with voices added some months later).
Argo ZRG 649 (stereo).

Façade, Suites Nos. 1 and 2
London Philharmonic Orchestra. Recorded No. 1 Studio, Abbey Road, 5 Mar. 1936. ('Noche espagnole' and 'Old Sir Faulk' recorded 25 Oct. 1938.)
HMV C 2836–7, C 3042 (78 r.p.m.), reissued in EMI ED 29 0715 1 (mono, 33 r.p.m.).
The Philharmonia Orchestra. Recorded Kingsway Hall, London, 20 Apr. 1955 and 26 Mar. 1957.
Columbia 33C 1054 (mono), reissued as EMI HQM 1006 (mono) and in EMI SLS 5246 (mono).

Portsmouth Point
The Philharmonia Orchestra. Recorded Kingsway Hall, 18 and 21 Mar. 1953.
Columbia 33C 1016 (mono), reissued as 45 Columbia SEG 8217 (mono). Reissued in EMI HQM 1006 (mono) and in EMI SLS 5246 (mono).

London Philharmonic Orchestra. Recorded Walthamstow Assembly Hall, London, 15 Apr. 1970.
Lyrita SRCS 47 (stereo).

Siesta
London Philharmonic Orchestra. Recorded No. 1 Studio, Abbey Road, 25 Oct. 1938.
HMV C 3042, reissued in EMI ED 29 0715 1 (mono, 33 r.p.m.).
London Philharmonic Orchestra. Recorded Walthamstow Assembly Hall, 15 Apr. 1970.
Lyrita SRCS 47 (stereo).

Sinfonia Concertante
Phyllis Sellick (piano), City of Birmingham Orchestra. Recorded Dudley Town Hall, 8 Aug. 1945.
HMV C 3478–80 (78 r.p.m.). Reissued in World Record Club SH 128 (mono, 33 r.p.m.) and on EMI EH29 12761 (mono, 33 r.p.m.).
Peter Katin (piano), London Symphony Orchestra. Recorded Walthamstow Assembly Hall, 14 Apr. 1970.
Lyrita SRCS 49 (stereo).

Viola Concerto
Frederick Riddle (viola), London Symphony Orchestra. Recorded Kingsway Hall, 6 Dec. 1937.
Decca X199–201 (78 r.p.m.).
William Primrose (viola), The Philharmonia Orchestra. Recorded Abbey Road, 22 and 23 July 1946.
HMV DB 6309–11 (78 r.p.m.), reissued in EMI EH 29 12761 (mono, 33 r.p.m.).
Yehudi Menuhin (viola), New Philharmonia Orchestra. Recorded Abbey Road, 9 and 11 Oct. 1968.
HMV ASD 2542 (stereo).

Belshazzar's Feast
Dennis Noble (bar.), Huddersfield Choral Society, Liverpool Philharmonic Orchestra. Recorded in Philharmonic Hall, Liverpool, 3 and 10 Jan. 1943.
HMV C 7572–6 (78 r.p.m.), reissued as HMV ALP 1089 (mono, 33 r.p.m.) and in EMI ED 29 0715 1 (mono).
Donald Bell (bar.), Philharmonia Chorus and Orchestra. Recorded Kingsway Hall, 2, 3, 4, and 5 Feb. 1959.
Columbia CX 1679 (mono, 33 r.p.m.), reissued as SAX 2319 (stereo) and in EMI SXLP 30236 and EMI SLS 5246.

Symphony No. 1 in B flat minor
The Philharmonia Orchestra. Recorded Abbey Road, 17, 18, and 19
   Oct. 1951.
   HMV ALP 1027 (mono, 33 r.p.m.), reissued in EMI SLS 5246.

Crown Imperial
The Philharmonia Orchestra. Recorded Kingsway Hall, 18 Mar. 1953.
   Columbia 33C 1016 (mono, 33 r.p.m.), reissued in EMI HQM 1006
   (mono) and in EMI SLS 5246 (mono).

 Violin Concerto
Jascha Heifetz (violin), The Philharmonia Orchestra. Recorded Abbey
   Road, 26 and 27 June 1950.
   HMV DB 21257–9 (78 r.p.m.), reissued as HMV BLP 1047 (mono, 33
   r.p.m.) and as RCA LSB 4102 and in RCA GD 87966 (compact disc).
Yehudi Menuhin (violin), London Symphony Orchestra. Recorded
   No. 1 Studio, Abbey Road, 12–15 July 1969.
   HMV ASD 2542 (stereo, 33 r.p.m.).

Suite, The Wise Virgins
The Sadler's Wells Orchestra. Recorded Abbey Road Studios, 24 July
   1940 (sides 1 and 2), 8 Aug. 1940 (sides 3 and 4).
   HMV C3178–9

'Sheep may safely graze' from *The Wise Virgins*
Philharmonia Orchestra. Recorded Kingsway Hall, 21 Mar. 1953.
   Columbia 33C 1016 (mono).

Music for Children
London Philharmonic Orchestra. Recorded Walthamstow Assembly
   Hall, 15 Apr. 1970.
   Lyrita SRCS 50 (stereo, 33 r.p.m.).

Scapino
The Philharmonia Orchestra. Recorded Abbey Road, 19 Oct. 1951.
   HMV DB 21499 (78 r.p.m.), reissued in EMI ED 29 0715 1 (mono,
   33 r.p.m.).
London Symphony Orchestra. Recorded Walthamstow Assembly
   Hall, 13 Apr. 1970.
   Lyrita SRCS 49 (stereo, 33 r.p.m.).

Spitfire Prelude and Fugue
The Hallé Orchestra. Recorded Houldsworth Hall, Manchester, 24
   June 1943.

HMV C 3359 (78 r.p.m.), reissued as HMV 7P 312 (mono, 45 r.p.m.) and in EMI ED 29 0715 1 (mono, 33 r.p.m.).
The Philharmonia Orchestra. Recorded Abbey Road, 16 Oct. 1963.
EMI SXLP 30139 (stereo, 33 r.p.m.) and in EMI SLS 5246 (stereo).

Suite, The Quest
London Symphony Orchestra. Recorded Walthamstow Assembly Hall, 14 Apr. 1970.
Lyrita SRCS 49 (stereo, 33 r.p.m.).

Henry V (film music)
Laurence Olivier with The Philharmonia Orchestra. Recorded Abbey Road, 27 and 28 Aug. and 12 and 13 Oct. 1946 (extra session on 13 Nov. 1946 conducted by Roy Douglas).
HMV C 3583–6 (78 r.p.m.), reissued as RCA RB 16144 (mono, 33 r.p.m.), in HMV ALP 1375 (mono, 33 r.p.m.), and in RCA LSB 4104 (mono, 33 r.p.m.).

Two Pieces for Strings ('Death of Falstaff' and 'Touch her soft lips and part')
The Philharmonia String Orchestra. Recorded Abbey Road, 12 Oct. 1945.
HMV C 3480 (78 r.p.m.), reissued in HMV ALP 1375 (mono, 33 r.p.m.) and in EMI ED 29 0715 1 (mono).

Suite from Henry V (arr. Mathieson)
The Philharmonia Orchestra. Recorded Abbey Road, 15 Oct. 1963.
Columbia 33CX 1883 (mono, 33 r.p.m.), reissued as Columbia SAX 2527 (stereo). Reissued in World Record Club ST 656 (mono, 33 r.p.m.), and in EMI SXLP 30139 (stereo, 33 r.p.m.) and in EMI SLS 5246 (stereo).

Hamlet: Funeral March
The Philharmonia Orchestra. Recorded Abbey Road, 15 Oct. 1963.
Columbia 33CX 1883 (mono, 33 r.p.m.), reissued as Columbia SAX 2527 (stereo). Reissued in World Record Club ST 656 (mono, 33 r.p.m.) and in EMI SXLP 30139 (stereo, 33 r.p.m.).

Orb and Sceptre
The Philharmonia Orchestra. Recorded Kingsway Hall, 18 Mar. 1953.
Columbia LX 1583 (mono, 78 r.p.m.), reissued as Columbia 33C 1016 (mono, 33 r.p.m.) and as Columbia SEL 1504 (mono, 45 r.p.m.). Reissued in EMI HQM 1006 (mono, 33 r.p.m.) and in EMI SLS 5246 (mono).

Troilus and Cressida

Excerpts: Act I: 'Is Cressida a slave?' 'Slowly it all comes back'. Act II: 'How can I sleep?' 'If one last doubt remains'. 'Now close your arms'. Act III: 'All's well! Diomede! Father!' Elisabeth Schwarzkopf (sop.), Richard Lewis (ten.), The Philharmonia Orchestra. Recorded Kingsway Hall, 18, 19, and 20 Apr. 1955.

Columbia 33CX 1313 (mono, 33 r.p.m.), reissued in World Record Club OH 217 (mono, 33 r.p.m.).

Excerpt: Act II: 'How can I sleep?' Marie Collier (sop.), Peter Pears (ten.), Covent Garden Opera Orchestra. Recorded Kingsway Hall (as part of Covent Garden anniversary set) in one of two sessions, Feb. and July 1968.

Decca SET 392–3 (stereo).

Richard III, Prelude and Suite

The Philharmonia Orchestra. Recorded Abbey Road, 15 and 16 Oct. 1963.

Columbia 33CX 1883 (mono, 33 r.p.m.), reissued as Columbia SAX 2527 (stereo). Reissued in World Record Club ST 656 (mono, 33 r.p.m.) and in EMI SXLP 30139 (stereo, 33 r.p.m.) and *Prelude* only in EMI SLS 5246 (stereo).

Johannesburg Festival Overture

The Philharmonia Orchestra. Recorded Kingsway Hall, 26 Mar. 1957.

Columbia 33C 1054 (mono, 33 r.p.m.), reissued in EMI HQM 1006 (mono, 33 r.p.m.) and in EMI SLS 5246 (stereo).

Partita for Orchestra

The Philharmonia Orchestra. Recorded Kingsway Hall, 6 and 16 Feb. 1959.

Columbia 33CX 1679 (mono, 33 r.p.m.), reissued as Columbia SAX 2319 (stereo). Reissued as EMI SXLP 30236 (stereo) and in EMI SLS 5246 (stereo).

Capriccio Burlesco

London Symphony Orchestra. Recorded Walthamstow Assembly Hall, 13 Apr. 1970.

Lyrita SRCS 49 (stereo, 33 r.p.m.).

APPENDIX IV

Select Bibliography

BOOKS ABOUT WALTON

CRAGGS, S. R., *William Walton, A Thematic Catalogue of his Musical Works*, with a critical appreciation by Michael Kennedy (London, 1977).
HOWES, F., *The Music of William Walton* (London, 1965; 2nd edn., London, 1974).
OTTAWAY, H., *Walton* (Novello Short Biographies) (Sevenoaks, 1972).
TIERNEY, N., *William Walton: His Life and Music* (London, 1984).
WALTON, S., *William Walton: Behind the Façade* (Oxford, 1987).

RELEVANT BOOKS AND ARTICLES

Aberconway, C., *A Wiser Woman* (London, 1966).
Aprahamian, F., 'Walton and his New Symphony', *The Listener*, 25 Aug. 1960.
—— 'Walton Retrospective', *Music and Musicians*, 20/11 (1972).
Ashton, F., 'The Ballet Façade' in limited edition of full score of *Façade* (Oxford, 1972).
Avery, K., 'William Walton', *Music and Letters*, 28 (1947).
Blom, E., 'The Later William Walton', *The Listener*, 20 Sept. 1945.
Brook, D., 'William Walton', in *Composers' Gallery: Biographical Sketches of Contemporary Composers* (London, 1946).
Campbell, R., *Light on a Dark Horse* (London, 1951).
Cooper, M., 'The Unpredictable Walton', *The Listener*, 25 July 1957.
Cox, D., 'Walton', in R. Simpson (ed.), *The Symphony, 2. Elgar to the Present Day* (Harmondsworth, 1967).
Evans, E., 'Walton and Lambert', *Modern Music*, 7/2 (1930).
—— 'Walton's Symphony', *Radio Times*, 1 Nov. 1935.
—— 'William Walton', *Musical Times*, 85 (1944).
Evans, P., 'Sir William Walton's Manner and Mannerism', *The Listener*, 20 Aug. 1959.
Foss, H. J., 'William Walton', *The Chesterian*, 11 (1930).
—— 'William Walton', *Musical Quarterly*, 26 (1940), repr. in D. Ewen (ed.),*The Book of Modern Composers* (New York, 1943).

—— 'The Music of William Walton', *The Listener*, 17 May 1945.

Frank, A., 'The Music of William Walton', *The Chesterian* 20, (1939).

Gilbert, G., 'Walton on Trends in Composition', *New York Times*, 4 June 1939.

Gray, C., *Musical Chairs, or Between Two Stools* (London, 1948).

Greenfield, E., 'William Walton', *The Guardian*, 29 Feb. 1972.

Hassall, C., 'Walton's Opera', *Music and Musicians*, 3/4 (1954).

Heath, E., *Music: A Joy For Life* (London, 1976).

Howes, F., *The English Musical Renaissance* (London, 1966).

Hughes, P. [Spike], *Opening Bars* (London, 1946).

—— 'Nobody Calls him Willie now', *High Fidelity*, Sept. 1960.

Hussey, D., 'William Turner Walton', in A. L. Bacharach (ed.), *The Music Masters*, iv (London, 1954).

—— 'Walton's *Troilus and Cressida*', *Music and Letters*, 36 (1955).

Hutchings, A., 'The Symphony and William Walton', *Musical Times*, 78 (1937).

Jacobs, A., 'William Walton', *Musical America*, Feb. 1952.

—— Interview with Sir William Walton at Savoy Hotel, 1 Nov. 1977, for BBC World Service.

Jefferson, A., 'Walton: Man and Music', *Music and Musicians*, 13/7 (1965).

Keller, H., and Walton, Sir W., 'Contemporary Music: Its Problems and its Future', *Composer*, No. 20 (1966) (transcript of BBC interview).

Kennedy, M., 'William Walton: A Critical Appreciation', in Stewart R. Craggs's *William Walton, A Thematic Catalogue* (London, 1977).

——'The Origins of Belshazzar's Feast', in full score of *Belshazzar's Feast* (Oxford, 1978).

—— 'Sir William Walton at Eighty', in *Keynote*, Mar. 1982.

—— 'Walton in Perspective', *The Listener*, 11 Aug. 1983.

Lambert, C., 'Fresh Hand; New Talent; Vital Touch; Brief Record of William Walton, Composer of *Portsmouth Point* and a Score or Two Besides', *Boston Evening Transcript*, 27 Nov. 1926.

—— 'Some Recent Works by William Walton', *The Dominant*, I/4 (1928).

—— 'Some Angles of the Compleat Walton', *Radio Times*, 7 Aug. 1936.

—— *Music, Ho! A Study of Music in Decline* (London, 1934; 3rd edn., 1966).

Lambert, J. W., 'Imp and Sceptre: conversation with Sir William Walton', *Sunday Times*, 25 Mar. 1962.

Layton, R., 'Walton and his Critics', *The Listener*, 29 Mar. 1962.

Lehmann, J., *A Nest of Tigers: Edith, Osbert and Sacheverell Sitwell in their Times* (London, 1968).

Lutyens, E., *A Goldfish Bowl* (London, 1972).

Maine, B., 'The Music of William Walton', *Musical Opinion*, No.60, 1937.

Mason, C., 'William Walton' in A. L. Bacharach (ed.), *British Music of our Time* (Harmondsworth, 1946).

Mellers, W., 'Sir William Walton and Twentieth-century Opera', *The Listener*, 2 Dec. 1954.

Merrick, F., 'Walton's Concerto for Violin and Orchestra', *Music Review*, 2 (1941).

Mitchell, D., 'Some Observations on William Walton', *The Chesterian*, 26 (1952).

—— 'The Modernity of William Walton', *The Listener*, 7 Feb. 1957.

Moorehead, C., 'Beyond the façade . . . the reluctant Grand Old Man', *The Times*, 29 Mar. 1982.

Morrison, A., 'Willie: The Young Walton and his Four Masterpieces', talk given at British Festival of Recorded Sound, 31 Jan. 1984. Repr. in *RCM Magazine*, 80/3 (1984), 119–127.

Murrill, H., 'Walton's Violin Sonata', *Music and Letters*, 31 (1950).

Newman, E., 'Façade', *Sunday Times*, 2 May 1926, repr. in E. Newman, *From the World of Music* (selected by Felix Aprahamian) (London, 1956).

Newton, I., *At the Piano: The World of an Accompanist* (London, 1966).

Ottaway, H., 'Walton and the Nineteen-Thirties', *Monthly Musical Record*, No. 81 (1951).

—— 'Walton and his Critics', *The Listener*, 25 June 1970.

—— 'Walton's First Symphony: the composition of the finale', *Musical Times*, 113 (1972).

—— 'Walton's First and its Composition', *Musical Times*, 114 (1973).

—— 'Walton Adapted', *Musical Times*, 115 (1974).

—— 'Walton', in *The New Grove Dictionary of Music and Musicians*, 20 (London, 1980).

Palmer, C., 'Walton's Film Music', *Musical Times*, 113 (1972).

—— 'Walton's Church Music', *Church Music*, 3/19 (1973).

—— 'Walton', *Musical Times*, 114 (1973).

Pearson, J., *Façades: Edith, Osbert and Sacheverell Sitwell* (London, 1978).

Pirie, P. J., 'Scapino: The Development of William Walton', *Musical Times*, 105 (1964).

—— 'Walton at Seventy', *Music and Musicians*, 20/3 (1972).

—— *The English Musical Renaissance* (London, 1979).

Poulton, A., *The Recorded Works of Sir William Walton: A Discography Celebrating Fifty Years of Recording History, 1929–79* (Kidderminster, 1980).

Reid, C., *Malcolm Sargent: A Biography* (London, 1968).

Reizenstein, F., 'Walton's *Troilus and Cressida*', *Tempo*, No. 34 (1954–5).

Routh, F., 'William Walton', in *Contemporary British Music: The 25 Years from 1945 to 1970* (London, 1972).

Rubbra, E., 'William Walton's 70th Birthday', *The Listener*, 23 Mar. 1972.

Rutland, H., 'Walton's New Cello Concerto', *Musical Times*, 98 (1957).

Schafer, R. M., 'William Walton', in *British Composers in Interview* (London, 1963).

Shawe-Taylor, D., 'The Challenge of Walton', *Sunday Times*, 26 Mar. 1972.

Shead, R., *Constant Lambert* (Lutterworth, 1973).

Shore, B., 'Walton's Symphony', in *Sixteen Symphonies* (London, 1949).

Sitwell, E., *Taken Care Of* (London, 1965).

—— *Façade*, sleeve-note written for Decca recording, LXT 2977, 1954, repr. in limited edition of full score of *Façade* (Oxford, 1972).

—— 'Young William Walton Comes to Town', *Sunday Times*, 18 Mar. 1962.

Sitwell, O., *Laughter in the Next Room* (London, 1949).

Sitwell, S., 'Façade', in limited edition of full score of *Façade* (Oxford, 1972).

Skelton, G., *Paul Hindemith: The Man behind the Music* (London, 1975).

Slonimsky, N., *Music Since 1900* (London, 1949).

Tertis, L., *My Viola and I* (London, 1974).

Tovey, D. F., *Essays in Musical Analysis*, 3. Concertos (London, 1936).

Vaughan, D., *Frederick Ashton and his Ballets* (London, 1977).

Walton, W. T., 'Contemporary Music: Its Problems and its Future', *Composer*, No. 20 (1966) (transcript of BBC interview with Hans Keller).

—— 'My Life in Music', *Sunday Telegraph*, 25 Mar. 1962.

—— 'Sir William Walton Talks about his First Symphony to Eric Roseberry', BBC interval talk, Music Programme, 3 Oct. 1965.

—— 'Preface' to Alan Poulton (ed.), *Alan Rawsthorne, 3. Essays on the Music* (Hindhead, 1986).

Warrack, J., 'Façade', in D. Drew (ed.), *The Decca Book of Ballet* (London, 1958).

—— 'Walton's Troilus and Cressida', *Musical Times*, 90 (1954).

—— 'Sir William Walton talks to John Warrack', *The Listener*, 8 Aug. 1968 (transcript of BBC interview).

Widdicombe, G., 'Walton's Troilus and Cressida', *Music and Musicians*, 25/3 (1976).

—— 'Walton at 70', *Financial Times*, 30 Mar. 1972.

Zoete, B. de, 'William Walton', *Monthly Musical Record*, No. 59 (1929).

Classified Index of Works

Arrangements and Transcriptions of Music by Other Composers

Bach, J. S.: *The Wise Virgins*, 108, 292, 301; see also Ballets and Orchestral
Choral Prelude for piano, 'Herzlich thut mich verlangen', 310

National Anthem (1953 arr.) 167, 296; (1955 arr.) 195, 296
Stafford Smith, J.: *The Star-Spangled Banner*, 195, 296

Ballets

Dr Syntax, 21, 290–1; see also Orchestral
Façade (ballets and suites), 62, 99, 167, 282, 291; see also Entertainment and Keyboard
First Shoot, The, 89–90, 243, 291; see also Brass Band and Orchestral

Quest, The, 121–3, 292; see also Orchestral
Siesta, 254, 291–2; see also Keyboard and Orchestral
Varii Capricci, 265, 292; see also Chamber Music and Orchestral
Wise Virgins, The, 108, 292; see also Arrangements and Orchestral

Brass Band

First Shoot, The, 90, 273, 311–2; see also Ballets

Chamber Music

Bagatelles, Five, 245–6, 254, 311; orch. as *Varii Capricci*, 264–5; see also Ballets and Orchestral
Duettino, 277, 310
Passacaglia, 274–6, 311
Pianoforte Quartet, 20, 309; composed, 13; discussed, 13–14; revised, 259
String Quartet (1919–22), 32, 39, 309; composed, 21; performed at Salzburg, 25
String Quartet in A minor, 137, 138,

212, 245, 247, 309; first ideas, 128; discussed, 134–5; see also Sonata for Strings under Orchestral
Theme (for Variations), 245, 311
Toccata in A minor, 21, 32, 309
Two Pieces for Violin and Piano, 145, 309–10; discussed, 147
Violin Sonata, 309; begun, 142; performed, 145, 150; discussed, 145–7

Choral Works

Belshazzar's Feast, 38, 52, 58, 59, 60, 69, 73, 93, 94, 133, 134, 138, 150, 159, 166, 167, 209, 223, 251, 254, 257, 277, 280, 282, 283, 289, 302; composed, 53–8; discussed, 60–1; revised, 142; in Rome, 209, 260; in Hoffnung concert, 218; W. on Previn recording, 257; on eightieth birthday, 276

Christopher Columbus (arr. Davis), 303; see also Orchestral and Radio

Coronation Te Deum, 277, 303; composed, 162–3; performed, 166; discussed, 166–7

Gloria, 303; composed, 215; discussed, 216–7

Henry V (arr. Palmer), 303; see also Film Music and Orchestral

In Honour of the City of London, 93, 302–3; discussed, 94–5

Church Music

Antiphon, 266, 268, 309
Jubilate Deo, 248–9, 251, 308
Magnificat and Nunc Dimittis, 262–3, 308

Missa Brevis, 226, 228, 308
Twelve, The, 227, 229, 251, 308

Entertainment

Façade, 3, 12, 15, 17, 20, 23, 24, 26, 33, 37, 39, 45, 47, 48, 89, 133, 162, 232, 234, 253, 254, 270, 282, 289, 304–5; composed, 27–8; first performed, 28–9; early performances, 29–31; in Siena, 33–4; definitive version, 34–5, 117; published, 151; at Aldeburgh, 254–5; fiftieth anniversary, 259; in Ischia, 277; see also Ballets, Keyboard, and Orchestral

Façade 2, 269–70, 305; discussed, 35–6

Fanfares and Marches

Anniversary Fanfare, 299
Birthday Fanfare, A, 300
Crown Imperial, 84, 163, 167, 240, 252, 298; composed, 93; see also Keyboard
Fanfare and March (Macbeth) (arr. Palmer), 300; see also Theatre
Fanfare for a Great Occasion, 139, 299; see also *Hamlet* under Film Music
Fanfare for the National, 266, 299
For the Red Army, 127, 299
Funeral March (Hamlet), 299; see also Film Music and Orchestral

History of the English-Speaking Peoples, A (March), 210, 298
Memorial Fanfare for Henry Wood, 127, 299
Orb and Sceptre, 165, 166, 167, 258, 298; composed and W.'s views on, 163
Queen's Fanfare, A, 299
Roaring Fanfare, 266, 300
Salute to Sir Robert Mayer (Introduction to the National Anthem), 272, 300

Film Music

As You Like It, 89, 312
Battle of Britain, The, 241, 243, 313; commissioned, 237; composed, 237–8; W.'s music rejected, 238–9; see also Orchestral

Dreaming Lips, 92, 99, 312
Escape Me Never, 76, 89, 312; see also Keyboard
First of the Few, The, 114, 116, 240, 312; composed, 117; see also Orchestral
Foreman Went To France, The, 114–5, 312
Hamlet, 138, 140, 313; discussed, 139; see also Orchestral
Henry V, 92, 255, 313; first mooted, 121; composed, 123–5; discussed, 125–6; see also Choral and Orchestral
Major Barbara, 109–10, 117, 312; see also Orchestral
Next of Kin, 112–3, 312
Richard III, 194–5, 237, 313; see also Keyboard and Orchestral
Stolen Life, 99, 100, 312
Three Sisters, 90, 243, 313
Went the Day Well?, 90, 119, 313

Keyboard

Ballet from *Escape Me Never*, arr. for piano, 310; see also Film Music
Crown Imperial, arr. for piano, 310; see also Fanfares and Marches
Duets for Children, for 2 pianos, 311; see also *Music for Children* under Orchestral
Portsmouth Point, arr. for 2 pianos, 310; see also Orchestral
Siesta, arr. for piano duet, 310; see also Ballets and Orchestral
Sinfonia Concertante, arr. for two pianos, 310; see also Orchestral
Three Pieces for organ, 195, 311; see also *Richard III* under Film Music, and Orchestral
Valse from *Façade*, for piano, 310; see also Ballets, Entertainments, and Orchestral
Valse in C minor, for piano, 310
'Wheatley', choral prelude for organ, 8, 9, 310

Operas

Bear, The, 132, 133, 229, 230, 231, 248, 254, 267, 290; suggested by Pears, 228; completed and performed, 232; discussed, 232–5; on TV, 245
Troilus and Cressida, 132, 139–40, 195, 197, 201, 202, 213, 233, 236, 253, 257, 290; synopsis, 314–9; decision on librettist, 136–7; Newman criticises libretto, 141; libretto 'ready', 145; work begins, 147–9; progress on, 150–3, 159–62, 164–6; Auden and sextet, 168–70; production problems, 171–8; completed, 178; rehearsal troubles, 179–80; first performances, 181–2; in Milan, 182–3; 1963 revival, 183–4; in Adelaide, 184, 223; hopes for revival, 184–6; revisions, 186–7; CG revival, 188–9; discussed, 189–93; revival mooted, 259; 1976 revival, 266–7; W.'s preoccupation with, 288–9; see also Orchestral and Songs

Orchestral (including Suites)

Battle of Britain, The (Suite arr. C. Matthews), 239, 301–2; see also Film Music
Capriccio Burlesco, 236, 283, 297
Cello Concerto, 212, 226, 236, 242, 282, 283, 288, 294; commissioned, 196; first performances, 197–8; new ending for, 198–9, 261; Heyworth on, 201–2; discussed, 203–4; W. and soloists, 244
Christopher Columbus (Suite arr. Palmer), 302; see also Choral Works and Radio
Façade (Suites Nos.1 and 2), 300–1;

see also Ballets, Entertainment, and Keyboard

First Shoot, The (arr. Palmer), 312; see also Ballets and Brass Band

Granada Prelude, Call Signs and End Music, 219, 297

Hamlet and Ophelia (arr. Mathieson), 139, 297; see also Fanfares and Film Music

Henry V (Suites), 126, 255, 301; see also Fanfares and Film Music

History of the English-Speaking Peoples, A, 210, 298; see also Fanfares and Marches

Improvisations on an Impromptu of Benjamin Britten, 132, 244, 251, 253, 297; composed, 241–2; discussed, 242–3; first British performance, 245

Johannesburg Festival Overture, 195–6, 295

Major Barbara (arr. Palmer), 297; see also Film Music

Music for Children, 109, 295–6; see also *Duets for Children* under Keyboard

Partita, 205, 208, 283; finished, 206; W.'s note on, 207; first performances, 207–8; discussed, 208

Passacaglia: Death of Falstaff (Two Pieces for Strings), 125, 296; see also *Henry V* under Film Music

Portsmouth Point, 9, 13, 21, 36, 44, 45, 52, 55n, 56, 83, 89, 110, 167, 208, 226, 283, 294–5; composed, 40–1; first performed, 41; discussed, 41–2; see also Keyboard

Prelude for Orchestra (from *Granada Call Signs*), 270, 297

Prelude, *Richard III,* 194, 301; see also Film Music and Keyboard

Prologo e Fantasia, 275–6, 298

Quest, The (Suite arr. Tausky), 123, 301; see also Ballets

Scapino, 100, 138, 214, 257, 283, 295; composed, 110; discussed, 111–2; first performance, 112

Shakespeare Suite, A (arr. Mathieson), 194–5, 301; see also *Richard III* under Film Music

Siesta, 42–3, 295; see also Ballets and Keyboard

Sinfonia Concertante, 44, 48, 49, 56, 167,

208, 252, 294; discussed, 45–6; revised, 124; see also Keyboard

Sonata for Strings, 205, 254, 297–8; transcription made, 247–8; discussed, 248; see also String Quartet in A minor under Chamber Music

Spitfire Prelude and Fugue, 117, 226, 239, 296; see also *First of the Few, The,* under Film Music

Symphony No.1, 9, 45, 89, 101, 104, 115, 169, 185, 201, 211, 212, 214, 230, 231, 238, 252, 280, 282, 293; first idea, 64; W. works on, 64–6, 68–9; W. is 'stuck', 69–70; progress of, 71–4; performed incomplete, 77–8, 80; completed, 80–1; first performance, 81–2; discussed, 82–4; success of, 85–6; recorded, 85

Symphony No.2, 205, 206, 209, 236, 253, 255, 293; commissioned, 196; first performed, 211; discussed, 211–4; in Rome, 258

Touch Her Soft Lips And Part (Two Pieces for Strings), 92, 125, 296; see also *Henry V* under Film Music

Troilus and Cressida (suite arr. Palmer), 302; see also Operas and Songs

Two Pieces for Strings (*Passacaglia: Death of Falstaff* and *Touch Her Soft Lips And Part*) from *Henry V,* 92, 125, 296; see also *Henry V* under Film Music

Variations on an Elizabethan Theme ('Sellinger's Round'), 163–4 and n, 167, 296

Variations on a Theme by Hindemith, 205, 226, 236, 241, 283, 297; composed, 219–20; performed, 220; discussed, 220–2

Varii Capricci, 265, 298; see also Ballets and Chamber Music

Viola Concerto, 9, 46, 49, 55n, 56, 57, 64, 84, 101, 103, 104, 133, 135, 146, 159, 196, 203, 219, 226, 251, 282, 286, 294; composed, 47–8; rejected by Tertis, 48; played by Hindemith, 49–50; discussed, 50–1, revised, 51–2, 92

Violin Concerto, 118, 130, 131, 146, 164, 169–70, 187n, 192, 196, 201,

212, 257, 280, 282, 294; commissioned, 92; composed, 97–101; discussed, 101–4; first performances, 105–6; revision, 124; in Rome, 260

Violoncello Concerto, see Cello Concerto
Wise Virgins, The, 108–9, 301; see also Arrangements and Ballets

Part-Songs, etc

All This Time, 307
Cantico del Sole, 258–9, 307–8
Drop, Drop, Slow Tears (*A Litany*), 8, 307; Mann on, 251
King Herod and the Cock, 268, 308
Litany, A, see Drop, Drop, Slow Tears
Make We Joy Now In This Fest, 65 and *n*, 307
Put Off The Serpent Girdle (*Troilus*

and Cressida), 187, 307; see also Operas and Orchestral
Set Me As A Seal Upon Thine Heart, 100, 253, 256, 307
Under The Greenwood Tree, 307; see also Songs
What Cheer?, 307
Where Does The Uttered Music Go?, 128, 307

Projected Works
(including those unpublished and withdrawn)

Adagio ed Allegro Festivo, 263–4
American Wind Band Piece (untitled), 271, 272
Antony and Cleopatra (film), 169
Brass Band Piece (untitled), 271, 272, 273
Bucolic Comedies, 29, 305
Clarinet Concerto, 217, 267
Double Concerto (violin and cello), 206, 237
Dr Syntax, 21, 294; see also Ballets
Fantasia Concertante, 39–40, 293
Flute Concerto, 205
Forsaken Merman, The, 302
Importance of Being Earnest, The (opera), 219, 233

Macbeth (ballet), 196; (film), 195
Moses and Pharaoh (oratorio), 209
Motet, 277
Orchestral Work for Szell (untitled), 224
Passionate Shepherd, The, 20–1, 304
Piano Concerto, 210
Romeo and Juliet (film), 165
Sinfonietta, 205
Stabat Mater, 278
Symphony No.3, 224, 255, 257–8, 262, 263, 264, 266, 271; first version destroyed, 259; new start on, 260; shape of work, 271–2; abandoned, 275
Upon This Rock (film), 243–4

Radio Incidental Music

Christopher Columbus, 115, 116–7, 119, 121, 313; discussed, 120; see also Choral and Orchestral

Songs and Song-Cycles

Anon. in Love, 210, 306; transcribed for orchestra, 244, 245
Beatriz's Song, 120, 305; see also *Christopher Columbus* under Radio

Contrast, The (No.5 of *Song for the Lord Mayor's Table*), 219, 306
Daphne (No.1 of *Three Songs*), 29, 65, 305

Diomede! (No.3 of *Three Solo Songs* from *Troilus and Cressida*), 306; see also Operas and Orchestral

Fain Would I Change That Note (No.1 of *Anon. in Love*), 306

Glide Gently (No. 2 of *Song for the Lord Mayor's Table*), 219, 306

Holy Thursday (No.4 of *Song for the Lord Mayor's Table*), 219, 306

How Can I Sleep? (No.2 of *Three Solo Songs* from *Troilus and Cressida*), 306; see also Operas and Orchestral

I Gave Her Cakes (No.5 of *Anon. in Love*), 306

Lady, When I Behold (No.3 of *Anon. in Love*), 306

Lord Mayor's Table, The (No.1 of *Song for the Lord Mayor's Table*), 219, 306

My Love In Her Attire (No.4 of *Anon. in Love*), 306

Old Sir Faulk (No.3 of *Three Songs*), 29, 65, 66, 305

O Stay, Sweet Love (No.2 of *Anon. in Love*), 306

Rhyme (No.6 of *Song for the Lord Mayor's Table*), 219, 306

Slowly It All Comes Back (No.1 of *Three Solo Songs* from *Troilus and Cressida*), 306; see also Operas and Orchestral

Song for the Lord Mayor's Table, A, 185, 306; performed, 218; discussed, 219

Tell Me Where Is Fancy Bred (1916), 7–8, 304; (1937), 91

Three Solo Songs from *Troilus and Cressida*, 306; see also Operas and Orchestral

Three Songs, 29, 305; composed, 65–6; see also under *Daphne, Old Sir Faulk* and *Through Gilded Trellises*

Through Gilded Trellises (No.2 of *Three Songs*), 29, 65, 305

To Couple Is A Custom (No.6 of *Anon. in Love*), 306

Tritons, 20, 304

Under the Greenwood Tree, 91, 305; see also Part-Songs

Wapping Old Stairs (No.3 of *Song for the Lord Mayor's Table*), 219, 306

Winds, The, 20, 21, 304

Theatre Incidental Music

Boy David, The, 89, 293

Macbeth, 113–4, 293; see also Projected Works

Son of Heaven, A, 40, 292

Title Music for TV Shakespeare, 268, 293

General Index

Aberconway, Lady (Mrs Henry McLaren), 42, 52, 54, 78; dedicatee of Viola Concerto, 49
Abraham, Gerald, 136
Academy of St Martin-in-the-Fields, 245, 248, 297, 298
Adler, Guido, 289
Aldeburgh Festival, 3, 14, 35, 131, 163–4, 210, 231, 245, 269, 275, 290, 296, 297, 305, 306; W. honoured at, 132–3; *The Bear* first staged at, 232; W. seventieth birthday concert, 254–5
Allchin, Basil, 8
Allegri Quartet, 247
Allen, Sir Hugh P., 8, 9, 10, 11, 12, 13n, 16, 21
Allen, Thomas, 276
Amalfi, Italy, 47, 48, 54–5, 56, 144
Amar Quartet, 49
Amsterdam, 60, 71, 74, 222, 224
Anderson, Hedli, 35, 119
Anderson, R. Kinloch, 14
Andreae, Volkmar, 41, 55 and n, 295
Ansermet, Ernest, 19, 23, 57n, 97, 294
Argentina, W.'s visit to 143–4
Argyle, Pearl, 292
Arnold, Cecily, 32
Arnold, Malcolm, 106, 218, 238, 250, 252, 263, 267, 284; W.'s friendship with, 226–7; scores some of W.'s *Battle of Britain*, 237; W. offers him quartet transcription, 247; letters from W. quoted, 216, 226, 237–8, 244–5, 246, 250, 254–5, 257, 258, 259, 260, 261–2, 263–4, 265, 266, 270, 272
Ascona, Switzerland, 57, 63, 67, 68
Ashby St Ledgers, Northants, 68n, 75, 93, 105, 113, 118, 122, 127, 128, 137, 140, 144; W. at work there, 106; wartime life at, 110
Ashton, Sir Frederick, 62, 76, 90, 108, 122, 130, 196, 253, 254, 265, 271, 291, 292
Atherton, David, 253
Atkins, Sir Ivor, 59n
Auden, W(ystan) H(ugh), 96, 160–1, 191, 196, 226, 249, 308; *Troilus* suggestions, 161–2, 168–9; writes text of *The Twelve*, 227
Auric, Georges, 23, 44
Ayrton, Michael, 113, 142

Bach, Johann Sebastian, 108, 109, 251, 275, 292
Baden-Baden, 49, 50
Bainbridge, Elizabeth, 188
Baker, Dame Janet, 184, 190, 218, 259, 260, 261, 290, 306; considered as Cressida, 185, 186; sings Cressida in revival, 188–9
Baker, Richard, 269, 305
Balcon, Sir Michael, 114
Barber, Samuel, 193
Barbirolli, Sir John, 99, 167, 207–8, 209, 245, 296, 298, 300
Barenboim, Daniel, 287
Barr, Herbert, 28
Barrie, Sir James M., 89, 90, 293
Barry, John, 239
Bartlett, Sir Basil, 112
Bartók, Béla, 8, 22, 24, 25, 32, 39, 92, 154, 199, 284
Bashford, Lt.-Col. R.B., 299
Bath Festival, 246, 248, 297, 311
Bax, Sir Arnold, 22, 24, 47, 48, 65, 70, 89, 92, 97 and n, 115, 129, 170, 200; meets W., 66
BBC (See *British Broadcasting Corporation*),
BBC Northern Singers, 259, 308
BBC Symphony Orchestra, 23, 74, 81, 83, 93, 115, 119, 197, 244, 293, 294, 295, 296, 297, 298, 299, 301, 302, 303, 313

BBC Welsh Symphony Orchestra, 265, 298, 302
Beaton, Sir Cecil, 90, 130
Beddington, Jack, 112
Bedford, Steuart, 269–70, 305
Beecham, Sir Thomas, 4, 13, 47, 58
Beethoven, Ludwig van, 39, 71, 96, 127, 199, 252
Bender, Charles, 28–9
Benjamin, Arthur, 96
Bennett, Arnold, 31, 40
Bennett, Richard Rodney, 252, 253
Berberian, Cathy, 35, 270
Berg, Alban, 22, 23, 24, 25, 199; meets W., 25–6
Bergner, Elisabeth, 76, 89, 90, 92, 99, 100
Berio, Luciano, 200, 284
Berkeley, Sir Lennox, 97, 164n, 167, 178n, 232, 234, 263
Berlin Philharmonic Orchestra, 263
Berlioz, Hector, 49, 206
Berners, Lord, 18 and n, 25, 41, 61, 64, 130, 133; W. scores ballet for, 43
Birch, John, 263, 308
Birtwistle, Sir Harrison, 200, 271, 285
Blacher, Boris, 258
Black, Andrew, 4
Blair, James, 297
Blech, Enid, 172
Blech, Harry, 306
Blech Quartet, 108, 135, 309
Bliss, Sir Arthur, 23, 24, 25, 70, 81, 97, 116, 129, 253, 264, 277; becomes Master of Queen's Music, 170
Bloch, Ernest, 24
Blom, Eric, 152, 200
Blyth, Alan, interviews W., 266–7
Bogarde, Dirk, 243
Boosey & Hawkes, 134, 148, 154–5, 179, 233, 269
Boosey, Leslie, 132, 143, 154
Borodin, Alexander, 45
Bosco, Mwenda Jean, 196
Boston Symphony Orchestra, 104, 294
Bottone, Robert, 227, 308
Boulez, Pierre, 200, 230, 284, 285n
Boult, Sir Adrian, 18, 43, 53, 56, 61, 73, 74, 83, 93, 119, 167, 298, 299, 302, 313; finds W. a job, 16

Bournemouth Symphony Orchestra, 239, 301
Bowen, York, 44, 294
Bower, Dallas, 76, 79, 114, 117, 119, 131, 134; suggests Christopher Columbus, 115; letters from W. quoted, 117, 119, 121
Boydell, Brian, 258
Brain, Dennis, 195
Brannigan, Owen, 303
Bream, Julian, 210, 279, 306, 311; commissions Bagatelles, 245–6
Bridge, Frank, 24, 63, 89
Brighton Evening Argus, 236
British Broadcasting Corporation, 49; commissions Belshazzar's Feast, 53; commissions Troilus and Cressida, 136; TV programme on W., 236–7
British Council, 97, 98, 132, 260, 270; sponsors recording, 121
Britten, Benjamin (Lord Britten of Aldeburgh), 3, 74n, 99n, 112, 115, 120, 128, 143, 151, 159, 160–1, 164 and n, 167, 181, 190, 200, 201, 210, 224, 229–30, 243, 245, 253, 260, 262, 269, 270, 274, 278, 279, 281, 282, 284, 285, 286, 287, 296; meets W., 96–7; W. speaks for, 116; wartime works, 129; W.'s attitude to, 130–3; admiration for W.'s works, 226; W. composes Improvisations on B.B. theme, 241–2; B.'s comment on, 255; W. on B.'s death, 267; W. on Cello Symphony, 288; letters from W. quoted, 132, 225, 226, 228–9, 230, 231, 241, 255
Brosa, Antonio, 100; plays through W.'s Violin Concerto, 99
Bruckner, Anton, 201, 252
Bryan, Gordon, 304, 309
Buck, Dr Percy, 11
Buckley, Jack, 260, 261, 270, 277
Burrows, Stuart, 186
Bush, Alan, 136
Busoni, Ferruccio, 24; verdict on young W., 19
Byrd, William, 166

Callas, Maria, 172, 182; W. suggests her as Cressida, 209

Callot, Jacques, 111, 295
Cameron, Basil, 109, 295
Cameron, John, 303
Campbell, Roy, 14, 130, 145, 279; assesses W., 15, 20
Capell, Richard, on W.'s First Symphony, 82
Cardus, Sir Neville, 60; on *Belshazzar's Feast*, 59, 134
Carlyle, Joan, 186
Cassilly, Richard, 188, 290
Casson, Sir Hugh, 174, 177, 179
Cavalcanti, Alberto, 114, 118
Charpentier, Gustave, 179
Chekhov, Anton, 228, 233, 290
Chicago Symphony Orchestra, 56, 110, 295; commissions *Scapino*, 100; W. conducts, 223
Christ Church Cathedral School, Oxford, 91, 217, 251, 256, 308; W. enters, 5–6
Christ Church College, Oxford, 60, 226
Chung, Kyung-Wha, 260, 276; records W.'s Violin Concerto, 257
Cincinnati Symphony Orchestra, 150
City of Birmingham (Symphony) Orchestra, 85
Clark, Edward, 49, 53, 55, 57, 133
Clark, Lord (Sir Kenneth Clark), 253, 254, 271
Cleveland Orchestra, 105, 205, 207, 214, 294, 296
Cleveland Plain Dealer, on *Partita*, 207
Coates, Eric, 113, 143, 210
Cochran, Sir Charles B., 33, 56, 243, 273, 291, 311; W. composes for revue, 89–90
Cohen, Harriet, 59, 70, 124, 310
Collegium Musicum Londinii, 296
Collier, Marie, 183, 184, 223, 290
Collinson, Frank, 291
Colvin, Lady, 19
Colvin, Sir Sidney, 19
Confalonieri, Giulio, on *Troilus* in Milan, 183
Cooper, Martin, 227
Cork Festival, W. choral work for, 258–9
Coulling, John, 294
Council for the Encouragement of

Music and the Arts (CEMA), 116
Courtauld, Samuel, opinion of W., 67–8
Courtauld, Mrs Samuel, 76; leaves W. annuity, 67–8
Covent Garden (See *Royal Opera House*)
Coward, Sir Noël, 24; parodies *Façade*, 30
Craft, Robert, 284
Craggs, Stewart, 6, 45, 62, 239, 277
Craighead, David, 309
Crain, Jon, 182
Crichton, Ronald, on W.'s 70th birthday concert, 251
Cross, Joan, 177
Culshaw, John, 236
Cummings, Douglas, 254
Curtin, Phyllis, 182
Czinner, Paul, 76, 90, 92, 99

Daily Dispatch, 65 and *n*
Daily Express, 53, 57, 238
Daily Mail, 58, 90
Daily Telegraph, 59*n*, 227
Dalberg, Frederick, 290
Darlington, Stephen, 308
Davis, Carl, 239, 301, 302, 303
Davis, Sir Colin, 297
Day Lewis, Cecil, 245
Debussy, Achille-Claude, 8, 22, 39, 40, 84, 285
Decca Records, 85, 86
Dehn, Paul, 230, 232, 233, 253, 266, 290, 307; librettist of *The Bear*, 229; death, 235
del Giudice, Filippo, 121
Delius, Frederick, 15, 24, 38, 40; W.'s dislike of, 31–2
Delius, Jelka (née Rosen), 32
Della Casa, Lisa, 172, 175–6
Del Mar, Norman, 284, 287
Dent, Edward J., 19, 24, 25, 38
Desert Island Discs, 285
Devine, George, 174, 177, 178
Diaghilev, Serge, 21, 22 and *n*, 40, 41, 49; rejects W. ballet, 44
Diamand, Peter, 128, 185
Dickinson, Thorold, 112
Doernberg, Baron Hans-Karl, 49
Doernberg, Baroness Imma, 57, 61,

63, 64, 68 and *n*, 78; meets W.,
    49; affair ends, 74
Domingo, Placido, 185, 266
Donat, Robert, 115
Donnelly, Mrs Nora (See *Walton,
    Nora*)
Dorati, Antal, 287
Dorfman, Dr Ralph, 243;
    commissions W. *Improvisations*,
    241
Douglas, Roy, 77, 110, 113, 119, 122–
    3, 142, 172*n*, 179, 198; advises
    W. on conducting, 106; Alice
    Wimborne's influence on W.,
    118; letters from W. quoted,
    114–5, 122–3, 124, 125, 128, 150,
    161, 171, 189, 197, 231–2, 234,
    235, 271, 272–3
Dow, Dorothy, 182
Dowd, Ronald, 303
Dowdall, Ann, 308
Dowell, Anthony, 292
Draper, Paul, 28, 35
Drogheda, Earl of, 254
Duke, Vernon (See *Dukelsky, Vladimir*)
Dukelsky, Vladimir, 40
Du Pré, Jacqueline, 287, 288
Durey, Louis, 23
Dyson, Sir George, 94, 166

Edinburgh Festival, 211, 293
Elgar, Sir Edward, 3, 4, 9, 18, 37–8,
    39, 45, 47, 50, 59, 69 and *n*, 76
    and *n*, 91, 93, 100, 101, 102, 125,
    200, 201, 209, 214, 219, 238, 247,
    265, 278, 280, 281, 283, 286, 289;
    W.'s admiration for, 38; meets
    W., 52
Elgar, William H., 4, 6
Eliot, T.S., 19
Ellington, Duke, 33
Ellis, Vivian, 76
EMI Records, 188, 211, 231, 239, 255,
    261, 289
English Bach Festival, 248, 269, 305
English Chamber Orchestra, 218,
    232, 290, 306
English, Gerald, 188, 290
English National Opera, 187
Evans, Dame Edith, 243
Evans, Lt.-Col. G.E., 300
Evening Standard, 12, 85

Falla, Manuel de, 22, 24
Farron, Julia, 122, 292
Fauré, Gabriel, 18
Fell, Sidney, 217
Ferraresi, Aldo, 169
Fetler, David, 309
Financial Times, 251, 252
Firbank, Ronald, 14, 130
Fleischmann, Aloys, 259
Fonteyn, Dame Margot, 76, 108, 122,
    196, 292
Foras, Odette de, 304
Forbes, Bryan, 254
Foss, Dora (Stevens, Dora), 29, 64, 65,
    73, 78, 99–100, 104, 109, 305; at
    Façade, 31; W. writes songs for,
    65; describes evening with W.,
    66; lunch with a Sitwell, 66–7;
    on Violin Concerto, 100–1; letter
    from W. quoted, 80–1
Foss, Hubert, 5, 7, 26, 31, 41, 42–3,
    52, 57, 58, 64, 65, 66, 70, 72, 74,
    76, 78, 83, 85, 86, 90, 91, 97, 98,
    154, 305; on W.'s First
    Symphony, 71, 81; letters from
    W. quoted, 69, 81*n*, 98
Foster, Lawrence, 188, 189, 290
Fox, Ian, 308
Frank, Alan, 6*n*, 84, 118, 154, 174,
    179, 202, 205, 211, 215, 232,
    247; on *Troilus* casting, 172–3;
    retires, 274; letters from W.
    quoted, 31–2, 43, 75, 78, 83, 90,
    92, 101, 120, 163, 169–70, 185,
    197, 198, 205, 206–7, 209, 215,
    218, 219, 220, 224, 227, 228, 230,
    231, 241, 243, 249, 258, 259,
    262–3, 263, 268, 269, 270, 271–2,
    273, 274–5, 275, 275–6, 277, 284,
    285, 287, 288, 288–9
Fricker, Peter Racine, W. on *Vision of
    Judgement*, 209
Fry, Roger, 18, 66
Furse, Roger, 171
Furtwängler, Wilhelm, 65, 86, 105

Gauntlett, Ambrose, 28
Gedda, Nicolai, 172–3, 177
Gentleman, David, 179
Georgiadis, John, 254
Gerhard, Roberto, W.'s opinion of,
    284
Gershwin, George, 24; W. meets, 40

Gesualdo, Carlo, 97, 113, 136
Gibbons, Orlando, 166
Gielgud, Sir John, 113–4
Glenconner, Lady, 15, 42, 304
Glenconner, Lord, 42
Glock, Clement, 179
Glock, Sir William, 179, 185 and n, 186, 200, 280, 287
Goehr, Alexander, 200
Gomez, Jill, 186, 257
Goodall, Sir Reginald, 181
Goodman, Benny, 92
Goodman, Arnold (Lord Goodman), 239, 253, 254, 261
Goodwin, Ron, 238, 239
Goossens, Sir Eugene, 21, 23, 24, 41, 150
Gore, Walter, 291
Gounod, Charles, 5
Graham, Colin, 188, 232, 255n
Grainger, Percy, 24
Gray, Cecil, 31, 39, 72, 83, 97, 136, 139, 142, 285; on W., 20, 32–3, 73; writes libretto for W., 113; on W.'s generosity, 133
Greenbaum, Hyam, 79, 86, 91
Greenfield, Edward, 34, 71, 82; on Battle of Britain, 238
Grey, Dame Beryl, 122, 292
Grimethorpe Colliery Band, 90, 271, 272, 312
Grinke, Frederick, 310
Groves, Sir Charles, 297
Guardian, The (Manchester), 71n, 94, 181, 238
Guest, Hon. Ivor, 100

Hába, Alois, 262
Hadow, Sir Henry, 13n
Hale, Una, 181
Hallé, Sir Charles, 4
Hallé Orchestra, 12, 53, 64n, 71, 117, 207, 296
Halsey, Louis, 308
Hamilton, Guy, 237
Hammond-Stroud, Derek, 259
Handel, George Frideric, 5, 166
Handley, Vernon, 222
Harbach, Otto, 56n
Harewood, George, 7th Earl of, 132, 160, 181; on Troilus, 187n
Harper, Heather, 186
Harty, Sir Hamilton, 53, 56, 67, 71, 73, 74, 77, 78, 81, 85, 86, 282,

293; asks W. for symphony, 64–5; on Belshazzar's Feast, 69; letter from W. quoted, 70
Harwood, Elizabeth, 186
Hassall, Christopher, 140, 141, 150, 172, 182, 189, 210, 228, 235, 290, 306, 307, 314–5; chosen as librettist, 136–8; visits Ischia, 160, 174; writes to Olivier about Troilus, 165–6; upset over Auden, 168–9; death, 183; letters from W. quoted, 137, 142–3, 147–9, 150–1, 151–5, 159–63, 164–5, 168–9, 170–2, 173–4, 175–8, 182
Hawkes, Ralph, 134
Haydn, Franz Joseph, 5, 207
Heath, Edward, 239, 250; host to W. at No.10, 253–4
Heifetz, Jascha, 97, 98, 99, 100–1, 103, 104, 130, 167, 197, 206, 237, 294; commissions W.'s Violin Concerto, 92; gives first performance, 105; records, 150; his first British performance of, 164
Helpmann, Sir Robert, 122, 292
Hely-Hutchinson, Victor, 53, 54, 136
Hemmings, Peter, 185
Hemsley, Thomas, 255n
Henze, Hans Werner, 196, 202, 209, 231, 246, 261–2, 271, 279, 284; W.'s opinion of, 285; H.'s view of W., 285–6
Herbage, Julian, 92n, 124, 125, 127
Herbert, Sir Alan P., 143
Herbert, George, 268, 309
Herbert, Jocelyn, 179
Heseltine, Philip (Warlock, Peter), 24, 31–2, 41, 113
Hess, Günter, 62, 291
Heward, Leslie, 79, 85, 101
Heyworth, Peter, 198, 206, 229; critical of W.'s music, 199–200; relationship with W., 200–3; on Second Symphony, 214; letters from W. quoted, 217–8, 230, 250, 273–4
Hill, Ralph, 79
Hindemith, Gertrud, 220
Hindemith, Paul, 24, 25, 50, 56, 66, 133, 205, 243, 284, 285, 294; plays W.'s Viola Concerto, 49–50; and W.'s Variations, 220

Holbrooke, Joseph, 24
Hollywood Quartet, 247
Holroyd, Michael, 40
Holst, Gustav, 12, 24, 76 and *n*
Holst, Henry, 105–6, 118, 124*n*, 294
Holt Ltd., Harold, 185, 235, 244, 269
Honegger, Arthur, 23, 24, 61, 99
Hope-Wallace, Philip, 181
Horner, David, 79
Houston Chronicle, 237
Houston Symphony Orchestra, 237
Howard, Leslie, 114, 119
Howard, Michael, 277
Howard, Trevor, 238
Howarth, Elgar, 90, 271, 273, 312
Howells, Herbert, 30, 166, 253, 258
Howes, Frank, 25, 28, 30, 181, 191,
    202, 212, 220
Howgill, R.J.F., 57
Huddersfield Choral Society, 38, 121,
    215, 278, 303; commissions
    Gloria, 207
Hughes, Owain Arwel, 265, 298; W.
    agrees to write work for, 277–8
Hughes, Patrick (Spike), 31, 34, 38,
    76–7, 79, 92, 107, 219
Hunter, Sir Ian, 206, 235, 239, 245
Hurst, George, 287
Hussey, Rt. Revd. Walter, 262
Hutchings, Arthur, on W.'s First
    Symphony, 134
Hyman, Prudence, 291

Ibert, Jacques, 234
Imai, Nobuko, 276
Incorporated Society of Musicians
    (ISM), award to W., 269
Ingpen, Joan, 172
International Society for
    Contemporary Music (ISCM),
    25, 50, 60, 71, 97, 270, 295
Ireland, John, 24, 47, 115, 129, 173,
    200, 226, 287; on W.'s First
    Symphony, 85
Irving, Ernest, 114, 115, 122, 293
Ischia, 62, 147, 184, 196, 197, 202,
    218, 223, 224, 226, 227, 229, 230,
    238, 243, 246, 247, 248, 249, 251,
    252, 257, 265, 267, 273, 277, 278,
    300; W. lives in, 144; moves
    house, 151; W.'s ashes in, 278;
    not isolated in, 283–4, 286
Italy, 54–5, 97; W. first visits, 15–16

Jackson, Gerald, on W. conducting,
    78
Jacobs, Arthur, 12, 72, 76, 112, 188–9,
    199
Janáček, Leoš, 22, 25
John, Augustus, 31, 89
Jones, Dame Gwyneth, 185
Jurinac, Sena, 175

Kabos, Ilona, 311
Kallman, Chester, 161, 168
Karajan, Herbert von, 169, 195, 263,
    264, 296
Karr, Gary, 237
Kennedy, Nigel, 50
Kentner, Louis, 139, 150, 210, 309,
    311
Kern, Jerome, 24, 56*n*
Kern, Patricia, 255*n*
King George V, 89, 90
King George VI, 93, 154, 298
Kipling, Rudyard, 3
Kirsten, Dorothy, 181
Klemperer, Otto, 65
Kochno, Boris, 44
Kodály, Zoltán, 24, 61
Koechlin, Charles, 284
Kostelanetz, André, 236, 297
Koussevitzky, Serge, 56, 210
Kraus, Otakar, 183, 290
Kreisler, Fritz, 100
Krenek, Ernst, 24
Krips, Josef, 241, 243, 297
Kurtz, Efrem, 91, 196, 205, 295

Lalandi, Lina, 248, 251, 269
Lambert, Constant, 27, 32, 34, 35, 40,
    41, 42, 44, 53, 55, 56–7, 73, 74,
    79, 80, 83, 86, 89, 96, 108, 117,
    128, 129, 162, 285, 286, 291, 292,
    305; first recites *Façade*, 31; on
    Façade, 33; on *Fantasia Concertante*,
    39–40; death, 151
Lambert, Isabel, 162, 171
La Mortella (W.'s home in Ischia),
    208–9; W.'s ashes in 278
Lang, Paul Henry, 218
Lanigan, John, 183, 290
Laski, Marghanita, 277
Laszlò, Magda, 176–7, 290; sings
    Cressida, 179
Latham-Koenig, Jan, 277

Lawrence, Ashley, 292
Ledger, Philip, 268, 308
Lee, Jennie (Baroness Lee), 254
Leeds Festival, 57, 58, 92, 93, 94, 209,
    302, 303; and *Belshazzar's Feast*,
    58–9
Legge, Walter, 94, 106, 149, 172, 174–
    5, 263
Leicestershire Schools Symphony
    Orchestra, 254
Leigh, Vivien (Lady Olivier), 119,
    121
Leinsdorf, Erich, 181
Lennard, Joan, 306, 313
Lewis, Peter, 58
Lewis, Richard, 175, 180, 181, 184,
    186, 189, 209, 223, 257, 290, 303
Lewis, P. Wyndham, 19, 291
Ley, Henry G., 6, 8, 9, 16
Lidell, Alvar, 253
Lidka, Maria, 147
Lifar, Serge, 44
Lill, John, 254
Lipp, Wilma, 172
Listener, The, 82, 199
Liverpool Philharmonic Orchestra
    (See *Royal Liverpool Philharmonic
    Orchestra*)
Lloyd-Jones, David, 293
Lockhart, James, 232, 290
Lockyer, James, 309
London, *passim*; musical life of early
    1920s, 21–4; *Aeolian Hall*, 29, 43,
    117, 259, 295, 305, 309; *Barbican*,
    276, 297, 302; *Coliseum*, 259, 262;
    National Theatre, 243, 266, 299;
    Queen Elizabeth Hall, 246, 308,
    311; *Queen's Hall*, 61, 73, 81, 109,
    293, 294, 295, 303, 310; *Royal
    Albert Hall*, 105, 115, 127, 260,
    273, 294, 295, 296, 299, 300, 301,
    302, 303; *Royal Festival Hall*, 123,
    162, 164n, 167, 201, 220, 251,
    265, 268, 272, 294, 295, 296, 297,
    298, 299, 300, 301, 303, 311;
    anniversary work abandoned,
    263–4; W.'s eightieth birthday
    concert, 276; *Wigmore Hall*, 23,
    135, 202, 305
London Mozart Players, 306
London Philharmonic Choir, 303,
    308
London Philharmonic Orchestra, 90–

1, 105, 109, 114, 244, 268, 276,
    293, 294, 295, 296, 297, 299, 303,
    308
London Sinfonietta, 253, 268
London Symphony Chorus, 260
London Symphony Orchestra, 71, 73,
    77, 185, 244, 251, 254, 255, 257,
    265, 268, 276, 293, 298, 299,
    302
Long Jr., Nick, 291
Lopokova, Lydia, 291
Lorrayne, Vyvyan, 292
Los Angeles Philharmonic Orchestra,
    261
Lowe, John, 136
Luce, Claire, 291
Lush, Ernest, 310
Lutyens, Elisabeth, 79; on W.'s
    generosity, 133
Luxon, Benjamin, 188, 290
Lysy, Antonio, 311

McCullagh Quartet, 25, 309
McEacharn, Capt. Neil, 68 and *n*
McGrath, Barry, 292
Mackerras, Sir Charles, 269, 305
McKie, Sir William, 162, 163, 166,
    303
McLaren, Christabel (See *Aberconway,
    Lady*)
MacNeice, Louis, 115, 119, 120, 313
McPherson, Aimée Semple, 57
Mahler, Gustav, 22, 54, 200, 201, 206,
    265, 281, 284, 285, 289
Maine, Revd. Basil, 52; on W.'s First
    Symphony, 134
Malcolm, George, 218, 306
Manchester Guardian (See *Guardian, The
    (Manchester)*)
Mangeot, André, 23
Mann, William, 191, 251
Markevitch, Igor, 49
Markova, Dame Alicia, 291
Marriner, Sir Neville, 245, 247, 248,
    254, 259, 297, 298
Marsh, Sir Edward, 136 and *n*, 164
Martin Neary Singers, 253
Masefield, John, 14, 127–8, 307
Mason, Colin, on W.'s Violin
    Concerto, 104; on W.'s Cello
    Concerto, 199
Mathias, Mrs Robert, 28–9

Mathieson, Muir, 126, 133, 139, 255, 297, 299, 301
Matthews, Colin, 239, 301
Maw, Nicholas, 252
Maxwell Davies, Sir Peter, 200, 252
Mayer, Dorothy (Lady Mayer), commissions *Cantico del Sole*, 258
Mayer, Sir Robert, 247, 258
Mehta, Zubin, 261, 288
Mendelssohn, Felix, 5
Mengelberg, Willem, 22, 85
Menuhin, Diana (Lady Menuhin), 139
Menuhin, Sir Yehudi, 50, 74*n*, 138–9, 150, 251, 309
Messiaen, Olivier, 265
Milhaud, Darius, 23, 24
Mills, Florence, 33
Minchinton, John, 296
Minty, Shirley, 308
Mitchell, Donald, 198, 206; on W.'s 'decline', 199
Mitchell, R. J., 114
Mitter, Sheila, 179
Moeran, E.J., 79
Moffo, Anna, 185
Montale, Eugenio, on *Troilus* in Milan, 182–3
Monthly Musical Record, 62
Moore, Doris Langley, 122, 292
Moore, Gerald, 218, 306
Moore, Henry, 151, 162, 171, 253, 262, 271
Mornys, William de, 40
Morrell, Lady Ottoline, 15
Morris, Wyn, 216 and *n*
Morrison, Angus, 13*n*, 31, 35, 39, 44, 48, 53, 54, 71, 147, 250, 309; meets W., 32; on First Symphony, 72; on W.'s attitude to criticism, 73
Mozart, Wolfgang Amadeus, 207
Munch, Charles, 197
Murchie, Robert, 29
Musgrave, Thea, 252
Musical Opinion, 62
Musical Times, 134, 232*n*

National Symphony Orchestra of Washington, 298; W. writes work for, 275–6
Neel, Boyd, 112, 214
Newman, Ernest, 89, 140, 143, 152 and *n*, 163, 165, 169; on *Façade*,

31; on *Sinfonia Concertante*, 45; on *Belshazzar's Feast*, 59; on *Troilus* draft libretto, 141; opinion of final libretto, 190
Newman, Nanette, 254
New Statesman, 135, 164, 183
Newton, Cyril Ramon, 39
Newton, Ivor, 196, 197*n*, 198
New York Philharmonic Orchestra, 236, 237, 297, 300
New York Post, 236
Nielsen, Carl, 24, 252
Nilsson, Raymond, 181
Noble, Dennis, 58, 121, 302
Nono, Luigi, 200, 284, 285
Northcott, Bayan, on *Troilus* revival, 188
Novello, Ivor, 138, 151, 184

Observer, The, 199, 200, 201, 214, 273
Oldham, Arthur, 164*n*, 167
Oldham, Lancashire, 4, 6, 7, 10, 17, 43, 144, 145, 278, 287, 289; W. born in, 5; W. freeman of, 217
Olivier, Edith, 49, 65 and *n*, 66, 108
Olivier, Lord (Sir Laurence), 115, 119, 121, 138, 151, 165, 171, 174, 194, 237, 239, 243, 244, 253, 254, 255, 271; first meets W., 91; on *Henry V* music, 125, 126; W. on working with, 194
Ormandy, Eugene, 84
Osborne, John, 200
Ottaway, Hugh, 6 and *n*, 73
Oxford University Press (OUP), 14, 120, 134, 142, 223, 254, 265, 274, 276, 283–4, 287; W.'s association with begins, 41; W. refuses to leave, 154–5; W. annoyed with, 179

Palestrina, Giovanni, W. studies, 277
Palmer, Christopher, 297, 300, 302, 303, 306, 312
Palmer, Tony, 140, 272, 277; TV profile of W., 273
Pannain, Guido, on *Troilus* in Milan, 183
Parratt, Sir Walter, 10
Parry, Sir Hubert, 8, 9, 12, 166; interest in W., 7
Pascal, Gabriel, 98, 109
Peake, Revd. Edward, 10
Pears, Sir Peter, 31, 115, 129, 130,

175, 180, 183, 225, 231, 244, 269, 290, 291, 305, 306; commissions song-cycle, 210; suggests *The Bear*, 228; in *Façade*, 254–5
Peasgood, Osborne, 303
Pennington String Quartet, 21, 309
Performing Right Society, 41, 143, 245, 274, 287
Peron, Juan, 143
Peyton, John, 250*n*
Philharmonia Orchestra, 77, 139, 164*n*, 165, 195, 276, 295, 296, 298
Philharmonic-Symphony Orchestra of New York (See *New York Philharmonic Orchestra*)
Piatigorsky, Gregor, 206, 237, 261, 288, 294; commissions Cello Concerto, 196–7; new ending for concerto, 198–9, 288
Picker, Arnold, 238
Picker, David, 238
Piper, John, 117, 122
Pirie, Peter J., 199
Pizzetti, Ildebrando, 24
Platts, Trevor, 300
Plomer, William, 160
Poleri, David, 182
Ponnelle, Jean-Pierre, 177
Ponsonby, Noel, 8
Ponsonby, Robert, 8
Pope, Michael, 120
Porter, Andrew, on *The Bear*, 232
Porter, Neil, 31, 305
Post, Joseph, 223
Poulenc, Francis, 23, 24, 45, 234
Pound, Ezra, 19
Preston, Simon, 251, 277, 308
Previn, André, 185, 186, 188, 189, 230, 234, 251, 252, 255, 257, 259, 264, 265, 275, 276, 282, 288, 298; W. on Rome concert, 260–1; on Third Symphony, 271
Price, Leontyne, 185
Pride, Malcolm, 179
Priestley, J. B., 114
Primrose, William, 50, 52, 92, 309
Prince Charles, Prince of Wales, W. writes piece for, 245; at W. birthday concert, 254–5
Pritchard, Sir John, 211, 293, 296
Prokofiev, Sergei, 8, 22, 46, 49, 51, 83, 103, 125, 206, 244, 281, 284, 288
Promenade Concerts, 273; *Troilus* Act II at, 185–6, 189, 257

Puccini, Giacomo, 4, 164, 173, 188, 190, 191, 201
Purcell, Henry, 278

Queen Elizabeth II, 94, 162, 266, 270, 298
Queen Elizabeth the Queen Mother, 197, 253, 298
Queen Mary, 93
Quidhampton, Wilts, 49, 65 and *n*, 66

Rabinowitz, Harry, 299
Radio Times, 33
RAI Orchestra, Italy, 260
Ralton, Bert, 39
Rassine, Alexis, 122, 292
Ravel, Maurice, 8, 12, 13, 21, 24, 39, 45, 84, 113, 280, 281, 284
Rawlings, Margaret, 119
Rawsthorne, Alan, 46, 79, 97, 115, 287
Raybould, Clarence, 93, 298
Reid, Charles, 58, 181; on Sargent rehearsing, 179–80
Reizenstein, Franz, 172 and *n*, 173, 174
Remedios, Alberto, 186, 188
Rennert, Günther, 182
Reynolds, Robert, 272
Richardson, Sir Ralph, 243
Rimsky-Korsakov, Nikolay, 45, 112
Rix, Diana, 185, 256, 274; friendship with W., 235; letters from W. quoted, 185, 186, 243–4, 244, 245, 258, 259, 261, 264, 266, 269, 275, 287
Robert Masters Quartet, 14
Robertson, Stuart, 302
Robinson, Forbes, 290
Robinson, Stanford, 117, 136, 145
Rodzinski, Artur, 105, 169–70, 294
Rootham, Helen, 15
Rosenstock, Joseph, 182
Rossini, Gioachino, 34, 112
Rostropovich, Mstislav, 244, 288, 298, 311; W. writes for, 274–6
Rouault, Georges, 49
Roussel, Albert, 205–6, 208, 280
Rowlandson, Thomas, 41, 208, 294
Royal Academy of Music, London, 19
Royal College of Music, London, 19, 21, 56, 309
Royal Liverpool Philharmonic Orchestra and Society, 121, 124*n*, 196, 293, 294, 295, 296,

297, 303; commissions Second Symphony, 196; gives first performance, 211

Royal Manchester College of Music, 4; W. made hon. member, 250–1

Royal Military School of Music, Kneller Hall, 299, 300, 303

Royal Opera House, Covent Garden, 145, 151, 160–1, 162, 167, 172, 173, 176, 184, 185, 186, 209, 228, 259, 261, 290, 296; *Troilus* rehearsals at, 179–81; *Troilus* revival at, 188–9, 266–7

Royal Philharmonic Orchestra, 297

Royal Philharmonic Society, 44, 57*n*, 136, 197, 294; commissions *Hindemith Variations*, 219

Rubbra, Edmund, 200

Rubinstein, Artur, 23, 167

Russell, Leslie, 6

Russian Ballet, 21–2, 42

Sackbut, The, 62

Sadie, Stanley, 252

Sadler's Wells Ballet, 108, 122–3, 196, 292

Saltzman, Harry, 237

Salzburg, 21, 96, 219, 261; W. quartet played in, 25

San Francisco Symphony Orchestra, 237, 241, 297

Sanzogno, Nino, 182, 201

Sargent, Sir Malcolm, 94, 117, 124*n*, 126, 139, 142, 150, 173, 174, 176, 177, 188, 196, 197, 201, 215, 231, 290, 294, 295, 296, 299, 301, 302, 303; conducts first *Belshazzar's Feast*, 58; conducts *Troilus*, 179–80; defends W., 199

Sassoon, Siegfried, 18, 37, 38, 42, 45, 55, 61, 78, 130; meets W., 14; his financial aid to W., 50, 56, 68; letters from W. quoted, 41, 47–8, 49, 56, 57–8, 63–5, 67, 69, 82, 89

Satie, Erik, 15, 22, 23

Sauguet, Henri, 44

Savage, Richard Temple, describes Sargent rehearsing, 180

Savoy Havana Band, 39

Savoy Orpheans, 33, 39

Schaffer, Elaine, 205

Schippers, Thomas, 209, 258

Schoenberg, Arnold, 12, 22–3, 24, 35,

39, 199, 200, 201, 284; meeting with W., 25–6; *Pierrot Lunaire* and *Façade*, 34–5; W.'s admiration for, 285

Schools Music Review, The, 62

Schubert, Franz, 253–4

Schuster, Frank, 18, 37–8, 52

Schwarzkopf, Elisabeth, 182, 218, 306; Cressida prototype, 149; not available for Cressida, 172–5; records Cressida extracts, 181

Scott, Cyril, 24

Searle, Humphrey, 164*n*, 167

Shawe-Taylor, Desmond, 183–4; on W.'s quartet, 135; on W.'s Cello Concerto, 199; on W. at seventy, 252

Shaw, George Bernard, 98; advice to W., 109–10

Shaw, John, 232

Shearer, Moira, 122, 292

Sheepshanks, Lilias (Mrs Robert Sheepshanks), 254

Shirley-Quirk, John, 255, 260, 261

Shore, Bernard, 21, 83

Shostakovich, Dmitri, 83, 284

Shuttleworth, F.G., 307

Sibelius, Jean, 22, 39, 71, 91, 252, 258, 281; and W.'s First Symphony, 82, 83; W. rediscovers, 285

Sibley, Antoinette, 292

Siena, Italy, 286, 305; *Façade* in, 33–4; fiftieth anniversary performance, 270

Simpson, Dr Cuthbert, 227

Simpson, Robert, 252

Sinclair, Monica, 232, 290

Sitwell, Dame Edith, 16, 17, 19, 30, 35, 45, 62, 81, 255, 304, 305; writes *Façade*, 27–8

Sitwell, Sir George, 14, 206

Sitwell, Sir Osbert, 14, 16, 17, 19, 21, 27, 30, 44, 45, 47, 55, 66, 81, 89, 130, 145, 206, 207, 244, 291, 302; describes W. at Oxford, 15; W. at Swan Walk, 20; on W. quartet at Salzburg, 25; first *Façade*, 28; on Elgar, 37–8; compiles *Belshazzar's Feast* libretto, 54–5; disapproves of Lady Wimborne, 78–9; writes oratorio libretto, 209

Sitwell, Sir Sacheverell, 19, 23, 27, 31, 43–4, 53, 66–7, 74, 80, 291; meets W., 14–5
Skelton, Geoffrey, 49
Skryabin, Alexander, 22
Smith, Cyril, 294
Solti, Sir Georg, 253, 268, 287
Somers, Debroy, 33, 39
Somes, Michael, 108, 292
Sorabji, Kaikhosru, 32
Spenser, Edmund, 122, 292
Speyer, Sir Edgar, 37
Speyer, Edward, 18
Spohr, Ludwig, 285
Stanford, Sir Charles Villiers, 166
Steber, Eleanor, 172
Stein, Erwin, 132
Stevens, Dora (See Foss, Dora)
Stevenson, Ronald, W.'s admiration for, 284
Still, Dr Karl-Friedrich, 300
Stock, Frederick, 100, 110, 112, 295
Stockhausen, Karlheinz, 284; W.'s views on, 262
Strachey, Lytton, 18, 40, 292
Strauss, Richard, 12, 22, 24, 38, 39, 59, 60, 74, 154, 159, 183, 207, 233, 279–80, 281, 282
Stravinsky, Igor, 8, 12, 15, 21–2, 23, 24, 39, 45, 59, 61, 154, 167, 175, 199, 201, 217–8, 229, 267, 274, 280, 281, 285; W. revalues, 284
Strong, Dr Thomas B., 7, 9; beneficence to W., 10–11, 12, 16, 18
Sunday Telegraph, 188, 239
Sunday Times, 31, 45, 140, 252
Szell, George, 86, 205, 235, 237, 245, 296; W.'s tributes to, 214–5, 222; W. scraps work for, 224
Szigeti, Joseph, 92, 100
Szymanowski, Karol, 24

Tailleferre, Germaine, 23,
Tas, Pierre, 309
Tausky, Vilem, 301
Taverner, John, 251,
Taylor, David, 255n, 291, 292
Tchaikovsky, Pyotr Ilyich, 12
Tear, Robert, 35, 245, 270, 306, 308
Te Kanawa, Dame Kiri, 186
Temple, Richmond, 39
Tennant, Stephen, 42

Tertis, Lionel, 47, 49, 50, 52, 55n, 57, 64n, 253; rejects Viola Concerto, 48
Thomas, Marjorie, 303
Thomas, Mary, 259
Thomson, Bryden, 302
Three Choirs Festivals, 3, 52, 59, 263
Tibbett, Lawrence, 76, 77
Tierney, Neil, 175
Times, The, 128, 181, 191, 202, 251, 252, 266–7; on W.'s early quartet, 25; on Belshazzar's Feast, 59; on Gloria, 215–6
Tippett, Sir Michael, 47, 112, 128, 129, 131, 164n, 167, 181, 279; W. on his operas, 284–5, 288–9
Tonhalle Orchestra, 295
Tooley, Sir John, 185 and n, 186, 260, 266
Toscanini, Arturo, 109–10
Tovey, Sir Donald, 52, 60, 159
Tudor, Anthony, 291
Turp, André, 183, 290

Uhde, Hermann, 173 and n

Valois, Dame Ninette de, 76
Van Allan, Richard, 188, 290
Vandervelde, Emil, 18
Vandervelde, Lalla, 18
Van Dieren, Bernard, 14, 15, 39; W.'s opinion of, 31–2
Vaughan Williams, Ralph, 3, 11, 13 and n, 22, 24, 38, 48, 59, 61, 80, 89, 96, 97 and n, 115, 116, 119, 120, 124 and n, 125, 129, 136, 151, 154, 164n, 166, 200, 201, 235, 278, 281, 284; and W.'s Henry V, 123
Veasey, Josephine, 183
Verdi, Giuseppe, 159, 183, 191, 271
Vickers, Jon, 183, 209
Vic-Wells Ballet, 62
Villa-Lobos, Heitor, 284
Vinter, Gilbert, 219, 270
Vlad, Roman, 258
Vogue, on first Façade, 30

Wagner, Richard, 191, 240
Walker, Dr Ernest, 9, 10
Walters, Jess, 181
Walton, Alexander (brother), 5, 44, 93, 167, 244, 253

Walton, Charles Alexander (father), 4–6, 10, 12, 17, 251
Walton, Elizabeth (niece), 55, 109
Walton, Louisa Maria (née Turner, mother), 4, 7, 10, 144, 217; takes W. to Oxford, 5–6; meets Susana Walton, 145; letters from W. quoted, 9, 11, 13, 14–5, 19, 21, 39, 43–4, 54–5, 56–7, 93
Walton, Michael (nephew), 55, 109
Walton, Noel (brother), 5, 6 and n, 7, 8, 109, 217
Walton, Nora (Mrs Nora Donnelly, sister), 5, 217, 223
Walton, Susana (Lady Walton, née Gil), 49, 63, 132, 147, 175, 179, 182, 198, 209, 223, 235, 270, 271, 274, 276, 278; marries W., 143–4; dedicatee of Troilus and Cressida, 178; on W.'s 1976 illness, 267; in Wagner film, 277; life with W., 286
Walton, William, 62–3
Walton, Sir William Turner
Career: birth, 5; choirboy at Oxford, 6–10; first compositions, 7–8; undergraduate, 10–15; sent down from Oxford, 11; sees operas, 13; meets Sassoon and Sitwells, 14; visits Italy, 15–6; with Sitwells, 16–21; meets Schoenberg, 25–6; composes Façade, 27–8; meets Delius, 32; at Savoy, 39; meets Gershwin, 40; visits Spain, 40–1; first OUP contract, 41; scores Berners ballet, 43; Diaghilev rejects W. ballet, 44; composes Viola Concerto, 47–8; Tertis rejects concerto, 48; first performance of concerto, 49; affair with Imma Doernberg begins, 49–50; meets Elgar, 52; begins Belshazzar's Feast, 53–5; living with Imma, 57; worry over international finance, 63–4; begins First Symphony, 64–6; receives annuity, 67–8; 'stuck' in

symphony, 69–70; break with Imma, 74; meets Alice Wimborne, 75; composes first film score, 76; symphony performed incomplete, 77–8; rift with Sitwells, 78–9; completes symphony, 80–2; writes for Cochran, 89–90; meets Olivier, 91; Violin Concerto commissioned, 92; writes 1937 Coronation music, 93–4; meets Britten, 96–7; hernia operation, 97; composes Violin Concerto 97–101; visit to USA, 101; arranges Bach ballet, 108–9; war work on films, 112; plans Gesualdo opera, 113; speaks for Britten, 116; composes First of the Few, 117; composes Christopher Columbus, 119–20; records Belshazzar's Feast, 121; composes The Quest, 121–3; composes Henry V, 123–5; declines 'Victory Anthem', 124; writes Wood memorial work, 127–8; begins quartet, 128; attitude to Britten, 130–3; post-war position, 134; awarded RPS Gold Medal, 136; contemplates Troilus and Cressida, 136–7; Alice Wimborne dies, 140; marriage, 143–4; moves to Ischia, 144; begins work on Troilus, 147–9; records with Heifetz, 150; further work on Troilus, 150–5; knighted, 150; involves Auden in Troilus, 161–2; composes 1953 Coronation works, 162–3; visits USA, 167; problems over Troilus, 168–78; advice on smuggling tobacco, 174; operation, 179; troubles with Sargent, 179–80; première of Troilus, 181; Troilus in Milan, 182–3; CG revival of Troilus, 188–9; composes Richard III,

194–5; begins Cello
Concerto, 196–7; in car
crash, 198; critical reaction
against, 199–202;
friendship with Peter
Heyworth, 200–3; work
schedule, 205; in séance,
206–7; builds 'La Mortella',
208–9; works on
symphony, 209; opera
commissioned, 209–10;
Second Symphony
performed, 211; visits New
York, 214; freeman of
Oldham, 217; at Hoffnung
concert, 218; conducts in
USA, New Zealand, and
Australia, 223–4; letters to
Britten, 224–6; friendship
with M. Arnold, 226–7;
lung cancer operation, 229;
radiation treatment, 230–1;
The Bear performed, 232;
appointed O.M., 235;
letters to Diana Rix, 235;
visits USA, 236; TV
programme, 236–7; Battle of
Britain film, 237–40;
musical tribute to Britten,
241–3; writes guitar pieces,
245–6; seventieth birthday,
250–1; party at No.10, 253–
4; begins third symphony,
255; conducts last Façade,
259; Troilus revision, 261;
writing work for Karajan,
263–4; zoo fanfare, 266;
Troilus revived, 266–7; his
plans, 267; Britten's death,
267; illness, 267–8; Façade
'rejects' revised, 269; lunch
for O.M.s, 270–1; writing
for brass band, 271; writes
for Rostropovich, 274–6;
abandons third symphony,
275; eightieth birthday, 276–7;
in hospital, 277; in Wagner
film, 277; plans Stabat Mater,
278; death and funeral, 278;
temperament and
technique, 279–82; interest
in contemporaries, 283–5;
business sense, 286–7;

preoccupation with Troilus,
288–9
Financial acumen: 12, 63–4, 67,
154, 274, 286–7
Health: 74, 78, 97, 117, 142, 179,
198, 229, 230–1, 250, 258,
264, 267–8, 272, 275, 278
Opinions on: Billy Budd, 151–2;
Burning Fiery Furnace, 231;
'causes', 112;
contemporaries, 284–5;
Cochran, 90; Delius, 31;
Elgar, 38; fluency, 279;
Foss, 72; Henze, 285;
Heseltine, 31–2; influence
of Sibelius, 83, 285;
Midsummer Marriage, 185–6,
288–9; Siesta, 43;
Stevenson, 284;
Stockhausen, 262;
Stravinsky, 284; Van
Dieren, 31–2; War Requiem,
225. Et passim
Ward, John, 215, 222
Warlock, Peter (See Heseltine, Philip)
Warrack, Guy, 21, 43
Warrack, John, 123, 143, 171, 236;
letters from W. quoted, 86, 122
Watson, Dr Sydney, 227, 308
Waugh, Evelyn, 29, 42
Webern, Anton, 22, 23, 24, 199
Webster, Sir David, 132, 134, 160 and
n, 171, 172, 173, 174, 177, 183
Welles, Orson, 243
Wellesz, Egon, 24
Whistler, Rex, 42, 47, 108
White, Bradbridge, 313
Widdicombe, Gillian, 120, 184, 274,
277; on W. at seventy, 252
Wilcox, Herbert, 76, 215
Wilkinson, Stephen, 259, 308
Williams, John, 210
Williams, Stephen, 12
Williamson, Malcolm, 46, 264n
Wilson, Sir Steuart, 150
Wimborne, Viscount, 75
Wimborne, Viscountess (Alice), 78–9,
93, 97, 100, 101, 106, 113, 118,
131, 136, 142, 143, 144, 145, 147,
178, 190, 230, 256, 278; meets W.,
75; and Violin Concerto, 102; Roy
Douglas describes, 110, 118;
letters to Hassall, 137–8, 139–40;

Wimborne, Viscountess (Alice), *cont'd*
    illness, 138–9; death, 140
Winbergh, Gösta, 188
Winn, Godfrey, 133*n*
Wood, Sir Henry J., 49, 66, 79, 99,
    124, 301; on W.'s First
    Symphony, 85; W.'s memorial
    to, 127–8
Wood, Jessie, 127, 128
Woodgate, Leslie, 128, 307
Woolf, Virginia, 18, 29–30
Wright, Kenneth, 92*n*, 108, 116;
    letters from W. quoted, 74, 116

Wyss, Sophie, 96

Xenakis, Iannis, 284

Yeomans, Walter, 86
Yorkshire Evening News, 54
Yorkshire Post, 58
Young, Alexander, 173

Zoete, Beryl de, 42
Zuckerman, Lord, 266, 271
Zukerman, Pinchas, 50, 287

OXFORD

MORE OXFORD PAPERBACKS

Details of a selection of other books follow. A complete list of Oxford Paperbacks, including The World's Classics, Twentieth-Century Classics, OPUS, Past Masters, Oxford Authors, Oxford Shakespeare, and Oxford Paperback Reference, is available in the UK from the General Publicity Department, Oxford University Press (JN), Walton Street, Oxford OX2 6DP.

In the USA, complete lists are available from the Paperbacks Marketing Manager, Oxford University Press, 200 Madison Avenue, New York, NY 10016.

Oxford Paperbacks are available from all good bookshops. In case of difficulty, customers in the UK can order direct from Oxford University Press Bookshop, 116 High Street, Oxford, Freepost, OX1 4BR, enclosing full payment. Please add 10 per cent of published price for postage and packing.

ISLAND CROSS-TALK

Tomás Ó'Crohan

Translated by Tim Enright

In these pages from his diary Ó'Crohan jotted down snatches of conversation, anecdotes, and descriptions of the landscape and the sea.

Island Cross-Talk, first published in 1928, was the first book to come out of the Blasket Islands, that remote, tiny community off the West Kerry coast speaking a dying language. It sowed the seeds of a rich and extraordinary flowering of literature: Maurice O'Sullivan's *Twenty Years A-Growing*, Peig Sayers's *An Old Woman's Reflections*, Ó'Crohan's later book, *The Islandman*, and many others.

THE ISLANDMAN

Tomás Ó'Crohan

Translated by Robin Flower

Tomás Ó'Crohan was born on the Great Blasket Island in 1856 and died there in 1937, a great master of his native Irish. He shared to the full the perilous life of a primitive community, yet possessed a shrewd and humorous detachment that enabled him to observe and describe his world. His book is a valuable description of a now vanished way of life; his sole purpose in writing it was, in his own words, 'to set down the character of the people about me so that some record of us might live after us, for the like of us will never be again'.

BRITISH WRITERS OF THE THIRTIES

Valentine Cunningham

This wide-ranging discussion of British writing and writers in that momentous and notorious decade offers interpretations of central texts of the period, not in linguistic isolation, but in the contexts—social, political, historical, ideological, personal—in which they were written.

'*British Writers of the Thirties* must be acknowledged a success. The immense strings of citations, marshalled with care and often with wit, eventually weave together an enormous design.' *Times Literary Supplement*

'This rich book . . . independent-minded, iconoclastic and articulate' *Sunday Times*

'*British Writers of the Thirties* is by far the best history of its kind published in recent years . . . it will become required reading for those who wish to look back as a society and a culture in which writers, for all their faults, were taken seriously.' Peter Ackroyd, *The Times*

'Cunningham's extraordinary assiduousness . . . makes it hard to see how *British Writers of the Thirties* could be surpassed as a reading of what has become—probably now more firmly than ever—the 'canonical' literature of the period.' *Times Higher Education Supplement*

CLASSIC SCOTTISH SHORT STORIES

Selected and edited by J. M. Reid

This suberb collection of twenty-two memorable stories from over 150 years of Scottish literature abounds in a strong sense of history, and a deep feeling for the eerie and supernatural. The writers include Sir Walter Scott, Robert Louis Stevenson, John Buchan, Sir James Barrie, Naomi Mitchison, Eric Linklater, Dorothy K. Haynes, and many more.

CLASSIC ENGLISH SHORT STORIES

Selected and introduced by Derek Hudson

A welcome collection of nineteen stories from the years 1930–1955, this book presents some of the best modern English short fiction from authors such as Somerset Maugham, Virginia Woolf, Evelyn Waugh, Elizabeth Bowen, Rosamond Lehmann, V. S. Pritchett, and Graham Greene.

'represents most of the best qualities to be found in this genre in England' *Spectator*

MARXISM AND LITERATURE

Raymond Williams

This book extends the theme of Raymond Williams's earlier work in literary and cultural analysis. He examines previous contributions to a Marxist theory of literature from Marx himself to Lukacs, Althusser, and Goldmann, and develops his own approach by outlining a theory of 'cultural materialism' which integrates Marxist theories of language with Marxist theories of literature.

'Williams has brought his authority and experience, established by his immense critical achievement, into the Marxist tradition.' Anthony Barnett, *New Society*

MORE LETTERS OF OSCAR WILDE

Edited by Rupert Hart-Davis

Sir Rupert Hart-Davis's edition of *The Letters of Oscar Wilde* received great acclaim when it was first published a quarter of a century ago. Since then, many new letters have come to light. Full of splendid Wildeisms, they are now presented for the first time in paperback.

'Almost every page contains something amusing or picturesque.' John Gross, *Observer*

SELECTED LETTERS OF OSCAR WILDE

Edited by Rupert Hart-Davis

When Sir Rupert Hart-Davis's magnificent edition was first published in 1962, Cyril Connolly called it 'a must for everyone who is seriously interested in the history of English literature— or European morals'. That edition of more than 1,000 letters is now out of print; from it Sir Rupert has culled a representative sample from each period of Wilde's life, 'giving preference', as he says in his Introduction to this selection, 'to those of literary interest, to the most amusing, and to those that throw light on his life and work'. The long letter to Lord Alfred Douglas, usually known as *De Profundis*, is again printed in its entirety.

'In Mr. Hart-Davis's *The Letters of Oscar Wilde*, the true Wilde emerges again for us, elegant, witty, paradoxical and touchingly kind . . . I urge all those who are interested in the contrasts between pride and humiliation, between agony and laughter, to acquire this truly remarkable book.' Harold Nicolson, *Observer*

THE FLOWER MASTER

Medbh McGuckian

In 1979 the young Ulster poet, Medbh McGuckian, won the annual Poetry Society competition, and in 1980 she received an Eric Gregory Award.

The vivid and brilliant surface attractiveness of her style beguiles the reader into untangling the evasive meanings of her mysterious and sensual poems. Anne Stevenson, reviewing the pamphlet 'Portrait of Joanna' in *The Times Literary Supplement*, wrote: 'she is as clever (probably) as Craig Raine, as perceptive (possibly) as Elizabeth Bishop . . . Reading these poems, one senses that thoughts and perceptions make mysterious connection with a hidden terror in the poet's mind.'

Oxford Poets

VENUS AND THE RAIN

Medbh McGuckian

This is the second collection of poems by Medbh McGuckian, one of the best known of the younger generation of Ulster poets. Her mysterious and erotic poems, sometimes puzzling, often beautiful, exercise a fascination over the reader who is willing to be beguiled as her subject-matter—domesticity, love, moving house, children—takes on an unnerving precariousness.

'a wonderfully original, bewitching collection' *London Magazine*

Oxford Poets

JAMES JOYCE
Richard Ellmann

Winner of the James Tait Black and the Duff Cooper Memorial Prizes

Professor Ellmann has thoroughly revised and expanded his classic biography to incorporate the considerable amount of new information that has come to light in the twenty-two years since it was first published. The new material deals with most aspects of Joyce's life: his literary aims, a failed love affair, domestic problems, and his political views.

'the greatest literary biography of the century' Anthony Burgess

'Richard Ellmann's superb biography . . . [is] a great feat of twentieth-century literary scholarship.' Christopher Ricks

'a superlatively good biography of Joyce' Frank Kermode, *Spectator*

Oxford Lives

CLASSIC AMERICAN SHORT STORIES
Selected and introduced by Douglas Grant

Professor Grant's selection of fourteen tales provide a first-rate introduction to a significant and influential branch of American literature—the short story.

The book begins with three classic examples from the nineteenth century, including Edgar Allan Poe's *The Black Cat*, and continues with works by Mark Twain, Jack London, Henry James, and Edith Wharton; finally, stories from Faulkner, Hemingway, and others make this a truly representative collection of all that is best in American short fiction.